Prospects and Challenges of Global Pilgrimage Tourism and Hospitality

S.K. Gupta
Centre for Mountain Tourism and Hospitality Management, HNB Garhwal University, India

Lilibeth C. Aragon
College of International Tourism and Hospitality Management, Lyceum of the Philippines University, Philippines

Pankaj Kumar
Department of Tourism and Hospitality Management, Mizoram University, India

Madhurima
Department of Environmental Science, Mizoram University, India

Rajesh Ramasamy
Department of Tourism and Hospitality Management, Mizoram University, India

A volume in the Advances in Hospitality, Tourism, and the Services Industry (AHTSI) Book Series

IGI Global
PUBLISHER of TIMELY KNOWLEDGE

Published in the United States of America by
IGI Global
Business Science Reference (an imprint of IGI Global)
701 E. Chocolate Avenue
Hershey PA, USA 17033
Tel: 717-533-8845
Fax: 717-533-8661
E-mail: cust@igi-global.com
Web site: http://www.igi-global.com

Library of Congress Cataloging-in-Publication Data

Names: Gupta, S.K., editor. | Aragon, Lilibeth C., editor. | Kumar, Pankaj, editor. | Madhurima, editor. | Ramasamy, Rajesh, editor.
Title: Prospects and challenges of global pilgrimage tourism and hospitality / edited by S.K. Gupta, Lilibeth C. Aragon, Pankaj Kumar, Madhurima, Rajesh Ramasamy.
Description: Hershey, PA : Business Science Reference, [2023] | Includes bibliographical references and index. | Summary: "This book helps readers identify, understand and recognize the changing facts and facets of pilgrimage tourism around the world and how to develop and promote pilgrimage tourism for community integration, faith-sharing, perseverance, tolerance, peace etc. for secular and sustainable futures"-- Provided by publisher.
Identifiers: LCCN 2022023443 (print) | LCCN 2022023444 (ebook) | ISBN 9781668448175 (hardcover) | ISBN 9781668448182 (paperback) | ISBN 9781668448199 (ebook)
Subjects: LCSH: Spiritual tourism. | Globalization and tourism. | Hospitality industry--Management.
Classification: LCC G156.5.S65 P76 2023 (print) | LCC G156.5.S65 (ebook) | DDC 203/.51--dc23/eng/20220801
LC record available at https://lccn.loc.gov/2022023443
LC ebook record available at https://lccn.loc.gov/2022023444

This book is published in the IGI Global book series Advances in Hospitality, Tourism, and the Services Industry (AHTSI) (ISSN: 2475-6547; eISSN: 2475-6555)

British Cataloguing in Publication Data
A Cataloguing in Publication record for this book is available from the British Library.

All work contributed to this book is new, previously-unpublished material. The views expressed in this book are those of the authors, but not necessarily of the publisher.

For electronic access to this publication, please contact: eresources@igi-global.com.

Advances in Hospitality, Tourism, and the Services Industry (AHTSI) Book Series

Maximiliano Korstanje
University of Palermo, Argentina

ISSN:2475-6547
EISSN:2475-6555

Mission

Globally, the hospitality, travel, tourism, and services industries generate a significant percentage of revenue and represent a large portion of the business world. Even in tough economic times, these industries thrive as individuals continue to spend on leisure and recreation activities as well as services.

The Advances in Hospitality, Tourism, and the Services Industry (AHTSI) book series offers diverse publications relating to the management, promotion, and profitability of the leisure, recreation, and services industries. Highlighting current research pertaining to various topics within the realm of hospitality, travel, tourism, and services management, the titles found within the AHTSI book series are pertinent to the research and professional needs of managers, business practitioners, researchers, and upper-level students studying in the field.

Coverage

- Health and Wellness Tourism
- Travel Agency Management
- Leisure & Business Travel
- Cruise Marketing and Sales
- Service Design
- Destination Marketing and Management
- Hotel Management
- Sustainable Tourism
- Casino Management
- Service Management

IGI Global is currently accepting manuscripts for publication within this series. To submit a proposal for a volume in this series, please contact our Acquisition Editors at Acquisitions@igi-global.com or visit: http://www.igi-global.com/publish/.

The Advances in Hospitality, Tourism, and the Services Industry (AHTSI) Book Series (ISSN 2475-6547) is published by IGI Global, 701 E. Chocolate Avenue, Hershey, PA 17033-1240, USA, www.igi-global.com. This series is composed of titles available for purchase individually; each title is edited to be contextually exclusive from any other title within the series. For pricing and ordering information please visit http://www.igi-global.com/book-series/advances-hospitality-tourism-services-industry/121014. Postmaster: Send all address changes to above address.

Titles in this Series

For a list of additional titles in this series, please visit: www.igi-global.com/book-series

Exploring Niche Tourism Business Models, Marketing, and Consumer Experience
Maria Antónia Rodrigues (CEOS.PP, ISCAP, Polytechnic of Porto, Portugal) and Maria Amélia Carvalho (ISCAP, Polytechnic of Porto, Porugal)
Business Science Reference • © 2023 • 397pp • H/C (ISBN: 9781668472422) • US $250.00

Cases on Traveler Preferences, Attitudes, and Behaviors Impact in the Hospitality Industry
Giuseppe Catenazzo (AUS American Institute of Applied Sciences, Switzerland & ICN Business School, Fance)
Business Science Reference • © 2023 • 348pp • H/C (ISBN: 9781668469194) • US $215.00

Women's Empowerment Within the Tourism Industry
Gül Erkol Bayram (Sinop University, Turkey) Syed Haider Ali Shah (Bahria University, Pakistan) and Muhammad Nawaz Tunio (The University of Sufism and Modern Sciences, Bhitshah, Pakistan)
Business Science Reference • © 2023 • 602pp • H/C (ISBN: 9781668484173) • US $235.00

Handbook of Research on Innovation, Differentiation, and New Technologies in Tourism, Hotels, and Food Service
Gonçalo Poeta Fernandes (CITUR, Polytechnic Institute of Guarda, Portugal & CICS, Universidade NOVA de Lisboa, Portugal) and António Silva Melo (CiTUR, Polytechnic Institute of Porto, Portugal & CIDTFF, University of Aveiro, Portugal)
Business Science Reference • © 2023 • 400pp • H/C (ISBN: 9781668469859) • US $295.00

Global Perspectives on Human Rights and the Impact of Tourism Consumption in the 21st Century
Maximiliano E Korstanje (University of Palermo, Argentina) and Vanessa G.B. Gowreesunkar (Anant National University, India)
Business Science Reference • © 2023 • 263pp • H/C (ISBN: 9781668487266) • US $240.00

Dark Gastronomy in Times of Tribulation
Demet Genceli (Istanbul Kent University, Turkey)
Business Science Reference • © 2023 • 283pp • H/C (ISBN: 9781668465059) • US $205.00

Inclusive Community Development Through Tourism and Hospitality Practices
Vipin Nadda (University of Sunderland in London, UK) Faithfull Gonzo (University of Sunderland in London, UK) Ravinder Batta (Shimla University, India) and Amit Sharma (Shimla University, India)
Business Science Reference • © 2023 • 361pp • H/C (ISBN: 9781668467961) • US $215.00

IGI Global
PUBLISHER of TIMELY KNOWLEDGE

701 East Chocolate Avenue, Hershey, PA 17033, USA
Tel: 717-533-8845 x100 • Fax: 717-533-8661
E-Mail: cust@igi-global.com • www.igi-global.com

Table of Contents

Detailed Table of Contents

This manuscript investigates the opinions of travellers in pilgrimage tourism, focusing on the three phases of the pilgrimage journey: pre, in-transit, and post. The study explores the different kinds of needs that emerge in each phase and examines their impact on shaping travellers' opinions. The research methodology employed SmartPLS (partial least squares) analysis to evaluate the relationships between independent variables and travellers' opinions. Data was collected through an online questionnaire. The findings indicate that all three phases significantly influence travellers' opinions, with the in-transit phase emerging as the most influential. Additionally, the study highlights the importance of pre-visit factors in framing travellers' opinions. The results of the study contribute to theoretical knowledge by identifying influential components in pilgrimage tourism and their impact on travellers' opinions. From a managerial perspective, the findings provide insights for destination managers to enhance the quality of experiences in the three phases and address pre-visit factors.

Pilgrimage, or religious, tourism is regarded to be one of the primaeval forms of tourism. Also, an extensive amount of research on pilgrimage and religious tourism reveals the vitality of the field of study. The chapter aims to conceptualise the notion of pilgrimage tourism and prepare the knowledge map of the pilgrimage tourism literature. The chapter uses the Scopus database and applies a systematic literature review as well as a bibliometric approach to uncover the most prolific studies, authors, organisations, countries, and themes of pilgrimage tourism research worldwide. Also, the growth and development of knowledge in the field has been demonstrated. For the analysis and data visualisation the Biblioshiny package of R and visualization of similarities (VOSviewer) software is used. The findings uncover that the United Kingdom, Spain, and China are the most prolific countries; the studies are centered around the themes of heritage tourism, sustainable development, and coronavirus in the pilgrimage tourism research domain from the last few years.

Calvaries are firmly rooted in Christianity and refer to the depiction of the scenes of the Passion of Christ. This chapter examines suburban calvaries and their potential role within the context of urban tourism, by widening the horizon of pilgrimage tourism while also challenging the notion of city tourism, by taking tourists to the fringe of the city. The research site is the capital of Hungary, Budapest that boasts some fine religious buildings which are inundated with tourists all-year-around, causing overcrowding, mobility challenges and conflicts with locals. The ethnographic account gives a descriptive analysis of six suburban calvaries that are dissected in terms of cultural, sustainable, and niche tourism, and explains how they can contribute to urban tourism. It does not deal with the inner city calvaries because they do not fit the purpose of making urban tourism more sustainable, thus, pushing tourists away from the crowded core of the city. The study concludes that including calvaries into urban tourism would be beneficial for locals, visitors, and the environment alike.

The work purports to present how the Shri Kashi Vishwanath Temple Corridor (SKVTC) Project has propositioned its roadmap of 'Extension and Beautification' of a densely-populated pilgrim hotspot (Kashi Vishwanath Mandir Visistikaran- Saundrikaran Yojana). The scope of this study includes the rediscovered historical temples of the region that were saved from encroachment as a result of this Corridor. The present study fills the gap of a systematic study of the Corridor in terms of relevance in pilgrimage tourism, particularly in the context of the rediscovered historical temples, focusing on heritage sustainability to support the revival of Varanasi's tangible and intangible heritage. The findings of the study, which discussed the rapidly changing landscape of the most important site of the city, the Kashi Vishwanath temple, enabled local communities, tourism stakeholders, and academicians worldwide to encapsulate the diverse complexities of the SKVTC Project and its impact on pilgrimage activities and tourism.

In the present book chapter, we have interrogated not only the nature of dark tourism but also the evolution of tourism since the former century to date. We have offered a model based on three stages of tourism, each one conditioned by a turning point or founding event. Classic tourism was originated in the WWII end but finalized as a result of Oil embargo in 1970. A new sustainable form of tourism supported by heritage finally came to stay until the attacks to the US. The surfacing of new morbid form of tourism witnessed the exhaustion of an industry chiefly worried to face unnumberable global risks.

on spiritual things and frequently visit sacred locations. Various ceremonies and offerings are conducted at pilgrim centers to atone for sins. Over the last three decades, a fresh focus on pilgrimage has evolved from tourism scholars who have investigated fascinating political, cultural, behavioral, economic, and geographical study themes. Pilgrimage tourism is important for socio-economic growth.

The travel and tourism industry is one of the fastest-growing industries in today's Indian global economy. The fast growth of the industry is not only generating jobs for millions, but also conducive on a large scale to the entire economy of the world. Pilgrimage tourism in Chidambaram is a popular form of spiritual tourism that attracts thousands of devotees every year. Chidambaram is a holy town located in the southern state of Tamil Nadu, India, and is considered one of the most important pilgrimage destinations in the country. The town is renowned for its ancient temple dedicated to Lord Nataraja, a form of Lord Shiva, and is also known for its rich cultural heritage, traditional arts, and dance forms. In Tamil Nādu, there are a number of tourist places but pilgrimage tourism in Chidambaram has its own importance. Therefore, it attracts a large number of pilgrims from various parts of the world. This chapter is an attempt to study the pilgrimage tourism on the overall development of the place and moreover aims at the challenges which the tourism industry is facing in Chidambaram.

This empirical investigation focuses on assessing the quality of water and sanitation services provided by hotels located on pilgrimage sites, especially after the COVID-19 pandemic. The study examines the factors that contribute to customers' perception of service quality and the associated hygiene practices implemented by the hotels. Data were collected through personal interaction using the SERVQUAL questionnaire which measures service quality in accordance with five dimensions. It highlights the importance of improving the tangibility of service and increasing empathy between hotel staff and guests, especially in the post-COVID-19 era. The study suggests recommendations for improving the quality of hospitality services and increasing tourist satisfaction. Overall, the study highlights the importance of providing high-quality water and sanitation services at pilgrimage sites, especially in the current context of the pandemic, to ensure that tourists have a positive experience and feel confident in using these facilities.

Tourism has been observed as one of the biggest economic sectors in the service industry, which has

Preface

Pilgrimage tourism is also known as religious tourism. It is an oldest form of tourism activity and motivation for travellers across the globe. Since ancient times, people used to travel to fulfill their quest for knowledge, their religious belief, peace, calm and wellbeing to the respective religious places which are also called as shrines or pilgrimage centres. There are huge number of such destinations and attractions, continuously attracting the visitors since time immemorial.

It has also been observed that, the existing modern practices of tourism motivations are the extension of the ancient pilgrimage tourism, where many destinations with other characteristics such as cultural tourism, urban tourism, rural tourism, adventure tourism, wellness tourism and others have been promoted and are being visited every year globally. The contribution of pilgrimage tourism and related activities have been acknowledged and appreciated due to its economic, social, cultural, and environmental contribution to society.

There are several government agencies, non-governmental organizations, and international organizations such as World Tourism Organization (UNWTO), United Nations Educational, Scientific and Cultural Organizations UNESCO, Archaeological Survey of India (ASI), Ministry of Tourism, Government of India and respective departments of the different countries across the globe are also working for the protection, conservation and promotion of such cultural-religious sites as our heritage. The visitors and destination communities are being offered financial assistance and awareness towards it.

With a view towards enhancing the positive effects of spiritual tourism on the economic and social advancement of communities and societies, the first UNWTO International Conference on Spiritual for Sustainable Development was organized in Ninh Binh Province, Vietnam on November, 21st and 22nd, 2013. The conference explored the multiple ways in which living culture, traditions and benefits can be integrated into tourism while respecting the four pillars of sustainability to draw the particular attention towards understanding religious values and assets in context of tourism, development, promotion and interpretation of spiritual tourism products, socio economic inclusion and community empowerment of local communities. Thereafter, an International Congress on Religious Tourism and Pilgrimage was organized by UNWTO in the year 2017, when the year was declared as "International Year of Sustainable Tourism". Along with the diverse benefits of pilgrimage tourism, this segment is facing several challenges at numerous destinations across the globe.

It is believed that the visitor of sacred places is characterized by a remarkable loyalty to the visited sites. Thus, a key principle to promote religious tourism or pilgrimage tourism is to preserve the spirit of destination's authenticity and integrity in the way that they are managed and promoted while making them accessible for all. On the other-hand these places of spirituality are also instrumental to drain part of their tourist flow to other tourist attractions in the region.

The governments, national, International organizations and industry stakeholders emphasized how tourism including the pilgrimage tourism can be an impressive tool for overcoming the inequalities among communities and cumulative destination development.. This book is an attempt to bring knowledge, information and ideas on various aspects of pilgrimage tourism and hospitality to the young academicians, researchers, tourism & hospitality industry facilitators, policy makers and destination community.

This book titled *Prospects and Challenges of Global Pilgrimage Tourism and Hospitality*, comprises 18 chapters discussing different dimensions of tourism and hospitality centred to pilgrims and pilgrimage tourism practices. The chapters in this book largely covers the concerns of pilgrims, pilgrimage, pilgrimage tourism, pilgrimage hospitality- nutritional needs of spiritual visitors, entrepreneurial aspects and sustainable development, destination management practices and case studies from different countries such as- India, Hungary, Malaysia, Argentina, Hong Kong, etc.

Each and all chapters are discussed with some valuable inputs. by the researchers, academicians, NGO practioners, and industry professional from different countries, which would be beneficial to the researchers, academicians, industry practioners, policy makers and communities to the society at large.

CHAPTER OVERVIEW

Chapter 1 explores the opinions of travellers engaged in pilgrimage tourism, focusing on the three distinct phases of their journey: pre-visit, in-transit, and post-visit. It investigates the different types of needs that arise during each phase and how these needs shape travellers' opinions. The study employs the SmartPLS (Partial Least Squares) analysis to establish the relationships between independent variables and travellers' opinions. Through an online questionnaire, data is collected, revealing that all three phases significantly influence travellers' opinions, with the in-transit phase emerging as the most influential. The chapter underscores the significance of pre-visit factors in shaping travellers' opinions. The findings contribute to both theoretical knowledge and practical application, aiding destination managers in enhancing the quality of experiences across the three phases and addressing pre-visit factors.

Chapter 2 focuses on conceptualizing pilgrimage tourism by mapping out the existing literature in the field. Using the Scopus database, a systematic literature review and bibliometric analysis are conducted to identify the prominent studies, authors, themes, and countries contributing to pilgrimage tourism research worldwide. The chapter emphasizes the growth and development of knowledge in this domain, spotlighting countries like the United Kingdom, Spain, and China as prolific contributors. Themes such as heritage tourism, sustainable development, and the impact of coronavirus on pilgrimage tourism are explored. Data visualization tools like R's Biblioshiny package and VOSviewer software aid in presenting the findings. This chapter contributes to understanding the landscape of pilgrimage tourism research and its trends.

Chapter 3 introduces a unique form of pilgrimage, suburban calvaries, and examines their potential within urban tourism contexts. It argues that these often-overlooked attractions can play a significant role in managing crowds and offering distinctive experiences. Departing from mainstream discourse, the chapter explores the characteristics of suburban calvaries and their relevance to cultural, sustainable, and niche tourism. The discussion is richly illustrated with examples from Budapest, Hungary. The chapter concludes that integrating calvaries into urban tourism can benefit local communities, visitors, and the environment, challenging traditional notions of pilgrimage sites.

Chapter 4 delves into the Shri Kashi Vishwanath Temple Corridor (SKVTC) Project, analyzing its roadmap for extending and beautifying the pilgrimage hotspot. The study fills a gap by systematically examining the Corridor's relevance in pilgrimage tourism, particularly in the context of rediscovered historical temples. Heritage sustainability is a focal point, contributing to the revival of tangible and intangible heritage in Varanasi. The findings uncover the project's impact on the rapidly changing landscape of the Kashi Vishwanath temple and provide insights for local communities, tourism stakeholders, and academics worldwide. This chapter sheds light on the complex dynamics of the SKVTC Project and its influence on pilgrimage activities and tourism.

In chapter 5, the authors explore the evolution of tourism since the previous century, with a specific focus on dark tourism. They propose a model based on three stages of tourism, each marked by significant turning points or events. The chapter examines how dark tourism emerged as a new form, encompassing morbid attractions and indicating shifts in the tourism industry's response to global risks. The chapter underlines the importance of understanding the interplay between tourism evolution and societal changes to anticipate and adapt to emerging trends.

Chapter 6 addresses the challenges posed by the COVID-19 pandemic on Hajj, a significant spiritual event in Islam involving mass global movement. It discusses threats and realities related to COVID-19 across various aspects of Hajj management, including health, logistics, and technology. The chapter analyzes potential strategies for the pre-departure, on-site management, and post-return phases of Hajj, focusing on health and resilience. The recommendations aim to guide governments and stakeholders in overcoming the current crisis and ensuring the sustainability of Hajj in future.

Chapter 7 shifts focus to non-secular destinations, investigating the motivations of tourists visiting local religious sites in cities not commonly promoted for tourism. Employing qualitative methods like participant observation and semi-structured interviews, the study uncovers insights into pilgrimage tourism motivations. By exploring this less-studied angle, the chapter contributes new perspectives to the field, shedding light on elements that could develop new tourism attractions in non-secular destinations.

Chapter 8 draws a comparative analysis between two Marian Pilgrim destinations in Tamil Nadu, India, and the Sanctuary of Our Lady of Lourdes in France. The authors examine how these destinations can learn and unlearn from Lourdes' construct to enhance their own reputation and global recognition. Through this comparative lens, the chapter explores strategies for improvement and growth, aiming to elevate the Indian destinations to global prominence.

Chapter 9 explores the potential of temple architecture to drive tourism. It discusses various styles of temple architecture from around the world and the ways in which tourism can be harnessed to showcase these cultural landmarks. The chapter emphasizes the economic, social, and cultural impact of temple tourism on local communities. Through case studies, the authors highlight how temple tourism can provide immersive experiences and contribute to local economies and cultural preservation.

Chapter 10 delves into the paradoxical relationship between animal cruelty, conservation, and pilgrimage tourism. The study reviews literature to explore the intricate connection between religious beliefs, pilgrimage practices, and animal welfare. The chapter highlights the tensions and complexities surrounding this issue, ultimately seeking to bridge ancient religious inscriptions with modern sustainable development goals. This chapter sheds light on an often-overlooked aspect of pilgrimage tourism.

Chapter 11 provides an overview of the multifaceted impact of pilgrimage tourism on political, social, cultural, and economic aspects of society. It highlights the integral connection between tourism and pilgrimage, underscoring how pilgrimage sites often attract local visitors who contribute to the tourism industry. The chapter explores the evolving landscape of pilgrimage tourism and its diverse implications for various study themes. It emphasizes the importance of pilgrimage tourism in driving socio-economic growth.

Chapter 12 focuses on the role of pilgrimage tourism in Chidambaram, Tamil Nadu, India. The chapter examines how this form of tourism contributes to the overall development of the region, both economically and culturally. Chidambaram's significance as a pilgrimage destination is explored, along with the challenges faced by the tourism industry in the area. The chapter aims to provide insights into the impact of pilgrimage tourism on the local community and the broader region.

Chapter 13 investigates the quality of water and sanitation services provided by hotels at pilgrimage sites, with a specific focus on the post-COVID-19 pandemic context. The study examines customers' perceptions of service quality and the hygiene practices implemented by hotels. Data collection utilizes the SERVQUAL questionnaire to measure service quality. The findings highlight the importance of improving tangible aspects of service and fostering empathy between hotel staff and guests, especially in the wake of the pandemic.

Chapter 14 assesses the potential of pilgrimage tourism for sustainable development in Punjab, India. It evaluates tourist areas, identifies obstacles hindering sustainable development, and provides recommendations for the growth of pilgrimage tourism in the region. Through a field-based survey, the chapter sheds light on the opportunities and challenges faced by pilgrimage tourism in Punjab, ultimately suggesting measures for its sustainable development.

Chapter 15 examines entrepreneurship's role in the growth of pilgrimage tourism in Jammu and Kashmir. The study assesses the current state of entrepreneurship in this sector, identifies opportunities and challenges, and explores future prospects. Through mixed methods, the chapter offers insights into the potential of entrepreneurship to drive economic development in the region. It also highlights obstacles that entrepreneurs encounter and suggests ways to overcome them.

Chapter 16 explores how pilgrimage tourism contributes to sustainable development in Ladakh, known for its unique landscape and spiritual attractions. The chapter discusses how pilgrimage tourism revives ancient traditions, uplifts local living standards, reduces unemployment, and promotes economic growth. By analyzing the socio-economic and cultural advantages, the chapter demonstrates the positive impact of pilgrimage tourism on the region's overall sustainable development.

Chapter 17 investigates the nutritional requirements of spiritual tourists visiting International Society for Krishna Consciousness (ISKCON) centers in India. The chapter analyzes diet menus, nutritional needs, and the impact of "Satvik Aahar" (ISKCON diet) on the physical and mental well-being of visitors. The study provides insights into how this dietary practice complements spiritual practices and contributes to the visitors' overall health and well-being.

Chapter 18 presents a theoretical exploration of the intersection between religious events and the tourism industry, with a focus on post-COVID-19 prospects. The chapter discusses the impacts of the pandemic on religious events, identifies emerging themes, trends, and challenges for the recovery of pilgrimage tourism. It offers a conceptual perspective on how these events can adapt and thrive in a post-pandemic world, contributing to the evolving discourse on pilgrimage tourism.

We take this opportunity to thank all the authors, research paper/article contributors, reviewers and dedicated and highly supportive team of IGI Global Publishers, who made possible to materialize this book. We also extend our sincere gratitude to all who contributed directly or indirectly.

S.K. Gupta
Centre for Mountain Tourism and Hospitality Management, HNB Garhwal University, India

Lilibeth C. Aragon
College of International Tourism and Hospitality Management, Lyceum of the Philippines University, Philippines

Pankaj Kumar
Department of Tourism and Hospitality Management, Mizoram University, India

Madhurima
Department of Environmental Science, Mizoram University, India

Rajesh Ramasamy
Department of Tourism and Hospitality Management, Mizoram University, India

Chapter 1
Pilgrimage Tourism:
A Look at Travelers' Opinions

Praveen Srivastava
https://orcid.org/0000-0001-5310-694X
Birla Institute of Technology, Mesra, Ranchi, India

Gautam Shandilya
https://orcid.org/0000-0002-9510-3787
Birla Institute of Technology, Mesra, Ranchi, India

ABSTRACT

This manuscript investigates the opinions of travellers in pilgrimage tourism, focusing on the three phases of the pilgrimage journey: pre, in-transit, and post. The study explores the different kinds of needs that emerge in each phase and examines their impact on shaping travellers' opinions. The research methodology employed SmartPLS (partial least squares) analysis to evaluate the relationships between independent variables and travellers' opinions. Data was collected through an online questionnaire. The findings indicate that all three phases significantly influence travellers' opinions, with the in-transit phase emerging as the most influential. Additionally, the study highlights the importance of pre-visit factors in framing travellers' opinions. The results of the study contribute to theoretical knowledge by identifying influential components in pilgrimage tourism and their impact on travellers' opinions. From a managerial perspective, the findings provide insights for destination managers to enhance the quality of experiences in the three phases and address pre-visit factors.

1. INTRODUCTION

Ever since the evolution of mankind on this planet people have moved from one place to another for different purposes. People travelled and migrated in search of food, livelihoods, trade, settlement, and better pursuit of life. This took a huge turn of event with the invention of wheel and the subsequent advancements in the means of transportation. People also started to travel to trail their roots and to find solace with nature and religious abodes. The movement of people took the shape of tourism got the industry status and has become one of the leading contributors to the world economy.

DOI: 10.4018/978-1-6684-4817-5.ch001

Pilgrimage tourism, characterized by journeys undertaken for religious or spiritual purposes, has a rich historical and cultural significance across the globe. These sacred journeys attract millions of travellers each year, fostering a unique intersection of faith, heritage, and exploration. It encompasses a myriad of needs that travellers seek to fulfil during their journeys, spanning the pre, during, and post phases of their pilgrimage experience. Understanding these diverse needs is crucial for developing effective strategies to enhance the overall pilgrimage experience and cater to the expectations and aspirations of pilgrims. This manuscript aims to delve into the various dimensions of pilgrims' needs throughout these distinct phases, shedding light on their preferences, challenges, and desired outcomes.

The pre-pilgrimage phase marks the beginning of the spiritual journey, where travellers make preparations, plan their itineraries, and engage in various activities to ready themselves for the pilgrimage ahead. During this phase, pilgrims may seek information and guidance about the religious significance of the destination, transportation logistics, accommodation options, and cultural norms of the host community. Exploring their opinions and perspectives during this stage offers valuable insights into the informational and emotional support required by pilgrims as they embark on their sacred voyage.

The In-Transit phase represents the heart of the pilgrim's journey, where they engage in religious rituals, visit sacred sites, and immerse themselves in the spiritual ambiance of their destination. Understanding the needs of pilgrims during this phase involves examining their preferences for accessibility to pilgrimage sites, availability of religious services, provision of facilities for accommodation, dining, and sanitary needs, as well as the importance of communal interactions and cultural experiences. By capturing the opinions and experiences of pilgrims in real-time, this study aims to illuminate the ways in which pilgrimage organizers and destination managers can enhance the pilgrim's experience and foster a deeper connection to the spiritual essence of the journey. Travelers seek a meaningful travel experience during pilgrimage which may bring them closer to nature, spiritual self, and the divine force (Cheer, 2023; Cheer et al., 2017).

The post-pilgrimage phase reflects the aftermath of the pilgrimage, as travellers return to their daily lives carrying the transformative experiences and insights gained during their sacred voyage. During this phase, pilgrims may seek avenues for reflection, community engagement, and continued spiritual growth. Understanding their needs in this phase provides valuable insights into the development of post-pilgrimage support systems, such as online communities, pilgrimage-related events, and resources for maintaining a spiritual connection even after leaving the sacred destination.

This manuscript aims to delve into the realm of pilgrimage tourism, exploring the multifaceted aspects of this specialized form of travel through the lens of the travellers in all three phases. By examining the opinions of pilgrims, we seek to provide a comprehensive analysis of the significance, challenges, and benefits associated with this unique type of journey.

2. REVIEW OF LITERATURE

2.1 Tourism

Let us first understand what the tourism is? The history of tourism is more ancient to the coinage of the term 'Tourist' in the 18th century. According to Encyclopaedia Britannica tourism can be defined as "an act of recreation, relaxation, and pleasure outside the home", and a tourist is a person who travels for pleasure. Tourism is a phenomenon of movement of people outside their usual environment (UNWTO,

2008). Tourism is not limited to entertainment and holidaying activities but any movement of people from their usual place of residence for more than 24 hours and less than a year for any purpose sans earning activities and employment (Fleetwood, 1995; Hunziker & Krapf, 1942; Spode, 1998).

Tourism is a big source of income for people in various countries, adds to the local economy in the form of payment and contributes to employment generation. This benefits the transportation services, lodging facilities, entertainment venues and other allied industries. There is a multiplier effect of tourism on social, cultural, educational, and economic aspects of any nation.

Tourism can be classified based on destination, size, and purpose. There are some common tourism categories like mass tourism, eco-tourism, medical tourism, educational tourism, cultural tourism, social tourism, business tourism, leisure tourism and religious tourism or pilgrimage. Whereas there are some niche categories or unusual categories like MICE, event tourism, sports tourism, space tourism, DNA tourism, doom tourism, dark tourism, experiential tourism, creative tourism, recession tourism, pro-poor tourism, volunteer tourism, winter tourism etc. These tourism categories may vary or at times overlap in their purposes, nature, and activities. Table 1 depicts the classification of tourism categories.

Tourism can be defined as a system of discretionary travel and temporary stay at a place away from home and the activities carried out during the movement from origin to destination and back to origin (Leiper, 1979). The traditional belief of tourist to be a leisure vocational and pilgrim to be a religious traveller shields the motive, intention and the opinion of a traveller (Smith, 1992), whereas in a broader perspective *peregrinus,* the Latin origin of pilgrimage, covers all kind of travellers including foreigners, person in exile, wanders, new comers, and strangers and *tornus,* the Latin origin of tourist define it as a traveller leisurely moving from one place to another and back to origin (Kliot & Collins-Kreiner, 2000).

2.2 Traveller

Tourism and travel have often been used interchangeably but have distinctive purpose. Travel developed in 19th and 20th century with the advancements in air, water, rail, and road transportation infrastructures. With these developments now everyone has the access to travel.

A traveller is a person who moves from one place to another for any purpose and at any time (UNWTO, 2008). A tourist is a person who is involved in the act of tourism with movement for more than 24 hours, whereas if the length of stay is less than 24 hours it will be called excursion and the person is an excursionist (UNWTO, 2008).

To be counted as tourist one's visit has to be rare and infrequent, thus the mere travellers are not considered as tourists (Yu et al., 2012). Tourists usually stay at a place for brief period, show interest

Table 1. Classifications

Tourism		
Based on Movement	**Based on Purpose**	**Based on Activities**
Domestic	Leisure	Social
International	Business	Cultural
		Religious
		Family

Source: Authors' Compilation 2023

in local sites, click photographs, buy souvenir but do not fathom much about local populace (Pearce, 1982, 1985). Nash (1981), defined tourist as a traveller having leisure time. Distinction between tourist and traveller can be understood from (Burkart & Medlik, 1974) proposal of dynamic element (journey) and static element (stay) of tourism where tourists spend on their accommodation, food, recreation, and other activities while staying at a place away from their usual habitat.

2.3 Religion

Religion is as old as human civilization. From being pagan to praying to nature or to a supreme figure in the form of some shapes and figurines to being monotheist, humans have always followed some or the other religion. It is the belief, trust or faith in the supreme body that enlightens, guides, nurtures, protects men in all their vices and virtues. Religion is a socio-cultural system of believing in spiritual self that addresses the nature of existence (James & Mandaville, 2010; Tylor, 1871).

Dictionary.com (*Religion*, 2023a) defined religion as "a set of beliefs concerning the cause, nature, and purpose of the universe, especially when considered as the creation of a superhuman agency or agencies, usually involving devotional and ritual observances, and often containing a moral code governing the conduct of human affairs". According to New world Encyclopaedia (*Religion*, 2023b), religion is a theistic belief system of relationship between humanity and the cosmos codified by prayers, scriptures, religious laws and rituals.

Religion has always been the need of the mankind and people chose to practice certain religion based on their value system; civilization; sustained culture; geographical boundaries and interfaith mingling. Some of the common religions followed are Christianity, Jewish, Hinduism, Islam, Buddhism, Shintoism, Sikhism, Bahai and Jainism etc. Figure 1 depicts the common religious symbols used worldwide.

There are also people who chose to be non-religious practitioners or atheists or to be called seculars. For the ease of understanding, religion may be classified under major religions and minor religions. Approximately 85% of world population practice some or the other major or minor religion (Pew Research Center Forum on Religion & Public Life, 2012). Table 2 depicts the major religious population estimated on 2010 baseline.

Figure 1. Common religious symbols
Source: New World Encyclopaedia (Religion, 2023b)

Table 2. Major religious population estimates

Religion	Population (in billion)	Percentage (approx.)	Major Location
Christianity	2.4	32%	Europe, America, Australia, Africa
Islam	1.9	25%	Asia (Middle East, Far East), North Africa
Hinduism	1.2	15%	India, Nepal
Buddhism	0.5	7%	Far East Asia
Folk Religion	0.4	5%	China, Africa
Unaffiliated	1.2	15%	China, all continents
Other	0.1	1%	All continents

Source adaptation: (Pew Research Center Forum on Religion & Public Life, 2012)

With this table it can also be understood that the majority followers of a specific religion mostly try to reside in a certain geographic location. Religious groups in minor segment can be found in sporadic locations due to migration, better settlement opportunities and sometimes conversion also.

2.4 Pilgrimage

Pilgrimage is a millenniums old practice, and no timeline can be framed for it. It all started with people traversing the rough geographic terrain on foot taking all the pain and hardship and at times sacrificing their lives too to meet the supreme lord or deity of their faith to seek blessings and find peace, solace, redemption, and salvation. With the taming of animals' people started riding pillion on the back of a horse or a mule and stayed at dharmshalas, sarais or inns with miniscule facilities. But with modern day's technology and infrastructural support system people reach to their destination in a jiffy by modern transport means and stay in the state-of-the art accommodation facilities with all modern amenities.

Pilgrimage is a journey to a sacred place or a shrine of one's faith in search of knowing oneself, nature and the super nature and people generally experience the personal transformation (Simon Coleman & Eade, 2004; Plate, 2009; Walter, 1993). Person on pilgrimage is known as pilgrim. According to S. M. Coleman (2023) in Encyclopaedia Britannica, pilgrimage is a religious journey which alienates pilgrims from their daily world and the materialism by abstaining from physical comforts and they may mark a new identity.

Some of the world-famous pilgrimage sites are showcased in table 3.

In all the religions what we have seen is that the very basic purpose or intention of the pilgrim towards the pilgrimage remains the same. It is just their culture, beliefs and the value system which differentiate in the mode of their offerings and rituals to establish relationships with the supreme power the almighty god.

With the core principles of pilgrimage tourism remaining intact and similar in nature across all the religions, travellers have different opinions on pilgrimage. There could be a question mark itself whether pilgrimage and tourism are same, or a pilgrim can be touted as tourist? Fahey (2002) distinguished pilgrims from tourists on the count of eight elements viz. faith; penance; community; sacred space; ritual; votive offering; celebration and perseverance and found that these elements are usually missing with tourists. Barber (2001) opined that tourism is more secular than pilgrimage. In fact, many researchers have an opinion that modern tourism has evolved from pilgrimage and the interchangeable use of tourism and pilgrimage is evident from the fact that many tourists may like to visit a sacred place while

Table 3. World-famous pilgrimage sites

Religion	Popular Pilgrimage Sites	Purpose/Intention
Bahá'í	Shrine of Bahá'u'lláh at Bahji, Israel	Nine days pilgrimage to holy places
Buddhism	Lumbini, Bodh Gaya, Sarnath, Kusinagar	Kaalchakra Puja
Christianity	Jerusalem, Rome, Vatican City	Visit to holy land
Hinduism	Kumbh Mela, Char Dham, Shakti Peeths, Mathura-Vrindavan, Ayodhya, Varanasi, Prayag Raj, Dwarka, Jagannath Puri, Vaishno Devi, Tirupati, Shirdi	Religious congregation, Holy bathe, Kanwar pilgrimage, Yatra, Moksha, Holy cities
Islam	Mecca, Madina, Karbala	Hajj, Umra, Arba'een
Judaism	Jerusalem, Bethlehem	Passover, Shavuot, Sukkoth, Mitzvah
Sikhism	Golden Temple	Spiritual, Cultural
Taoism	Hong Kong, Macau, Taiwan	Mazu (Matsu)
Zoroastrianism	Fire Temple, Yazdi Shrines in Iran and India	Udvada

Source: Authors' Compilation 2023

pilgrims can behave like a tourist (Davie, 2000; Margry, 2008; Schramm, 2004). It can be understood that tourism is a mere epicurean situation.

Cohen; Turnbull (1992) found that pilgrimage has a binding character which follows norms and moves towards the centre (existential quest) with spiritual reason. A religious practitioner intends to connect with the community, explore practices of religion and search for inspiring things (Stibich, 2022).

Pilgrims of the Abrahamic religions like Judaism, Christianity and Islam intends to visit their holy land to sense the physical manifestations of their faith (Metti, 2011). Muslim pilgrims travel in search of knowledge to *ihla li-talab al-'ilm* (tombs) and *ziyara* (sacred sites) or *hijra* to live in accordance with Islamic fundamentals (Eickelman & Piscatori, 1990). Buddhist pilgrims travel to places which have the footprints of Lord Gautam Buddha's travel trails (Gurung, 2016; Hall, 2006; Rodriguez, 2017).

Hindu pilgrims do pilgrimage for their wellbeing, spirituality, and holy atmosphere (Dar, 2020). The shrine's status; reach and accessibility; safety and security; cleanliness, hygiene and spiritual ambience; traffic to the place; seasonality; temple service pricing, donations and total expenditure; local culture and host community's acceptance and hospitability; hedonistic and holistic tourism experience all matters for a Hindu pilgrim (Jyotsna & Prakash Sai, 2022).

Pilgrimage makes travellers more tolerant; more sociable; more acceptable towards multicultural, multiethnic and multi-confessional aspects; accepting odd and tough situations and enjoying environment and cultural landscape and heritage (Lopez, 2014). It also derives certain emotions like humility, spirituality, solidarity, faith, personal understanding, sense of loneliness, reflection, happiness, spiritual enrichment, and charm. Cultural attraction and religious heritage of the pilgrimage destination are the important factors apart from the core religious value in pilgrimage (Ascaniis & Cantoni, 2017).

The modern trend shows that individuals work for acquiring life achievements and then seek to find the sacred meaning of life, hence often go for pilgrimage (Rodriguez, 2017; Shao Zhen et al., 2019). This corroborates the self-actualization need of the man (Maslow, 1943). This is more evident after the outbreak of COVID 19 pandemic as people are turning to be more religious to thank God for saving their lives whilst devastation across the world.

The increased accessibility to pilgrimage sites due to modern transport facilities helps in profit making for international tourist market (Olsen & Timothy, 2006). Advancement in technology is also imperative in pilgrimage and performance of religious services. These days people even prefer to use online services for their religious expeditions and about 21% American adults even use apps to read religious texts (Faverio et al., 2023). Collins-Kreiner (2020), emphasized the need of Pilgrimage to be reidentified as a mix of tourism subsets like religious tourism, secular tourism, heritage tourism, dark tourism, and spiritual tourism altogether in quest of deep and enduring transformation of humanity.

The research incorporated the Delphi method, involving consultation with experts in the field of pilgrimage tourism. The expert identified, three major phase of pilgrimage tourism as under:

Pre-Phase: This phase is characterised by the quest to discover inner self and hence, the tourist start planning for the journey toward the pilgrimage site. These experts collectively identified that religious value, Self-actualization, Search for inspiration, and Mental health were the most significant components for individuals on a quest to discover their inner selves through pilgrimage. Pilgrimage, often rooted in faith and devotion, allows travellers to connect with their religious heritage, sacred sites, and rituals. The Pre-Phase enables pilgrims to reflect on their religious values, fostering a deepened sense of purpose and devotion. assessment. The decision to undertake a pilgrimage is often motivated by a desire for personal growth and transformation. During the planning stage, individuals may reflect on their aspirations, values, and life goals. This self-evaluation paves the way for self-actualization, as pilgrims align their intentions with the forthcoming journey, setting the stage for a meaningful exploration of their identity and purpose. Furthermore, the act of planning a pilgrimage can be a therapeutic process, providing a positive focus and a sense of anticipation. Researching and organizing the trip can offer a sense of control and purpose, positively impacting one's mental state.

In-Transit Phase: This phase pertains to the period when the tourist is en-route to the chosen destination and engaged in various travel-related activities. In relation to the in-transit phase, experts emphasized that Easy access, Comfortable stay, Connectivity, and Hassle-free performance of rituals were the most crucial determinants for a fulfilling pilgrimage experience. Well-established transportation networks, clear signage, and efficient routes enhance the accessibility of these sacred sites. Offering comfortable and suitable accommodation options is essential for pilgrims to rest and rejuvenate during their visit. In today's interconnected world, pilgrims often seek to stay connected with their families, friends, and broader communities during their pilgrimage. Providing reliable connectivity, including internet access and mobile networks, enables pilgrims to maintain communication, share their experiences, and seek guidance if needed.

Post-Phase: The post-phase takes place after the tourist has returned from their trip. The experts highlighted that Reflection, Community Engagement, and Continued Spiritual Growth were the most important components for the post-phase of the pilgrimage journey. This phase offers pilgrims a valuable opportunity for reflection and contemplation. Pilgrimage experiences often create a sense of camaraderie and shared purpose among fellow travellers. The Post-Phase provides a platform for pilgrims to engage with their community, share stories, and build connections with like-minded individuals who have undergone similar transformative journeys. It also serves as a continuation of the pilgrim's spiritual journey. Having encountered sacred sites, rituals, and profound moments during the pilgrimage, individuals are inspired to further nurture their spiritual growth.

Based on these expert insights, the identified items were incorporated into the conceptual model to guide the research analysis and understanding of pilgrims' opinions and experiences. Accordingly, Table 4 illustrates the opinions of travellers regarding the various types of needs during the pre, in-transit, and post phases of pilgrimage.

The conceptual framework is given in Figure 2

Table 4. Factor influencing travellers opinion

Quest to Discover Inner Self Religious value Self-actualization Search for inspiration Mental health	**Travellers' Opinion**
Modern Facilities Easy access Comfortable stay Connectivity Hassle-free performance of rituals	**Travellers' Opinion**
Reinforcement in religious beliefs Reflection Community engagement Continued spiritual growth	**Travellers' Opinion**

Source: Authors' Compilation 2023

Figure 2. Conceptual model
Source: The Authors (2023)

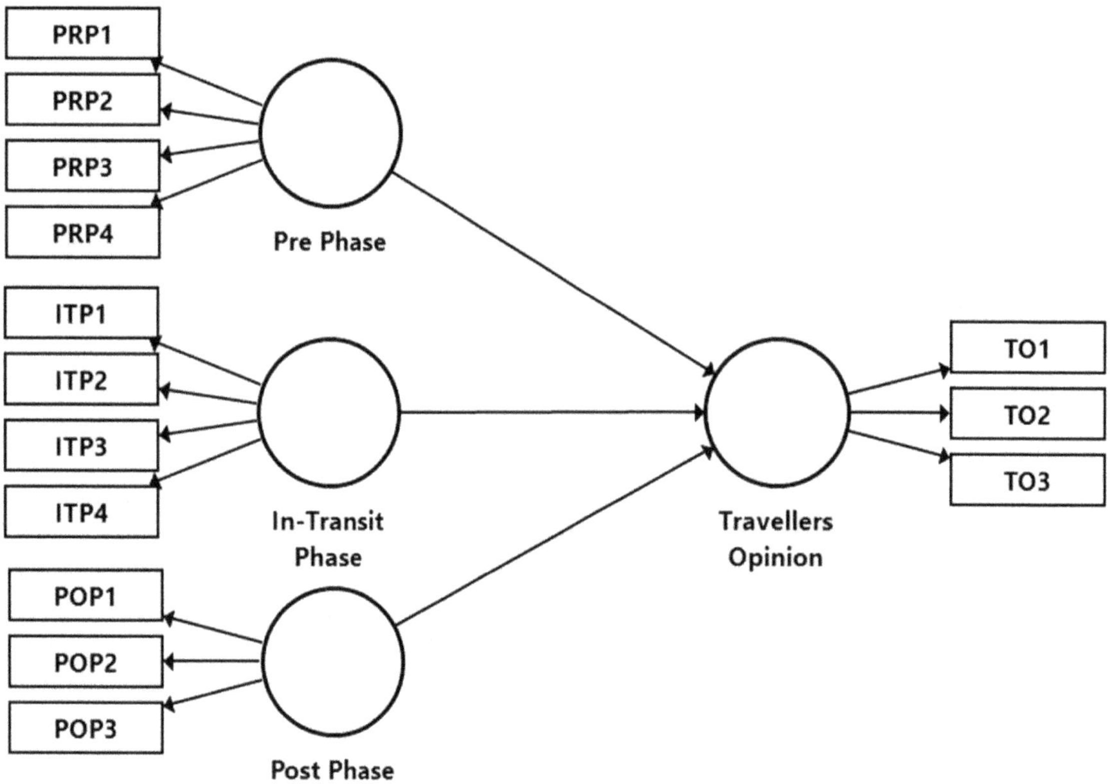

Hence, it can be affirmed that the entirety of a traveller's journey can be categorized into three principal stages. The perception of the traveller can be influenced by the encounters within these stages. Hence, following hypotheses can be drawn upon reviewing the existing literature:

H_1. *Quest to discover inner self (Pre-phase) has a significant impact on travellers' opinion.*
H_2. *Modern facilities availed during pilgrimage (In-Transit Phase) has a significant impact on travellers' opinion.*
H_3. *Reinforcement (Post Phase) in religious beliefs has a significant impact on travellers' opinion.*

3. RESEARCH METHODOLOGY

The research employed the SmartPLS (Partial Least Squares) to investigate the impact of diverse independent variables, obtained from the Delphi method, on the viewpoints of travellers. The data was obtained by means of an online survey that was developed using Google Forms. The survey was extensively distributed across various social media platforms to guarantee a comprehensive and diverse sample. The survey instrument comprised of two distinct sections.

The participants consisted of individuals who had visited to a minimum of two pilgrimage destinations within the past one year. The survey was disseminated through online channels, stipulating a prerequisite of visiting at least two such sites. Out of a total of 286 received responses, only 232 were deemed accurate following a screening process. The initial segment of the study collected demographic information from the participants, encompassing variables such as age, gender, income, and education, among other factors. The subsequent segment comprised of various components pertaining to distinct variables, specifically Pre-Phase (PRP-I to PRP-IV), In-Transit Phase (ITP-I to ITP-IV), Post-Phase (POP-I to POP-III), and Travellers' Opinion (TO-I to TO-III).

The participants employed a Likert scale consisting of five points to express their degree of concurrence, which spanned from 1 (strongly disagree) to 5 (strongly agree). The data that was gathered underwent a thorough and exhaustive analysis.

The measurement model was subjected to evaluation using SmartPLS to establish the validity and reliability of the constructs. The study proceeded to examine the connections between the independent variables and Traveller's Opinion to verify the proposed hypotheses.

4. RESULT

4.1 Measurement Model

The empirical results demonstrate compelling support for the reliability and validity of the constructs employed in the investigation. The study findings indicate that the factor loadings, Cronbach's alpha (α), Composite Reliability (CR), and rho_A all surpassed the recommended threshold of 0.70, as proposed by Hair et al., (2017). This suggests the reliability of the constructs. Furthermore, in accordance with the Fornell & Larcker, (1981) guidelines, the Average Variance Extracted (AVE) values for all constructs exceeded the recommended threshold of 0.5, thereby affirming adequate convergent validity (*See Table 5*).

To evaluate the discriminant validity, which pertains to the differentiation between constructs, all inter-dimensional correlations were observed to be below the square root of the AVE, as presented in Table 5. This provides evidence for the differentiation of the constructs. In addition, in order to address the problem of collinearity, the Variance Inflation Factors (VIFs) were examined, and it was observed that they were less than 5 for all the variables (*See Table 6*).

Table 5. Reliability and validity

		Loading	Reliability				Validity			
Constructs	Item	Outer Loading	α	rho_A	CR	AVE	PPP	ITP	POP	TO
Pre-Phase	PRP1	0.892	0.8	0.84	0.9	0.775	**0.88**			
	PRP2	0.886								
	PRP3	0.887								
	PRP4	0.856								
In-Transit- Phase	ITP1	0.922	0.8	0.81	0.9	0.774	0.83	**0.88**		
	ITP2	0.874								
	ITP3	0.887								
	ITP4	0.832								
Post Phase	POP1	0.864	0.8	0.85	0.9	0.804	0.72	0.69	**0.9**	
	POP2	0.904								
	POP3	0.921								
Travellers Opinion	TO1	0.873	0.9	0.87	0.9	0.812	0.88	0.87	0.8	**0.9**
	TO2	0.912								
	TO3	0.917								

Source: Authors (2023)

Table 6. VIF

	VIF		VIF
ITP1	2.64	PRP1	2.752
ITP2	2.587	PRP2	2.683
ITP3	2.765	PRP3	2.658
ITP4	2.102	PRP4	2.346
POP1	2.088	TO1	2.139
POP2	2.603	TO2	2.791
POP3	2.86	TO3	2.889

Source: The Authors (2023)

4.2 Structural Model

An analysis was performed to evaluate the structural model by scrutinising the statistically significant associations between the exogenous and endogenous variables. The researchers conducted a bootstrapping analysis using 5000 subsamples to establish the statistical significance of the path coefficients, as recommended by Hair et al., (2014). The findings of the assessment of the structural model are exhibited in Figure 3, which demonstrates that all three hypotheses exhibited noteworthy correlations. Furthermore, the comprehensive evaluation of the model's appropriateness, ascertained via bootstrapping, produced statistically significant outcomes for these propositions, thereby substantiating their soundness.

Figure 3. Structural model
Source: The Authors (2023)

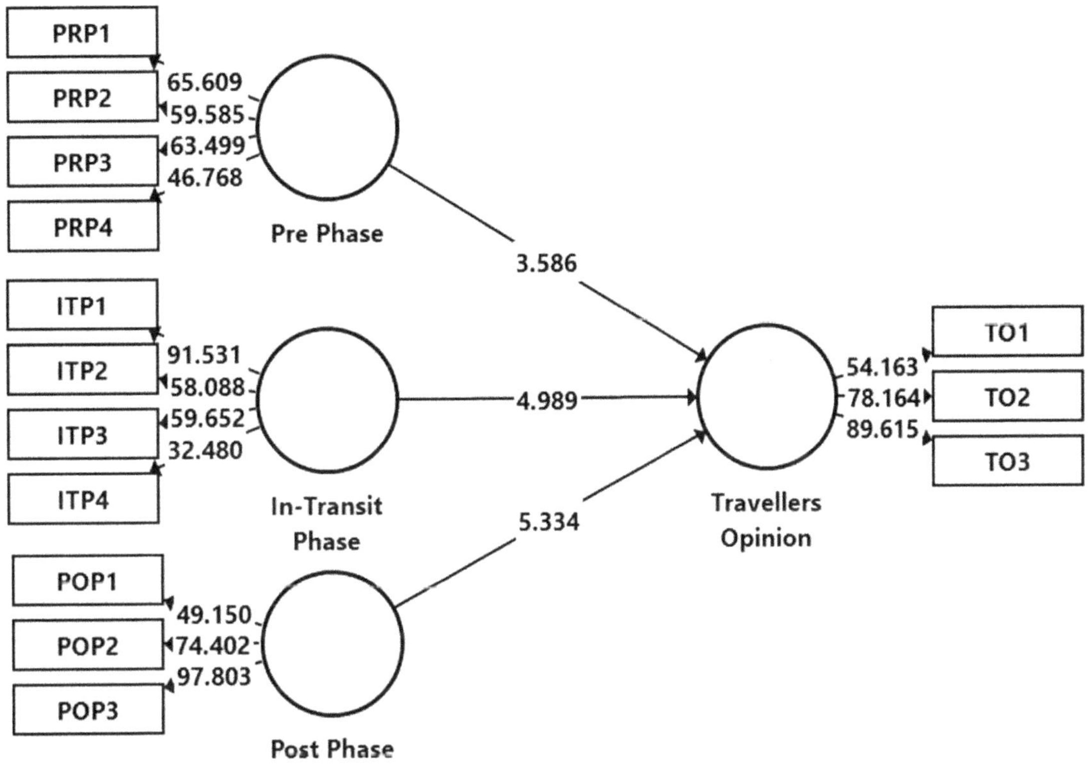

The findings of the structural model analysis indicate that all three phases of the pilgrimage journey, namely the pre-phase ($\beta = 0.307, p<0.05$), in-transit phase ($\beta = 0.378, p<0.05$), and post-phase ($\beta = 0.322, p<0.05$), have a significant impact on travellers' opinions. Hence, supporting all the framed hypotheses. This suggests that each phase plays a crucial role in shaping the overall perception of pilgrimage tourism. However, it is noteworthy that the in-transit phase, where travellers actually experience the pilgrimage activities and visit sacred sites, emerges as the most influential phase ($\beta = 0.378, p<0.05$). This is followed by the post-phase ($\beta = 0.322, p<0.05$), indicating that the reflection, community engagement, and continued spiritual growth after the pilgrimage journey also hold significant importance. These findings highlight the significance of both the in-transit and post phases in shaping travellers' opinions and emphasize the transformative nature of the pilgrimage experience. The same is provided in Table 7.

Table 7. Hypothesis testing

Hypo.	Relation	Original Sample (O)	Sample Mean (M)	Standard Deviation (STDEV)	T Statistics (|O/STDEV|)	P Values	Result
H1	PRP -> TO	0.307	0.292	0.085	3.586	0	Supported
H2	ITP -> TO	0.378	0.379	0.076	4.989	0	Supported
H3	POP -> TO	0.322	0.336	0.06	5.334	0	Supported

Source: The Authors (2023)

Table 8. R^2

	R Square	R Square Adjusted
Travellers Opinion	0.866	0.865

Source: The Authors (2023)

The analysis of the structural model revealed a high coefficient of determination (R-squared) value exceeding 0.80, indicating that over 80% of the variability in travellers' opinions can be explained by the three factors encompassing the pre-phase, in-transit phase, and post-phase of pilgrimage. This substantial R-squared value signifies a strong dependence of travellers' opinions on these three influential factors. It suggests that the combined impact of the pre, in-transit, and post-phases significantly shapes and influences the overall perception and experience of pilgrims, highlighting the importance of these factors in understanding and interpreting travellers' opinions regarding pilgrimage tourism.

5. CONCLUSION

The findings of our study highlight the importance of all three phases—pre, in-transit, and post—in shaping travellers' opinions in pilgrimage tourism. The in-transit phase emerged as a significant factor, as it strongly influences travellers' overall perception of the pilgrimage journey. A positive in-transit experience can lead to repeat customers and positive word-of-mouth recommendations, while a negative experience may deter future visits. The post-visit phase also holds considerable importance, providing an opportunity for reflection, community engagement, and continued spiritual growth. Positive experiences during this phase contribute to favourable opinions, while negative experiences may diminish travellers' perceptions. Furthermore, the pre-visit factors play an equally vital role, as they set the stage for the entire pilgrimage journey. A positive pre-visit experience, characterized by efficient planning, accessible information, and suitable facilities, contributes to favourable opinions from the outset. Conversely, challenges in the pre-visit phase may negatively impact travellers' opinions.

Overall, our study emphasizes the multifaceted nature of pilgrimage tourism and highlights the interplay between the three phases. It underscores the need for destination managers and stakeholders to focus on enhancing the in-transit experience, facilitating a positive post-visit phase, and addressing pre-visit factors to shape favourable opinions among pilgrimage tourists. By understanding the significance of each phase and addressing the associated factors, destinations can provide meaningful and transformative experiences, leading to increased visitor satisfaction, loyalty, and positive destination perceptions.

The result from this study hence shows that the religious value, personal and spiritual growth. and the need for social interaction have significant impact on travellers' opinions. At the same time, todays' travellers are highly inquisitive, knowledgeable, well equipped with information and technology and give strong opinion about products and services offered to them. Modern day pilgrims or tourists have access to resources and hence do not like to settle for mediocre services and voice strong opinions about them.

6. NOVELTY

Unlike many previous studies that might have focused on isolated aspects of pilgrimage tourism, our research adopts a holistic and phased approach, meticulously examining the interplay of the pre, in-transit, and post phases. This comprehensive lens provides a more nuanced understanding of how travellers' opinions evolve throughout the entire pilgrimage journey. The study brings to the forefront the post-visit phase's significance, highlighting its potential as a realm of reflection, community engagement, and ongoing spiritual development. This distinct focus enriches the understanding of how post-travel experiences shape opinions and contribute to long-term visitor satisfaction. As far as the author's comprehension goes, there has been no prior research that has underscored the significance of all three phases, particularly within the context of pilgrimage tourism.

7. THEORETICAL AND MANAGERIAL IMPLICATION

This study on the opinions of travellers in pilgrimage tourism has several theoretical and managerial implications. Theoretically, the findings contribute to the existing body of knowledge by providing insights into the significant factors that shape travellers' opinions in the three phases of pilgrimage. The identification of religious value, self-actualization, search for inspiration, mental health, easy access, comfortable stay, connectivity, hassle-free performance of rituals, reflection, community engagement, and continued spiritual growth as influential components enhances our understanding of the pilgrimage tourism experience. This study also validates the importance of pre, in-transit, and post phases in influencing travellers' opinions and emphasizes their interdependence.

From a managerial perspective, the findings offer practical implications for destination managers and stakeholders in the pilgrimage tourism industry. Recognizing the significance of the in-transit phase, managers can focus on enhancing the quality of experiences during this phase by providing well-maintained religious sites, efficient logistical support, and engaging activities that facilitate personal transformation. Attention should also be given to the post-visit phase, where opportunities for reflection, community engagement, and spiritual growth can be created through organized events, discussion platforms, and community involvement initiatives. Additionally, addressing the pre-visit factors, such as providing accurate and accessible information, facilitating ease of planning, and ensuring suitable accommodation options, can enhance travellers' initial perception and set a positive tone for their overall pilgrimage experience.

By understanding the travellers' opinions, the stakeholders will be benefited while participating in pilgrimage tourism operations. Governments and other agencies need to understand and recognize the economic potential of pilgrimage tourism. Proper product augmentation and promotion of holy places to the tourists need to be carried out. Best of the accessibility means to reach the pilgrimage sites, hotels, and resorts at pilgrimage sites for comfortable stay, uninterrupted communication means to relate to rest of the world during pilgrimage and a check and control over traffic and the behaviour of the clergy men are some of the desired requirements which need to be provided to boost pilgrimage tourism. Reputed broadcasters and documentary makers must be taken onboard to make online visits to the shrines, a good experience to the pilgrims.

Even the government appointed agencies like various shrine boards, waquf boards and committees need to be tourist oriented and work for their better pilgrimage experience.

8. LIMITATION AND FUTURE STUDY

While this study provides valuable insights, it is important to acknowledge its limitations and identify avenues for future research. Firstly, the study relied on self-reported data collected through an online questionnaire, which may be subject to respondent bias and limited generalizability. Future research could incorporate qualitative methods, such as interviews or focus groups, to gain a deeper understanding of travellers' opinions and experiences in pilgrimage tourism.

Secondly, the study primarily examined the impact of the three phases on travellers' opinions. Further research could explore additional factors, such as cultural influences, social interactions, and technological advancements, that may shape travellers' opinions in pilgrimage tourism.

Lastly, the study did not explore the long-term effects of travellers' opinions on destination loyalty. Future research could investigate the relationship between travellers' opinions, repeat visitation, and destination advocacy to provide insights into the long-term impacts of pilgrimage tourism experiences.

Overall, this study lays the foundation for future research on the opinions of travellers in pilgrimage tourism. By addressing the identified limitations and exploring new avenues, researchers can deepen our understanding of the factors that influence travellers' opinions, thereby informing effective strategies for destination management and enhancing the overall pilgrimage tourism experience.

REFERENCES

Barber, P. (2001). Mito, religione e conoscenza: la mappa del mondo medievale. In C. Pirovano (Ed.), Segni e sogni della terra. Il disegno del mondo dal mito di Atlante alla geografia delle reti (pp. 48–79). De Agostini.

Burkart, A. J., & Medlik, S. (1974). *Tourism: past, present and future* (2nd ed.). William Heinemann Ltd.

Cheer, J. M. (2023). Host Communities and Pilgrimage Tourism. In R. N. Progano, J. M. Cheer, & X. M. Santos (Eds.), *Perspectives on Asian Tourism* (pp. 229–233). Springer.

Cheer, J. M., Belhassen, Y., & Kujawa, J. (2017). The search for spirituality in tourism: Toward a conceptual framework for spiritual tourism. *Tourism Management Perspectives*, *24*, 252–256. doi:10.1016/j.tmp.2017.07.018

Cohen, E. (1992). Pilgrimage and Tourism: Convergence and Divergence. In A. Morinis (Ed.), *Sacred Journeys: The Anthropology of Pilgrimage* (pp. 47–61). Greenwood Press.

Coleman, Simon, & Eade, J. (Eds.). (2004). *Reframing Pilgrimage Cultures in Motion* (1st ed.). Routledge.

Coleman, S. M. (2023). Pilgrimage Religion. In *Britannica*. Encyclopedia Britannica.

Collins-Kreiner, N. (2020). Pilgrimage tourism-past, present and future rejuvenation: A perspective article. *Tourism Review*, *75*(1), 145–148. doi:10.1108/TR-04-2019-0130

Dar, H. (2020). Hindu Religious Motivations in Kashmir Valley. *International Journal of Religious Tourism and Pilgrimage*, *8*(3), 1–14.

Davie, G. (2000). *Religion in Modern Europe: A Memory Mutates*. Oxford University Press.

De Ascaniis, S., & Cantoni, L. (2017). Online visit opinions about attractions of the religious heritage: An argumentative approach. *CHURCH, COMMUNICATION AND CULTURE*, *2*(2), 179–202. doi:10.1080/23753234.2017.1350585

Eickelman, D. F., & Piscatori, J. P. (1990). *Muslim Travellers: Pilgrimage, Migration, and the Religious Imagination* (D. F. Eickelman & J. P. Piscatori, Eds.). Routledge.

Fahey, F. (2002). Pilgrims or Tourists? *The Furrow*, *53*(4), 213–218.

Faverio, M., Nortay, J., Diamant, J., & Smith, G. A. (2023). *Online Religious Services Appeal to Many Americans, but Going in Person Remains More Popular.*

Fleetwood, S. (1995). *Collection of Tourism Expenditure Statistics.*

Fornell, C., & Larcker, D. (1981). Evaluating Structural Equation Models with Unobservable Variables and Measurement Error. *JMR, Journal of Marketing Research*, *18*(1), 39–50. doi:10.1177/002224378101800104

Gurung, R. (2016). *Marketing Religious Tourism Destination: Case study: The birth place of Gautama Buddha*. Centria University of Applied Sciences.

Hair, J., Hollingsworth, C. L., Randolph, A. B., & Chong, A. (2017). An updated and expanded assessment of PLS-SEM in information systems research. *Industrial Management & Data Systems*, *117*(3), 442–458. doi:10.1108/IMDS-04-2016-0130

Hair, J. F., Sarstedt, M., Hopkins, L., & Kuppelwieser, V. G. (2014). Partial least squares structural equation modeling (PLS-SEM): An emerging tool in business research. *European Business Review*, *26*(2), 106–121. doi:10.1108/EBR-10-2013-0128

Hall, C. M. (2006). Buddhism, tourism and the middle way. In *Tourism, religion and spiritual journeys* (pp. 172–185). Routledge.

Hunziker, W., & Krapf, K. (1942). Grundriss der allgemeinen Fremdenverkehrslehre. *Polygraph.*

James, P., & Mandaville, P. (2010). Globalization and Culture, Vol. 2: Globalizing Religions. Sage.

Jyotsna, J. H., & Prakash Sai, L. (2022). Modelling pilgrim-tourist experience in Hindu religious destinations: An Interactive Qualitative Analysis. *Journal of Tourism and Cultural Change*, 1–19. doi:10.1080/14766825.2022.2095914

Kliot, N., & Collins-Kreiner, N. (2000). Pilgrimage tourism in the Holy Land: The behavioural characteristics of Christian pilgrims. *GeoJournal*, *50*(1), 55–67. https://link.springer.com/article/10.1023/A:1007154929681. doi:10.1023/A:1007154929681

Leiper, N. (1979). The framework of tourism: Towards a definition of tourism, tourist, and the tourist industry. *Annals of Tourism Research*, *6*(4), 390–407. doi:10.1016/0160-7383(79)90003-3

Lopez, L. (2014). How Long Does the Pilgrimage Tourism Experience to Santiago de Compostela Last? *International Journal of Religious Tourism and Pilgrimage*, *1*(1), 1–14. doi:10.21427/D7C133

Margry, P. J. (2008). Secular Pilgrimage: A contradiction in Terms? In P. J. Margry (Ed.), *Shrines and Pilgrimages in the Modern Word.* Amsterdam University Press.

Maslow, A. H. (1943). A theory of human motivation. *Psychological Review, 50*(4), 370–396. doi:10.1037/h0054346

Metti, M. S. (2011). *Jerusalem - the most powerful brand in history.*

Nash, D., Akeroyd, A. V., Bodine, J. J., Cohen, E., Dann, G., Hermans, D., Jafari, J., Kemper, R. V., LaFlamme, A. G., Manning, F., Noronha, R., Pi-Sunyer, O., Smith, V. L., Stoffle, R. W., Thurot, J. M., Watson-Gegeo, K. A., & Wilson, D. (1981). Tourism as an anthropological subject. *Current Anthropology, 22*(5), 461–468. doi:10.1086/202722

Olsen, D., & Timothy, D. (2006). Tourism and religious journeys. In *Tourism, religion and spiritual journeys* (1st ed., pp. 1–21). Routledge.

Pearce, P. L. (1982). *The social psychology of tourist behavior: International Series in Experimental Social Psychology* (1st ed.). Pergamon Press.

Pearce, P. L. (1985). A Systematic Comparison of Travel-Related Roles. *Human Relations, 38*(11), 1001–1011. doi:10.1177/001872678503801101

Pew Research Center Forum on Religion & Public Life. (2012). *The Global Religious Landscape.*

Plate, S. B. (2009). The Varieties of Contemporary Pilgrimage. *Cross Currents, 59*(3), 260–267. doi:10.1353/cro.2009.a782452

Rodriguez, J. (2017). Cleaning Up Bodhgaya: Conflicts over Development and the Worlding of Buddhism. *City & Society, 29*(1), 59–81. doi:10.1111/ciso.12112

Schramm, K. (2004). Coming Home to the Motherland: Pilgrimage Tourism in Ghana. In S. Coleman & J. Eade (Eds.), Reframing Pilgrimage: Cultures in Motion (pp. 133–149). Routledge.

Shao Zhen, T., Hassan, S. H., & Mohamad, O. (2019). Buddhists' Motivations for Revisiting Religious Temples in Malaysia. *European Journal of Business and Management Research, 4*(4). doi:10.24018/ejbmr.2019.4.4.81

Smith, V. L. (1992). Introduction: The quest in Guest. *Annals of Tourism Research, 19*(1), 1–17. doi:10.1016/0160-7383(92)90103-V

Spode, H. (1998). Geschichte der Tourismuswissenschaft. In G. Haedrich (Ed.), Tourismus-management: Tourismus-marketing Und Fremdenverkehrsplanung (German). de Druyter. doi:10.1515/9783110815993.911

Stibich, M. (2022). *What Is Religion? The Psychology of Why People Believe.* Verywell Mind.

Turnbull, C. (1992). Postscript: Anthropology as Pilgrimage: Anthropologist as Pilgrim. In A. Morinis (Ed.), *Sacred Journeys: The Anthropology of Pilgrimage* (pp. 257–274). Greenwood Press.

Tylor, E. B. (1871). *Primitive Culture: Researches Into the Development of Mythology, Philosophy, Religion, Art, and Custom* (Vol. 1). John Murray.

UNWTO. (2008). *GLOSSARY OF TOURISM TERMS.* UN, WTO.

Walter, T. (1993). *Pilgrimage in Popular Culture* (I. Reader & T. Walter, Eds.; 1st ed.). Palgrave Macmillan.

Yu, X., Kim, N., Chen, C.-C., & Schwartz, Z. (2012). ARE YOU A TOURIST? TOURISM DEFINITION FROM THE TOURIST PERSPECTIVE. *Tourism Analysis*, *17*(4), 445–457. doi:10.3727/108354212X13473157390687

Chapter 2
Mapping Pilgrimage Tourism Research Through Systematic Literature Review and Bibliometric Analysis

Aditya Sharma
Mizoram University, India

Pankaj Kumar
https://orcid.org/0000-0002-1479-9787
Mizoram University, India

ABSTRACT

Pilgrimage, or religious, tourism is regarded to be one of the primaeval forms of tourism. Also, an extensive amount of research on pilgrimage and religious tourism reveals the vitality of the field of study. The chapter aims to conceptualise the notion of pilgrimage tourism and prepare the knowledge map of the pilgrimage tourism literature. The chapter uses the Scopus database and applies a systematic literature review as well as a bibliometric approach to uncover the most prolific studies, authors, organisations, countries, and themes of pilgrimage tourism research worldwide. Also, the growth and development of knowledge in the field has been demonstrated. For the analysis and data visualisation the Biblioshiny package of R and visualization of similarities (VOSviewer) software is used. The findings uncover that the United Kingdom, Spain, and China are the most prolific countries; the studies are centered around the themes of heritage tourism, sustainable development, and coronavirus in the pilgrimage tourism research domain from the last few years.

INTRODUCTION

The world is home to various religions and each religion has its own history, attractions and landmarks (Koshim et al., 2020). Pilgrimage tourism or Religious tourism is the driver for the movement of people

DOI: 10.4018/978-1-6684-4817-5.ch002

worldwide (Nkwanyana & Njiyela, 2021). It is one of the earliest tourism forms (Collins-Kreiner, 2019; Nkwanyana & Njiyela, 2021) as travelling to pilgrimage and sacred sites has been practised by people for thousands of years (Collins-Kreiner, 2019; Olsen & Esplin, 2020) which continues till present days. Pilgrimage and tourism coexist with one another as the roots of tourism lies in the antiquated practice of pilgrimage. Therefore while studying tourism, the interconnected nature of tourism and pilgrimage cannot be disregarded (Singh, 2004; Collins-Kreiner, 2019). The term pilgrimage denotes "a journey towards ethical or religious implication" (Vidhya & Selvam, 2019, p. 1254). Pilgrimage is interwoven with religion and is the "oldest non-economic forms of migration"(Judák et al., 2022, p. 615). Pilgrimage sites are the locations pinned by gods, they act as junctions for the union of gods and humans (Judák et al., 2022). Stretching from a few days to weeks, pilgrimage journeys are characterised by a sense of sacrifice and effort (Judák et al., 2022). Religious tourism has developed as a result of the practice of pilgrimage, with religious devotion continuing to flourish the sector (Shinde, 2012) and strengthening the economies of several countries around the world (Abdulhanaa, 2021). Also, a nation's economy and spiritual well-being are both enhanced by religious tourism (Chistyakov, 2020). Religious sites, pilgrim centres and shrines which become significant tourist attractions are a part of every religion worldwide (Prasetio et al., 2022). Pilgrimage not only makes an essential element of religious traditions but also contributes to the spiritual development of people (Shinde & Olsen, 2023), leading to purification, penitence and spiritual renewal (Judák et al., 2022; Raj, 2012). Religious as well as ethnic and cultural tourism are strongly related (Chistyakov, 2020). Due to the intercultural and inter-religious conversation that occurs when a person travels to another location, the exchange of spiritual values is inevitable (Chistyakova, 2020). Tourists travel to distinct nations to experience new, unexpected, and novel things that are distinct from their everyday routines (Chistyakov, 2020; Smørvik, 2020), and the surge for experiencing religious tourism, in particular, is growing (Smørvik, 2020). As posited by projections, religious tourism is poised to play a more prevalent role in international tourism (Griffin & Raj, 2017). According to the Global Religion Landscape, 84% of the world's 6.9 billion inhabitants, or more than eight out of ten individuals, identified themselves as religious in 2010 (The Pew Forum on Religion & Public Life, 2012) the statistics portray the prodigious potentiality of this segment as along with religious people segment the non-religious people too frequently explore religious sites for a variety of compelling reasons. Moreover, besides religious motives, secular desires also entice people for visiting religious sites (Nkwanyana & Njiyela, 2021) which increases the popularity of religious-themed tourism (Wang, 2022).

BACKGROUND

Understanding Pilgrimage Tourism Literature

Religious tourism has been defined by scholars in different ways and perspectives, but there exists no single unified definition of the concept (Tozser, 2020). The FICCI in its report defines religious tourism as "travel with the core motive of experiencing religious forms, or the products they induce, like art, culture, traditions and architecture" (FICCI, 2012, p.2). Similarly, it has also been a tedious task for the researchers to categorise "tourists" and "pilgrims", due to the overlapping characteristics of these groups (Wang et al., 2020). Though, a close and integrated relationship exists between pilgrimage tourism, religious tourism and spiritual tourism but the forms also carry elementary dissimilarities which can be discerned based on the tourists' intentions (Hermawan et al., 2019). The literature entails numerous studies which have attempted to examine the motivation, experiences and challenges concerning pilgrimages and religious sites such as the motivation of Muslim pilgrims for visiting religious destinations (Raj, 2012; Almuhrzi & Alsawafi, 2017). Studies pointed out that worshipping, learning (Shinde & Olsen, 2023), religious motivation (AlMakhammad, 2020), performing rituals (Almuhrzi & Alsawafi, 2017; Olsen & Esplin, 2020), curiosity and other reasons are the drivers for motivating people to take part in pilgrimage (Olsen & Esplin, 2020). Also, Nyupane et al. (2015) studied the motivation of people of different faiths to visit other faiths' sacred sites. On the other hand, access to religious sites is hindered globally by religious privacy (Al-Makhadmah, 2020).

Religious tourism propels positive as well as negative economic, socio-cultural and environmental implications on the destinations (Olsen & Esplin, 2020). Therefore sustainability of religious tourism has been studied by various researchers, as per Narayanan (2013) religion and sustainable development mutually aims to accomplish "humanitarian welfare" and "ecological preservation", therefore they both must coexist together (p. 138). Shinde (2012) pointed out the essentiality of a sound religious tourism policy for the site's sustainability and mitigating its negative effects. Narayanan (2013) emphasised the significance of religion for sustainable development and suggests three ways in which religion can contribute to sustainable development namely through its values, through its potential for action in the areas of ecology, society, and politics, and through its ability to foster self-development. Importantly, academicians are sceptical about the sustainability of the modern pilgrimage (Shinde & Olsen, 2023). Few studies have also attempted to explore attachment. For instance, Majumdar and Majumdar (2004) proposed the typology for place attachment by classifying the attachment into two forms, namely secular and sacred. Patwardhan et al. (2020) evaluated the destination loyalty of visitors through place attachment, emotional solidarity, and perceived safety level concerning religious sites. Sestino et al. (2023) investigated the fervid attachment of native people to religious ceremonies. Latently studies with respect to coronavirus pandemic can be traced in the literature such as Prasetio et al. (2022) examined the impact of the coronavirus pandemic on religious tourism in the context of Iraq.

Though, the association between religion and tourism continues for a long time (Collins-Kreiner & Wall, 2015). Despite this, religious tourism has been losing popularity over the last decade due to the rise of other forms of travel (Collins-Kreiner, 2020). Similarly, Griffin and Raj (2017) note a paucity of literature on religious tourism globally. Studies have used the bibliometric approach in the context of religious tourism studies, namely Martinho and Nunes (2020) explored and highlighted the pilgrim routes of Camino de Santiago through a literature review and bibliometric analysis using the Web of Science database. Durán-Sánchez et al. (2018) performed a bibliometric analysis on religious tourism using the Web of Science database from 1968 to 2017. The analysis unveiled the most productive articles, authors, magazines, collaboration index and thematic areas of religious tourism research and recommended the application of other databases for upcoming studies. Therefore the present chapter attempts to address the gap and explore the pilgrimage tourism literature through a systematic literature review and bibliometric approach.

METHODOLOGY

Through a thorough assessment of the literature and bibliometric analysis, the chapter seeks to improve understanding and map the intellectual research structure in the field of pilgrimage tourism by unpacking the most significant publications, authors, journals, organisations, and nations that have contributed to the field of pilgrimage tourism research. The chapter makes an effort to consider the following research queries.

RQ: What is the current state and future outlook for research on pilgrimage tourism?

RQ: What are the most well-known publications and authors in the field of pilgrimage tourism?

RQ: What are the leading nations and organisations in the sector of pilgrimage tourism?

RQ: What are the prevailing topics and scholarly trends in pilgrimage tourism research?

Scopus database has been preferred for retrieving the data as it has broader coverage in comparison to the other databases as well as a majority of similar studies have used the same database such as (Bhowmik, 2021; Habibi et al., 2022; Hasana et al., 2022). For developing the Boolean search the keywords were chosen from the study of Durán-Sánchez et al. (2018) and were modified accordingly. The documents were searched in the "Article title, Abstract, Keywords" field on 23 March 2023. The keywords used in the Boolean search were "Pilgrim* Tour*" OR "Relig* Tour*" OR "Faith* Tour*". The search was not limited to specific period and was restricted to only open-access articles. The application of the Boolean search resulted in 994 documents. After limiting the language to English, 925 documents were available. After applying other inclusion and exclusion criteria documents were identified. Lastly, the manual check for the relevancy of studies was done which led to no elimination of any documents as all documents were found to be relevant to the study. The final number of documents used for the analysis are 155. The study uses "Scientific Procedures and Rationales for Systematic Literature Reviews" ("SPAR-4-SLR") i.e. Figure 1 for the systematic literature review. SPAR 4 SLR has been recommended by Paul et al. (2021) and is considered to be superior than the widely used PRISMA protocol (Paul et al., 2021). For the analysis and data visualisation, the Biblioshiny package of the R software (Aria & Cuccurullo, 2017) and VOSviewer which refers to "visualization of similarities" was utilised (Van Eck & Waltman, 2010).

RESULTS AND DISCUSSION

Trends and Patterns of Pilgrimage Tourism Research

Information about the 155 documents that were taken from the Scopus database and used for the bibliometric study is shown in Table 1. The number of journals which entailed these papers were 81. The calculated yearly growth rate is 10.5%. "Keyword plus" has been explained as "the total number of keywords that frequently appear in the title of the article" (Abbas et al., 2022, p. 5). The no of keywords plus was 313 which was twice the number of articles. 42 documents i.e. 27.09% of the total documents were written by a single author out of the 388 total authors. Figure 2 enumerates the annual scientific production for the period of 2001 to 2023. Only the publications up through May 2023 have been taken into account for the study. According to the trends, the first article entitled "The Tourist Product and its Evolution" authored by Kaczmarek et al. (2001) was released by Turyzm/Tourism. The articles shed light on the evolution of tourism products historically and discuss future tourism development. The scientific output was low from 2001 to 2011, with only one or no publication per year, for a whole ten years. From 2012 to 2016, the number of articles somewhat rose. With 28 publications each, 2021 and 2022 were the most productive years, which saw a large increase in studies. With nine studies published so far until May 2023, the trends in pilgrimage tourism research continue to indicate an increased tendency.

Figure 1. "SPAR 4 SLR" flowchart

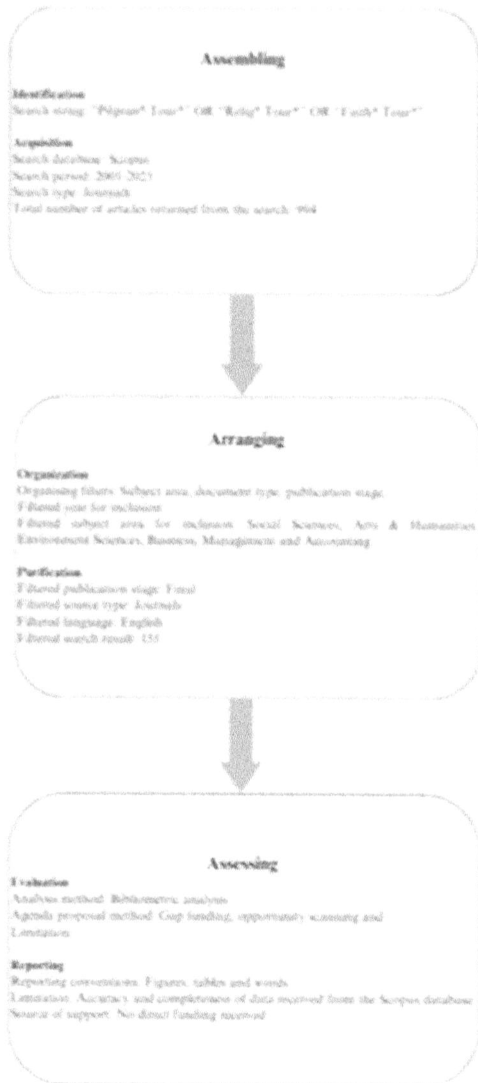

Most Productive Articles, Sources, Authors, Countries, and Organisations

The most influential articles as per the total citations have been depicted in Table 2. "A closer look at Santiago de Compostela's pilgrims through the lens of motivations" by Amaro et al. published in 2018 is the most cited document. The study investigated the modern pilgrim's drive for taking part in the Camino de Santiago (Amaro et al., 2018). "The complexities of religious tourism motivations: Sacred places, vows and visions" by Terzidou et al. published in 2018 is the second most cited article which focused on understanding the complexities of pilgrim travel motivations (Terzidou et al., 2018). "Motivation and personality traits for choosing religious tourism. A research on the case of Medjugorje" by Abbate & Di Nouvo published in 2013 mentions pilgrimage travel as the most conventional noneconomic reason

Table 1. Summary of the main information

Description	Results
Timespan	2001:2023
Sources (Journals, Books, etc)	81
Documents	155
Annual Growth Rate %	10.5
Document Average Age	3.79
Average citations per doc	9.026
References	8154
Keywords Plus (ID)	313
Author's Keywords (DE)	590
Authors	388
Authors of single-authored docs	42
Single-authored collaboration docs	44
Collaboration of Co-Authors per Doc	2.72
International co-authorships %	18.06

Source: Biblioshiny

Figure 2. Annual scientific production of articles

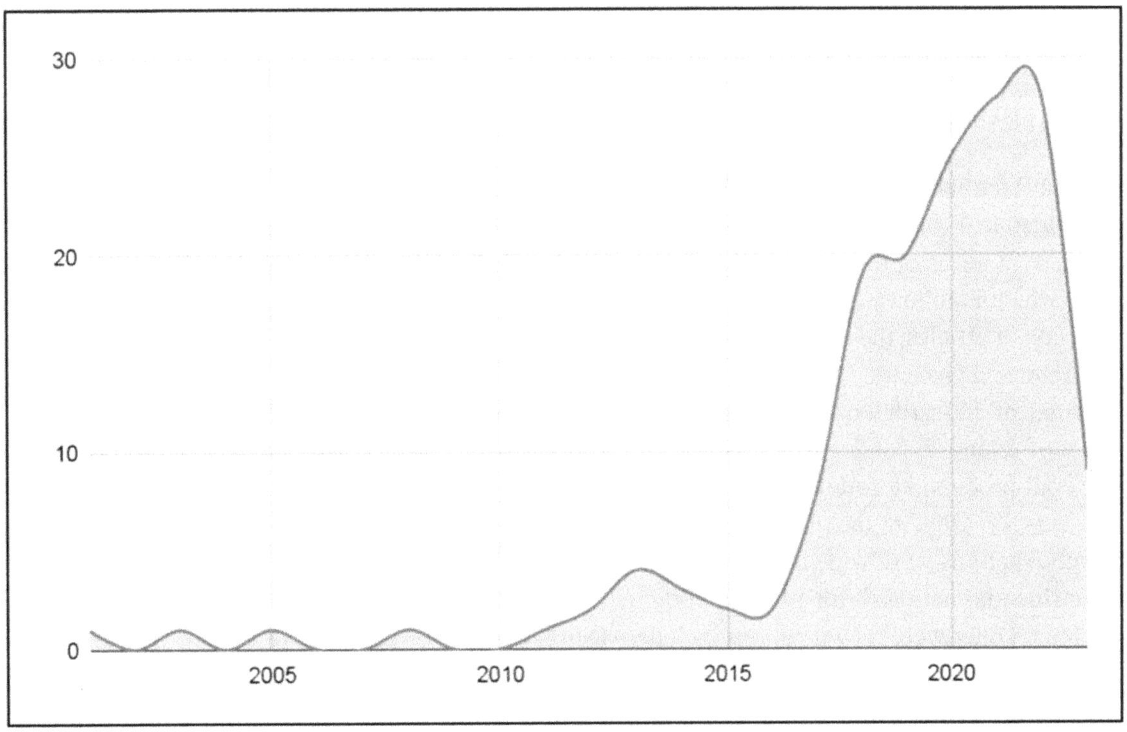

Table 2. Influential articles

Title	Author	Journal	Publication Year	TC	AAC
A closer look at Santiago de Compostela's pilgrims through the lens of motivations	Amaro et al.	*Tourism Management*	2018	83	20.7
The complexities of religious tourism motivations: Sacred places, vows and visions	Terzidou et al.	*Annals of Tourism Research*	2018	78	19.5
Motivation and personality traits for choosing religious tourism. A research on the case of Medjugorje	Abbate and Di Nouvo	*Current Issues in Tourism*	2013	59	6.5
The role of visual media in religious tourists' destination image, choice, and on-site experience: the case of Tinos, Greece	Terzidou et al.	*Journal of Travel & Tourism Marketing*	2018	49	12.2
Moving from Pilgrimage to "Dark" Tourism: Leveraging Tourism in Palestine	Isaac and Ashworth	*Tourism Culture & Communication*	2012	49	4.9
Exploring religious tourist experiences in Jerusalem: The intersection of Abrahamic religions	Albayrak et al.	*Tourism Management*	2018	48	12
Religiousness as tourist performances: A case study of Greek Orthodox pilgrimage	Terzidou et al.	*Annals of Tourism Research*	2017	47	9.4
Compassion as a neglected motivator for sustainable tourism	Weaver and Jin	*Journal of Sustainable Tourism*	2016	46	7.6
Religious Tourism and Pilgrimage: Bibliometric Overview	Durán-sánchez et al.	*Religions*	2018	44	11
Contesting the Commercialization and Sanctity of Religious Tourism in the Shaolin Monastery, China	Hung et al.	*International Journal of Tourism Research*	2017	43	8.6

Source: The Authors
TC: Total citations; AAC: Average Annual citations

for travel. The study analysed the motivation for religious travel in the context of the personality trait of the travellers (Abbate & Di Nouvo, 2013). The works are prominent in the domain of religious tourism literature and most cited literature revolves around examining motivation for visiting pilgrim sites or holy places.

The influential journals that are well-known for publishing in the literature on religious tourism are included in Table 3. The dataset comprised 81 journals in which the scientific production has been published. In terms of the total citations, the journal with the most citations is Annals of Tourism Research, which is followed by Tourism Management and Sustainability (Switzerland). In terms of total production of articles, the top three journals are Sustainability (Switzerland), Religions and GeoJournal of Tourism and Geosites.

A total of 155 articles were published by 388 authors from 55 countries. According to the total number of scientific publications produced, Poland takes the top spot with 13 documents, or 8.38% of the overall production, followed by the UK, Spain, and Indonesia, who each have 12 documents. 11 documents, or 7.09% of the total production, are from Italy. Figure 3 illustrates the ranking of the most influential nations worldwide in terms of total citations. With 246 citations, the United Kingdom is the most influential nation. With 144 and 100 citations in second and third place, respectively, are Spain and China. Other well-known nations include Australia, Italy, Portugal, and others. It can be noted that developed countries are dominating and contributing more to scientific production in the field of pilgrimage tourism research worldwide.

The total number of identified organizations in the data set was 300. The world's top organisations that have contributed to research on pilgrimage tourism are depicted in Figure 4. In terms of the total

Table 3. Influential journals

Journal	TC	TP	Publisher
Annals of Tourism Research	169	4	Elsevier
Tourism Management	147	3	Elsevier
Sustainability (Switzerland)	122	19	MDPI
Religions	95	17	Taylor & Francis
International Journal of Tourism Research	79	3	Wiley-Blackwell
GeoJournal of Tourism and Geosites	54	13	University of Oradea, Romania
Church, Communication and Culture	43	5	Taylor & Francis
Journal of Religion and Health	43	2	Springer
European Countryside	29	3	SCIENDO
International Journal of Environmental Research and Public Health	28	2	MDPI

Source: Author
TC: Total citations; TP: Total production

Figure 3. Most influential countries

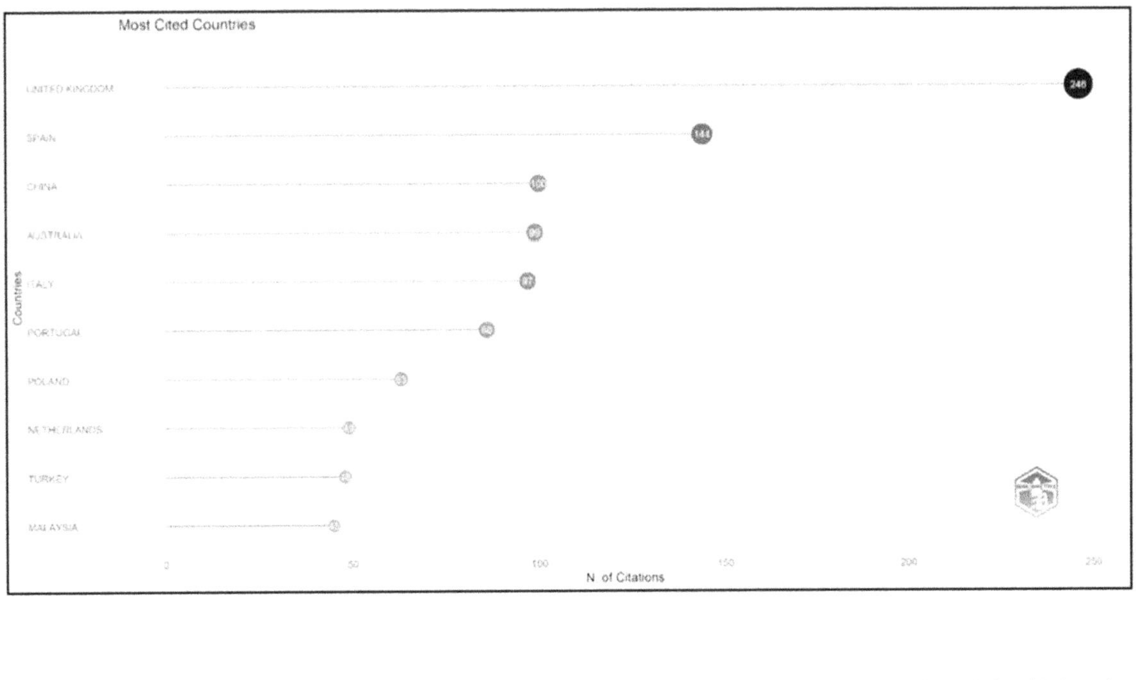

production of articles, King Faisal University, Saudi Arabia and Heriot-Watt University, United Kingdom earn the top spot with the scientific production of 12 and 7 articles respectively. The University of Venda, University of Cantabria, Sunway University, The Hong Kong Polytechnic University, and Alexandru Ioan Cuza University of IASI all take third place with 5 publications each.

From the data corpus of 151 authors, the top 10 most productive authors with their scientific production and citations have been depicted from 2011 to 2021 in Figure 5.

Figure 4. Most productive affiliations

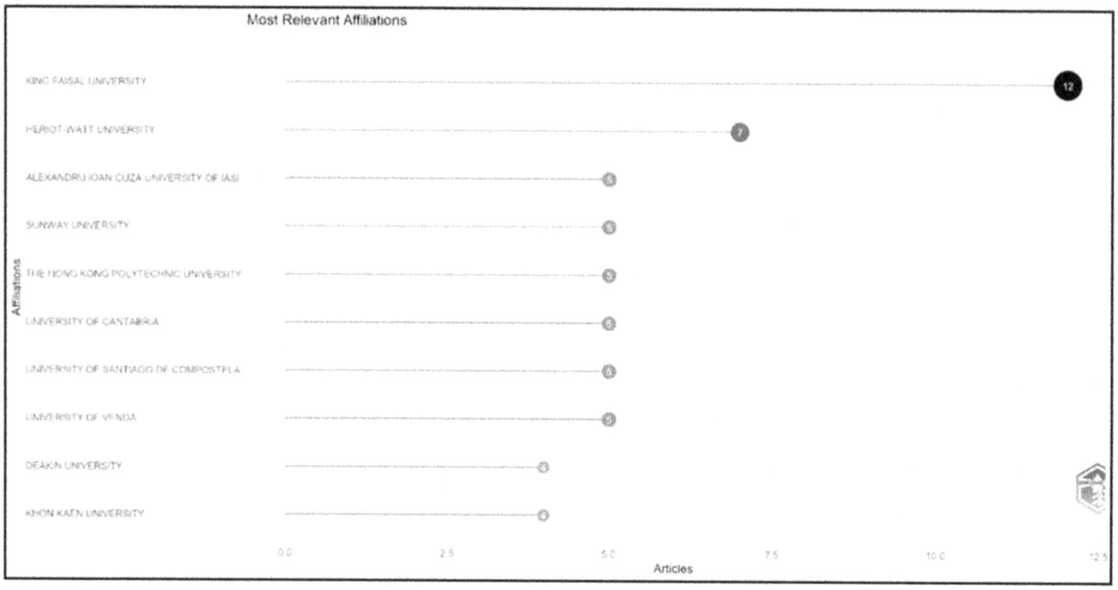

Figure 5. Most productive authors

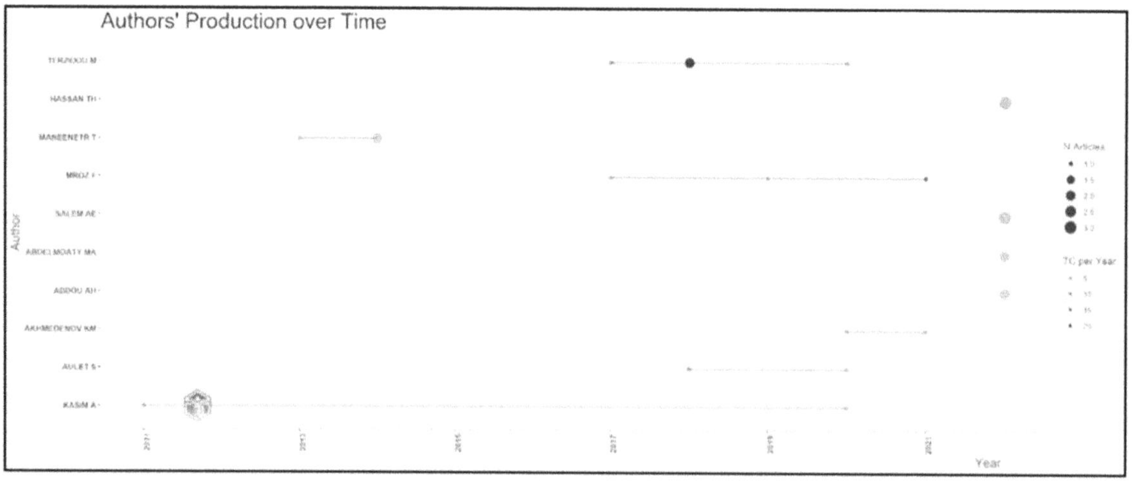

The treemap in Figure 6 denotes the keywords in Pilgrimage tourism literature. The larger the boxes appear, the more frequency in the literature. Words namely tourism (9%) and religion (8%) are the most popular terms followed by tourist destination (6%), sustainable development (5%), heritage tourism (5%) and tourism development (5%). It reveals that pilgrimage tourism literature has been prominent in the context of tourism development, destination, sustainable development, heritage etc.

The total keywords which were identified in the dataset were 851, after fixing the number of occurrences to a minimum of 4 times, the keywords resulted in 44. Figure 7 illustrates the keywords in

Figure 6. Tree map

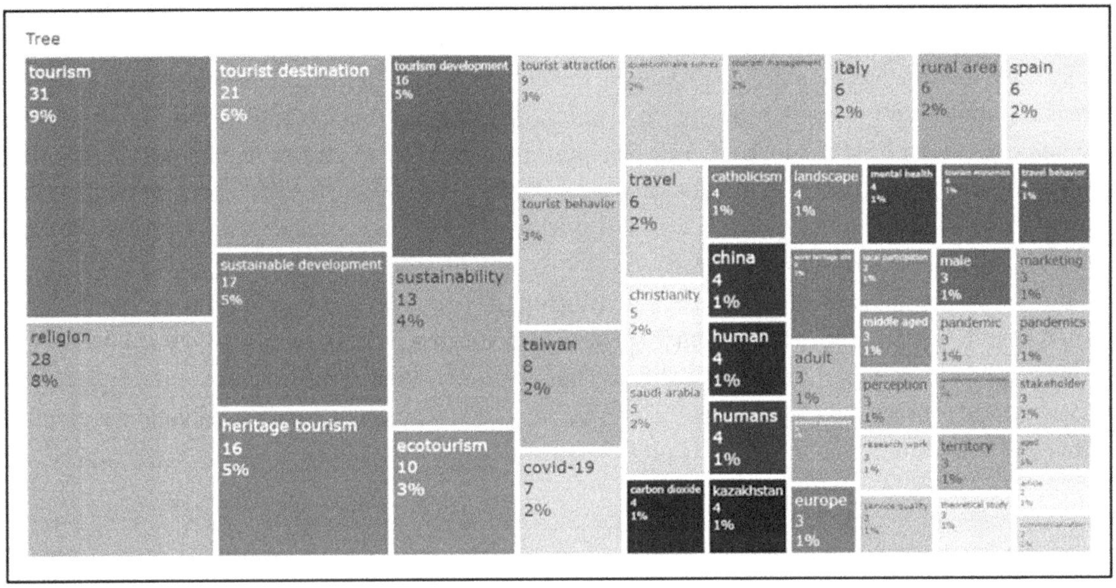

Figure 7. Keyword co-occurrence

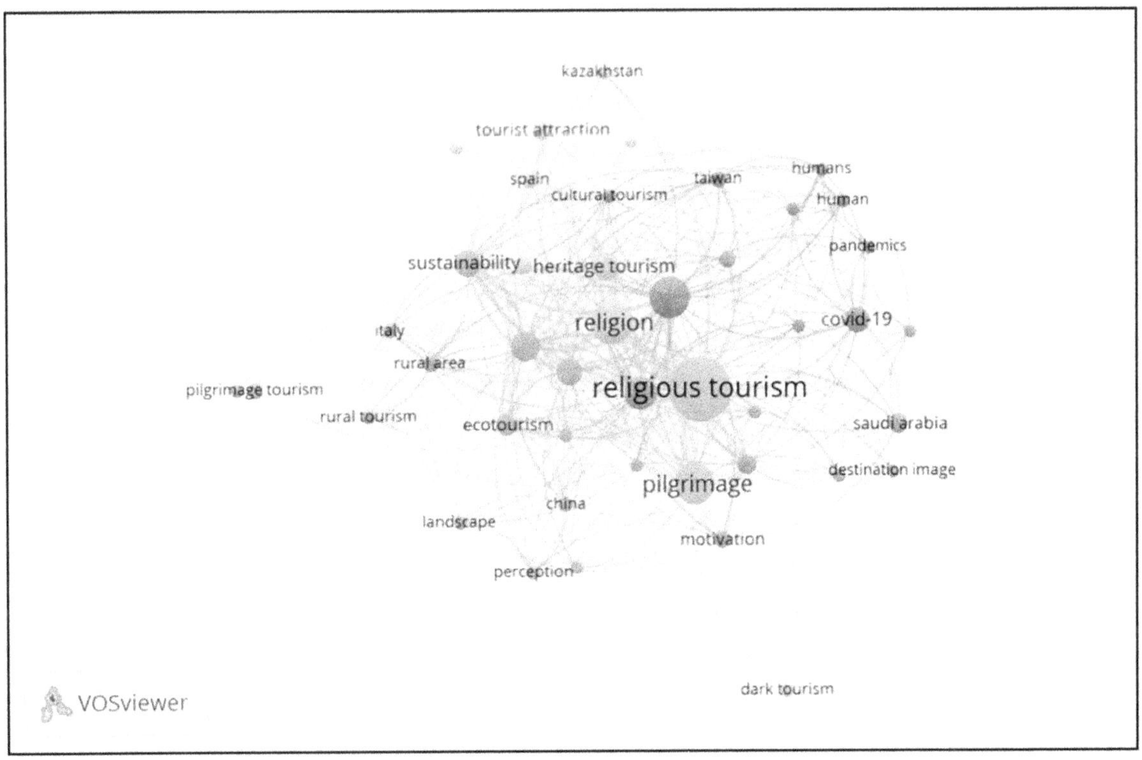

six clusters displayed in different hues. Cluster one which is denoted by a red hue is composed of nine keywords namely Christianity, cultural tourism, covid-19, humans, pandemic, pilgrimages, tourism and Taiwan. The second cluster is denoted by green hue which has a presence of terms like China, ecotourism, landscape, perception, sustainability, sustainable development and sustainable tourism. The blue hued cluster is composed of destination image, motivation, satisfaction, Saudi Arabia, tourism economics, tourist behavior and tourist destination. Yellow colour is for the fourth cluster which mentions heritage tourism, religion, Spain, tourist attraction and world heritage site. Cluster fifth is composed of Italy, pilgrimage tourism, rural tourism, rural area and spirituality. The sixth cluster has the presence of dark tourism, Kazakhstan and religious tourism.

Figure 8 lists the hot subjects in research on pilgrimage tourism. The trending topics in pilgrimage tourism research have been depicted from 2018 to 2022. Starting in 2018, research started focusing on Spain, heritage tourism and tourism destinations. The researchers looked into the topics of religion, travel destinations, and tourism development in 2020. The areas of sustainability and sustainable development were investigated starting in 2021 with COVID-19, Taiwan, and Italy as the main research targets in 2022.

FUTURE RESEARCH DIRECTIONS

While the chapter makes an effort to add to the body of pilgrimage tourism literature by highlighting the scientific production trends, notable organisations, nations, topics, and future directions for studies on pilgrimage tourism, the study is limited by the fact that it is based on a single database. Therefore, future researchers may think about employing multiple databases and may validate the findings of the study.

Figure 8. Trend topics

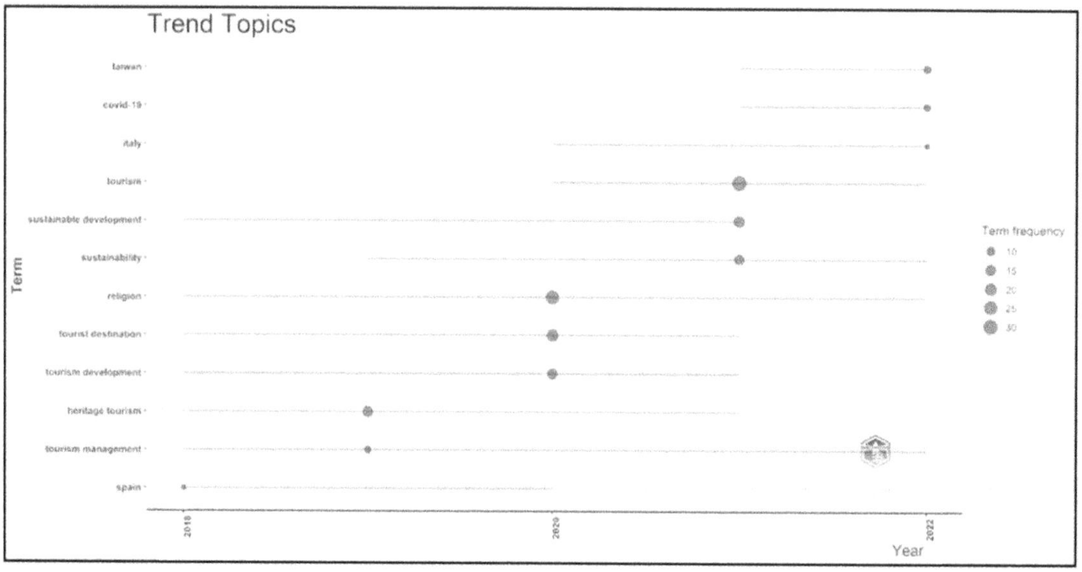

CONCLUSION

From the existing literature and scientific output, it is evident that religious or pilgrimage tourism has been well-studied in the tourism literature. There have been innumerable studies on the motivations, challenges, opportunities, growth, sustainability, and other aspects of pilgrimage tourism. While the issues of crowding at religious locations, accommodation crunch, unregulated carrying capacity worries, cultural deterioration, environmental degradation, inefficient waste management and commercialisation of religious tourist products continue to be the face of well-known holy places around the world. Future researchers can therefore concentrate on addressing and reflecting on these concerns in more depth. Future studies might also look at how technology, virtual reality, and artificial intelligence can help pilgrimage sites overcome obstacles and shape the future of pilgrimages. Also, themes regarding gastronomy tourism, risk and crisis management concerning religious sites can be explored in forthcoming articles.

REFERENCES

Abbas, A. F., Jusoh, A., Mas'od, A., Alsharif, A. H., & Ali, J. (2022). Bibliometrix analysis of information sharing in social media. *Cogent Business and Management*, *9*(1), 2016556. doi:10.1080/23311975.2021.2016556

Abbate, C. S., & Di Nuovo, S. (2013). Motivation and personality traits for choosing religious tourism: A research on the case of Medjugorje. *Current Issues in Tourism*, *16*(5), 501–506. doi:10.1080/13683500.2012.749844

Abdulhanaa. (2021). A Review of Islamic Economic Law on Religious Tourism Arrangements in South Sulawesi. *Samarah: Jurnal Hukum Keluarga dan Hukum Islam*, *5*(1), 450-470. http://dx.doi.org/doi:10.22373/sjhk.v5i1.9088

Al-Makhadmah, I. M. (2020). The role of virtual museum in promoting religious tourism in Jordan. *Geo Journal of Tourism and Geosites*, *28*(1), 268–274. doi:10.30892/gtg.28121-468

Almuhrzi, H. M., & Alsawafi, A. M. (2017). Muslim perspectives on spiritual and religious travel beyond Hajj: Toward understanding motivations for Umrah travel in Oman. *Tourism Management Perspectives*, *24*(1), 235–242. doi:10.1016/j.tmp.2017.07.016

Amaro, S., Antunes, A., & Henriques, C. (2018). A closer look at Santiago de Compostela's pilgrims through the lens of motivations. *Tourism Management*, *64*, 271–280. doi:10.1016/j.tourman.2017.09.007

Aria, M., & Cuccurullo, C. (2017). bibliometrix: An R-tool for comprehensive science mapping analysis. *Journal of Informetrics*, *11*(4), 959–975. doi:10.1016/j.joi.2017.08.007

Bhowmik, P. (2021). Heritage Tourism: A bibliometric review. *Anatolia*, *32*(3), 387–403. doi:10.1080/13032917.2021.1875250

Chistyakov, D. (2020). The influence of Mass Media on consumer culture: Religious tourism. *Logos*, 140–147. doi:10.24101/logos.2020.58

Chistyakova, O. V. (2020). Ethnic Tourism as Knowing Other. *RUDN Journal of Philosophy*, *24*(4), 720–729. doi:10.22363/2313-2302-2020-24-4-720-729

Collins-Kreiner, N. (2020). Pilgrimage tourism-past, present and future rejuvenation: A perspective article. *Tourism Review*, *75*(1), 145–148. doi:10.1108/TR-04-2019-0130

Collins-Kreiner, N., & Wall, G. (2015). Tourism and Religion: Spiritual Journeys and Their Consequences. In Business Media Dordrecht. doi:10.1007/978-94-017-9376-6_34

Durán-Sánchez, A., Álvarez-García, J., del Río-Rama, M. D. L. C., & Oliveira, C. (2018). Religious tourism and pilgrimage: Bibliometric overview. *Religions*, *9*(9), 1–15. doi:10.3390/rel9090249

FICCI. (2012). Diverse Beliefs: Tourism of Faith Religious tourism gains ground. FICCI. https://ficci.in/spdocument/20207/diverse-beliefs-tourism-of-faith.pdf

Griffin, K., & Raj, R. (2018). The Importance of Religious Tourism and Pilgrimage: Reflecting on definitions, motives and data. *International Journal of Religious Tourism and Pilgrimage*, *5*(3), ii–ix. doi:10.21427/D7242Z

Habibi, A., Mousavi, M., Jamali, S. M., & Ale Ebrahim, N. (2022). A bibliometric study of medical tourism. *Anatolia*, *33*(3), 415–425. doi:10.1080/13032917.2021.1954042

Hasana, U., Swain, S. K., & George, B. (2022). A bibliometric analysis of ecotourism: A safeguard strategy in protected areas. *Regional Sustainability*, *3*(1), 27–40. doi:10.1016/j.regsus.2022.03.001

Hermawan, B., Salim, U., Rohman, F., & Rahayu, M. (2019). Making Borobudur a Buddhist Religious Tourist Destination: An Effort to Preserve Buddhist Temples in Indonesia. *International Review of Social Research*, *9*(1), 71–77. doi:10.2478/irsr-2019-0008

Judák, V., Petrikovičová, L., & Akimjak, A. (2022). Religious Tourism on the Example of Nation Pilgrim Places in Slovakia (Patronages of the Virgin Mary). *Journal of Education Culture and Society*, *13*(2), 615–630. doi:10.15503/jecs2022.2.615.630

Kaczmarek, J., Stasiak, A., & Włodarczyk, B. (2001). The Tourist Product and its Evolution, *Turyzm/ Tourism, 11*(1), 7-23. doi:10.18778/0867-5856.11.1.02

Kim, H., Yilmaz, S., & Ahn, S. (2019). Motivational Landscape and Evolving Identity of a Route-Based Religious Tourism Space: A Case of Camino de Santiago. *Sustainability, 11*(13), 3547, 1-20. doi:10.3390/su11133547

Koshim, A. G., Sergeyeva, A. M., Saparov, K. T., Berdibayeva, S. K., & Assylbekova, A. A. (2020). Underground Mosques of Mangystau as the objects of Religious Tourism. *Geo Journal of Tourism and Geosites*, *34*(1), 33–41. doi:10.30892/gtg.34105-616

Martinho, V. J., & Nunes, J. (2020). Camino de Santiago: The routes in the region of Viseu. *Pasos (El Sauzal), 18*(3), 357–370. doi:10.25145/j.pasos.2020.18.025

Mazumdar, S., & Mazumdar, S. (2004). Religion and Place Attachment: A Study of Sacred Places. *Journal of Environmental Psychology*, *24*(3), 385–397. doi:10.1016/j.jenvp.2004.08.005

Narayanan, Y. (2013). Religion and Sustainable Development: Analysing the Connections. *Sustainable Development (Bradford)*, *21*(2), 131–139. doi:10.1002d.1557

Nkwanyana, S., & Njiyela, T. (2021). The Sanctity of Emmaus Pilgrimage Heritage Site: The Case of Abbot Francis Pfanner. *Pharos Journal of Theology*, *6*(102), 1–12. doi:10.46222/pharosjot.102.024

Nyaupane, G. P., Timothy, D. J., & Poudel, S. (2015). Understanding tourists in religious destinations: A social distance perspective. *Tourism Management*, *48*(1), 343–353. doi:10.1016/j.tourman.2014.12.009

Olsen, D. H., & Esplin, S. C. (2020). The Role of Religious Leaders in Religious Heritage Tourism Development: The Case of the Church of Jesus Christ of Latter-Day Saints. *Religions*, *11*(256), 1–22. doi:10.3390/rel11050256

Patwardhan, V., Ribeiro, M. A., Payini, V., Woosnam, K. M., Mallya, J., & Gopalakrishnan, P. (2020). Visitors' place attachment and destination loyalty: Examining the roles of emotional solidarity and perceived safety. *Journal of Travel Research*, *59*(1), 3–21. doi:10.1177/0047287518824157

Paul, J., Lim, W. M., O'Cass, A., Hao, A. W., & Bresciani, S. (2021). Scientific procedures and rationales for systematic literature reviews (SPAR-4-SLR). *International Journal of Consumer Studies*, *45*(4), 1–16. doi:10.1111/ijcs.12695

Prasetio, A. P., Tai, T. D., Opulencia, M. J. C., Abbas, M., El-Ebiary, Y. A. B., Abbas, S. F., Bykanova, O., Samal, A., & Iswanto, A. H. (2022). Impact of the COVID-19 pandemic on religious tourism amongst Muslims in Iraq. *Hervormde Teologiese Studies*, *78*(4), 1–6. doi:10.4102/hts.v78i4.7565

Raj, R. (2012). Religious tourist's motivation for visiting religious sites'. *International Journal of Tourism Policy*, *4*(2), 95–105. doi:10.1504/IJTP.2012.048997

Sestino, A., Pino, G., & Guido, G. (2023). The effect of fervid attachment to religious rites on tourism: Evidence from the Holy Week in Southern Italy. *Journal of Cultural Heritage Management and Sustainable Development*, *13*(2), 201–218. doi:10.1108/JCHMSD-01-2021-0010

Shinde, K. (2012). Policy, planning, and management for religious tourism in Indian pilgrimage sites. *Journal of Policy Research in Tourism, Leisure & Events*, *4*(3), 277–301. doi:10.1080/19407963.2012.726107

Shinde, K. A., & Olsen, D. H. (2023). Reframing the Intersections of Pilgrimage, Religious Tourism, and Sustainability. *Sustainability, 15*(1), 461, 1-13. doi:10.3390/su15010461

Singh, S. (2004). Religion, Heritage and Travel: Case References from the Indian Himalayas. *Current Issues in Tourism*, *7*(1), 44–65. doi:10.1080/13683500408667972

Smørvik, K. K. (2020). Why enter the church on holiday? Tourist encounters with the Basilica of Santa Maria in Trastevere, Rome. *Journal of Heritage Tourism*, *16*(3), 337–348. doi:10.1080/1743873X.2020.1807557

Terzidou, M., Scarles, C., & Saunders, M. N. K. (2018). The complexities of religious tourism motivations: Sacred places, vows and visions. *Annals of Tourism Research*, *70*, 54–65. doi:10.1016/j.annals.2018.02.011

The Pew Forum on Religion & Public Life. (2012). *Global Religion Landscape*. Pew Research. https://assets.pewresearch.org/wp-content/uploads/sites/11/2014/01/global-religion-full.pdf

Tozser, A. (2020). An examination of the Csíksomlyó shrine based on the tourist life cycle curve. *Geo Journal of Tourism and Geosites*, *32*(4), 1304–1309. doi:10.30892/gtg.32417-573

Van Eck, N. J., & Waltman, L. (2010). Software survey: VOSviewer, a computer program for bibliometric mapping. *Scientometrics*, *84*(2), 523–538. doi:10.100711192-009-0146-3 PMID:20585380

Vidhya, K., & Selvam, V. (2019). Factors affecting the perceptions of pilgrim tourists in selection of Accommodation and Transportation: A Case Study of Golden temple, Vellore city, Tamil Nadu, India. *Humanities & Social Sciences Reviews*, *7*(5), 1254–1261. doi:10.18510/hssr.2019.75164

Wang, K.-Y., Karim, A., & Yu, J. (2020). Religious Festival Marketing: Distinguishing between Devout Believers and Tourists. *Religions*, *11*(8), 413. doi:10.3390/rel11080413

Chapter 3
Sustainable Mini Pilgrimage in Urban Spaces:
Suburban Calvaries in Budapest

Brigitta Pécsek
Eszterházy Károly Catholic University, Hungary

ABSTRACT

Calvaries are firmly rooted in Christianity and refer to the depiction of the scenes of the Passion of Christ. This chapter examines suburban calvaries and their potential role within the context of urban tourism, by widening the horizon of pilgrimage tourism while also challenging the notion of city tourism, by taking tourists to the fringe of the city. The research site is the capital of Hungary, Budapest that boasts some fine religious buildings which are inundated with tourists all-year-around, causing overcrowding, mobility challenges and conflicts with locals. The ethnographic account gives a descriptive analysis of six suburban calvaries that are dissected in terms of cultural, sustainable, and niche tourism, and explains how they can contribute to urban tourism. It does not deal with the inner city calvaries because they do not fit the purpose of making urban tourism more sustainable, thus, pushing tourists away from the crowded core of the city. The study concludes that including calvaries into urban tourism would be beneficial for locals, visitors, and the environment alike.

INTRODUCTION

Religious heritage is way more common than most people would imagine. Sacred places in Europe represent an important part of the European heritage, and many of them are related to religious buildings (Nolan, 1992), which count over half a million buildings on the continent. No wonder that approximately 20 percent of the properties inscribed on the World Heritage List managed by UNESCO have a distinctive religious or spiritual connection (UNWTO, 2023). Each year, nearly 300 million people visit religious sites all over the world (40% taking place in Europe) and journeys mixing sightseeing with religious or spiritual experiences are far the most popular (Cohen & Cohen, 2014). So, religious tourism has always been more than a niche type of tourism. In spite of that healthy demand, there are still some types

DOI: 10.4018/978-1-6684-4817-5.ch003

of religious sites which have been overlooked by both scholars and tourists. For instance, no specific research was carried out on calvary visitations, in fact, they might be one of the most underresearched areas within the context of tourism. This hiatus is somewhat surprising since calvaries are very complex in terms of history, architecture and symbolism.

Many popular religious sites located in urban environment such as churches and temples are suffering from overtourism. The reason being is that they are often one of the most recognisable landmarks in many European cities such as the cathedrals in Paris, Barcelona, Vienna, or Milan. COVID-19 was an intermezzo in terms of tourist flow, providing some years of relief, however, sustainability should be readdressed, otherwise, cities will become even more unmanageable than before. Surely, the problem was raised in the academic arena, scholars like Bouchon and Rauscher (2019) investigated European cities finding that cities regardless of their sizes have been suffering from touristification due to the easiness to travel and the ability for tourists to blend in. The conclusion of their research was a call for holistic destination management. Similarly, Amore et. al. (2019) was also concerned with the degree of overtourism in European cities. Jover and Diaz-Para (2020) argue that tourism unsustainability from a social perspective should be investigated as well, as local communities are slowly physically blocked from accessing vital local infrastructure and services. Vetitnev et. al. (2021) explores the situation, taking the example of Sochi, signalling already some mild problems between residents and visitors. Abbasian et. al. (2020) gives a thorough account of the overcrowding situation at the Croatian costal destination, Dubrovnik.

Overtourism has been dissected by the ample volume of literature, however, easily implemented practical solutions are few and far in between. One of the solutions implemented in several cities is charging entry fees, for instance 3-10 euros depending on the season are levied for day-trippers in Venice. This might contribute healthily to the city budget, but a couple of euros will not deter tourists from flocking to the destination of their choice. Tourism scholars, planners and practitioners have to resign to the fact that city breakers will not change their motivation easily. Furthermore, the travel restrictions during COVID-19 have also proved that even overcrowded cities need tourist spending to prosper. As international tourist arrivals dropped by 73% in 2020 and by 72% in 2021, compared to 2019 levels (Atout France, 2022), most countries with a substantial tourism sector felt the consequences economically.

However, even at the most visited destinations tourist-free areas can be found. Taking Venice as an example again, a couple of neighbourhoods on the northern part of the city have stayed surprisingly low-key. So, in many cases similarly to Venice, the most serious problem is the spatial and temporal concentration of tourists, including on water surfaces due to cruise ships heavily investigated by several researchers including Butler (2008) and Settis (2014).

A number of scholars have already realised that religious tourism and pilgrimages help to boost sustainable tourism development along with the social and economic growth of sacred destinations (Romanelli et. al., 2021). Furthermore, religious tourism and pilgrimage contribute to delivering a high level of tourist satisfaction by providing meaningful and unique experiences (Lopez, 2013). They help preserve natural resources and cultural heritage, as well as the traditional values of communities, thus ensuring long-term economic and social benefits.

As previously noted, suburban calvaries are outside the tourist bubble, so in this light it is quite striking that they have not been used as examples of sustainable attractions to alleviate tourist concentration. Beside their beneficial role in the management of the tourist flow, they also diversify the religious and spiritual offers in a city. Apart from urban cemeteries (Pécsek 2015, 2021), no other religious sites feature in scholarly works related to the above notions. Since popular city churches and shrines are often on the receiving end of overcrowding, it is high time to rethink religious urban tourism as a whole.

This chapter guides the reader through the concept and history of calvaries to make a solid foundation for the argument of including suburban calvaries into the urban tourism offerings. Several international examples that were accepted into the UNESCO world heritage list are also discussed, proving that they have universal value worth cherishing, protecting, and visiting. Suburban cavalries are dissected from cultural, sustainability and niche points of view, analysing all the components that can contribute to urban tourism.

BACKGROUND

The Concept of Calvaries

Calvary is firmly rooted in Christianity and refers to the depiction of the scenes of the Passion of Christ. The English word Calvary and Golgotha derive from the Vulgate Latin Calvariae, Calvariae locus and locum (all meaning "place of the Skull" or "a Skull"), and Golgotha used by Jerome in his translations of Matthew 27:33, Mark 15:22, Luke 23:33, and John 19:17. The usually fourteen stations of the Cross are placed on the way up to the top of the pilgrimage hill where there is a chapel. The first calvary can be traced back to the 15th century in Spain (Bilska-Wodecka, 2000). The traditions of building of calvaries came to Hungary through Germany and spread around the country.

In literature the number of people attended the crucifixion of our Lord slightly differ, therefore, there is no evidence related to the size of the audience. In the Gospel of John, the mother of our Lord, Mary and her sister, Mary of Clopas, Mary Magdalene (John 19:25), and Mary, the mother of James and John (Matthew 27:56) are mentioned. All four gospels state the two criminals crucified on either side of Jesus. The crowd included the chief priests, scribes, and elders and many other women "were looking on from afar" (Grace Fellowship Church, 2020). The early set of seven scenes was usually number 2, 3, 4, 6, 7, 11 and 14 from the list below. From the late 16th century to the present, the standard Way of the Cross has consisted of 14 pictures or sculptures depicting the following scenes.

The reason why calvaries started spreading in the Middle Age is that Jerusalem was very far for most devoted potential pilgrims. Back at that time it must have taken weeks even months for people to reach the city of Jerusalem. St. Leonard of Port Maurice, the Franciscan missionary was an extensive traveller who were preaching and promoting special devotion to the Stations of the Cross. During his travel, he sensed the void and started erecting replicas of the stations so that Catholics could follow Our Lord's path to Golgotha. He built 572 Stations of the Cross in Italy, the most famous is the Stations in the Colosseum, where the Vatican's Good Friday service has been held for 300 years (Catholic Company, 2023). All in all, it is commonly believed that these calvaries were built to provide the same spiritual experience, symbolically as close to the original as possible without having to embark on a long journey.

Calvaries in Hungary

The first calvaries appeared in Hungary at the end of the 17th century through Austrian mediation, but their wider spread dates back to the 18th century, when the Counter-Reformation took hold. In the beginning, calvaries were mainly built in cities, later also in villages (Szilágyi, 1974).

Table 1. Original and revised stations

Traditional	*Revised from 1975*
Jesus is condemned to death	*The Last Supper*
Jesus carries his cross	*The Garden of Gethsemane*
Jesus falls the first time	*Jesus before the Sanhedrin*
Jesus meets his mother	*Jesus before Pilate*
Simon of Cyrene helps Jesus carry the cross	*Jesus is whipped and crowned with thorns*
Veronica wipes the face of Jesus	*Jesus carries his cross*
Jesus falls the second time	*Jesus is helped by the Cyrenean*
Jesus meets the women of Jerusalem	*Jesus speaks to the women of Jerusalem*
Jesus falls the third time	*Jesus is stripped and nailed to the cross*
Jesus is stripped of his garments	*Jesus and the Good Thief*
Crucifixion: Jesus is nailed to the cross	*Jesus speaks to Mary and John*
Jesus dies on the cross	*Jesus dies on the cross*
Jesus is taken down from the cross	*Jesus is buried*
Jesus is laid in the tomb.	*Jesus rises from the dead*

Source: https://www.catholic.au/s/article/Stations-of-the-Cross

When establishing a calvary, the environment had to be carefully considered, and usually they were set up on a hill, and the main group of statues was erected on the top. All of the characteristic components of calvaries are found in the calvaries in Budapest: the stations, the calvary scene and the chapel. The usual hermitages in the 18th century are no longer characteristic in the 19th century. Several types of calvaries are distinguished by Szilágyi (1974), based on the distance between the stations, calvaries show a very varied picture in the Hungarian practice. There are some where the distance reaches sixty to seventy metres between two stations, but in other cases this distance can be reduced to six to eight metres. However, the distance between the individual stations is usually constant, therefore, some calvaries need less strenuous effort to walk through while others make quite an athletic achievement next to the spiritual one.

Regarding the layout, the linear type of arrangement of the stations is more common in Hungary, for example, the ones that are lined along a road or path. Within this category there are ones with stations on only one side of the road or with stations on both sides. In addition, the closed type calvary layout should also be mentioned, where the stations are placed on the wall, or in niches. It is also possible that the entire calvary forms a single building (e.g., Strawberry Orchard Calvary, Budapest).

CALVARIES AS TOURIST ATTRACTIONS AND THEIR CHARACTERISTICS

While religious sites are featured in abundance on the UNESCO world heritage list, calvaries are greatly underrepresented. However, both urban locations and country examples can be found on the list, so it is not unprecedented.

Figure 1. Via Dolorosa in Jerusalem
Source: https://www.touristisrael.com/via-dolorosa/28670/

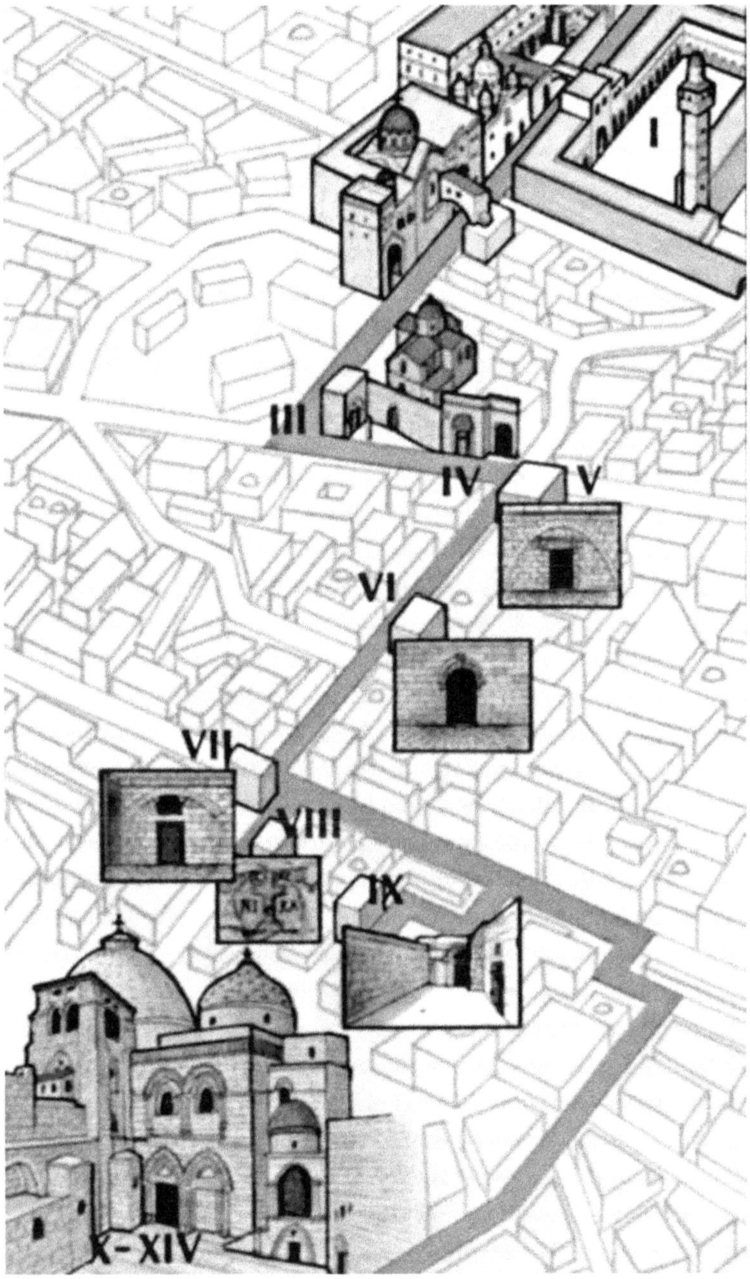

When it comes to urban location, the shining example is the Sanctuary of Bom Jesus do Congonhas located in Brazil. Congonhas is a city with a population of 55 000 residents and mainly famous for its baroque architecture. This sanctuary was built in the second half of the 18th century, consisting of a Rococo church, an outdoor stairway decorated with statues of the prophets, and seven chapels illustrating the Stations of the Cross, accessed by steep, cobbled streets (UNESCO, 2023).

The other example is a more spread-out site, called Kalwaria Zebrzydowska Park, located near the small town of the same name. Officially it is called the Architectural and Park Landscape Complex and Pilgrimage Park and was the first of the large-scale calvaries built in Poland, consisting of a monastery as well as a number of churches, chapels, and other architectural structures. The architecture and landscape in the park provide the ideal setting for reenacting the mysteries of the Way of the Cross and for celebrating the feast of the Assumption of the Blessed Virgin Mary. These events have been held there regularly for over 400 years and are attended by thousands of pilgrims and tourists (UNESCO, 2023).

Apart from the UNESCO heritage, modern calvaries are also created for religious tourism purposes. One of the most unique examples was created in Croatia under the sea to offer divers - a unique experience (figure 2). This modern Vida Dolorosa was specifically planned for the tourists in mind, in Trogir on the Dalmatian coast. The 50 statues are found 4 metre deep, 1.8 metres in height and weigh about 1,000 kilograms, which means they will be able to withstand the elements (Balu Nautica, 2023). Trogir was inscribed on the UNESCO World Heritage List in 1977, full of historic churches and buildings of different styles. Therefore, they offer tourists a myriad of experiences mixing cultural, spiritual and active elements. By attracting divers Trogir can extend the season with months thanks to the active tourists who are more resilient to cooler weather conditions.

The below table (table 2) lists a set of important characteristics of calvaries organised in three different categories, which make them ideal religious attractions within the city tourism context.

Cultural Characteristics

First of all, calvaries are man-made religious sites. They are not standardised at all, but highly varied religious sites embedded into the local history of the community due to their specific circumstances under which they were erected. In most cases even the choice of location and the background of their creation are unique and closely linked to the place. In the case of Hungary, the Stations of the Cross sometimes are placed along a road where certain groups of people were deported. Individual narratives in any shape and form can develop place attachments because they trigger emotional response. When it comes to heritage, storytelling is essential for proper interpretation (Park, 2013) and for calvaries even more crucial, without it, there will not be any point of difference.

The tangible element of calvaries is the architecture. Since calvaries started to be erected in the Middle Age, they appear in all different styles, reflecting the dominant architectural style of the era. The layout of the stations, the shape and size of them, and the chapel with the statues on top of the mountain combined with the green backdrop create a unique, non-replicable experience and urge visitors to see other calvaries for similar, yet different experience.

Customs and traditions play a key role in nurturing the local culture and identity, and they belong to the intangible part of the calvaries. The intangible elements give the opportunity to enhance the appeal of the site, by putting the abstract layer on them, making them a unique, unforgettable experience in the eye of the beholder. The most common custom related to calvaries is walking of the Cross during Lent, especially on Good Friday, where at each station specific prayers are recited. There are also traditions to reenact the crucifixion of Jesus Christ involving the local populace, which tradition is practised in Máriabesnyő, close to Budapest. Newer traditions include to commemorate for those deported in the history of Hungary and walking the Stations of the Cross has a symbolic meaning in that case, relating the suffering of people to that of Christ. Participating as a guest in such an event might forge a lifetime

Figure 2. Station of the Cross under water, Trogir
Source:https://dynamic-media-cdn.tripadvisor.com/media/photo-o/1b/d5/11/53/img-20200820-wa0005-largejpg.jpg?w=1100&h=-1&s=1

bond with the local community, which can foster return visitations, strong "word of mouth" advocacy and co-creation experiences.

Sustainable Characteristics

Calvaries are one of the most sustainable religious sites. They always blend into the natural environment because their placement was carefully selected in the first place and over the years the fusion of the landscape and the calvary naturally takes place. Green environment in big cities is scarcer than in the countryside, therefore, they can be visitor attractions in their own right, coupled with unobtrusive, architectural site laden with history and symbolism it should be enough to provide a short getaway from the centre. Since calvaries usually can be accessed either via a set of stairs or an ascending path, getting

Table 2. Characteristics of Calvaries

Tourism type	Characteristics	Touristic value
Cultural		
	Local history	Place attachment
	Architecture	Aesthetic benefit
	Customs and traditions	Interaction with locals, Common experiences
Sustainable		
	Green environment	Physical activities
	Easily accessible	Day trip option
	No commodification needed	Authenticity
	Time-tested	Appeal across historical time
	Slow attraction	Tourist management
Niche		
	Complex attraction	Body, mind and spirit
	Unique	No attraction has the same composition
	Specific marketing	Finding USP

Source: own research, 2023

there is a physical - exercise, but the surroundings offer way more opportunities such as hiking, nordic walking, or sleighing in winter.

In Europe suburbs are usually easily accessible from the centre, since many of them used to be independent villages or small towns, merging recently with the city. Some of these suburbs are more residential, others are less densely populated with generous green spaces and serve as a destination for city dwellers going on a short excursion. Nonetheless, each suburb has its unique landscape and culture evolved around the place over centuries. This is the reason why most calvaries differ to some extent and worth visiting more than one. Most city breakers are short of time but including a couple of interesting suburbs with calvaries into the itinerary provides an unexpected angle for visitors.

These days when tourists are constantly looking for new tourism attractions and even non-places (like airports) can become tourist sites, it is always useful to look around what is available yet has been overlooked in the cacophony of tourism media. Since suburban calvaries never stand alone, embedded in the local environment, no extra infrastructural or hospitality services are to be developed. Protection is needed but not development. They are truly authentic, which is highly valued in an urban environment that is highly globalised and uniformised.

Calvaries have been around for centuries, so they are time-proof heritage sites, most of them are an integral part of the local vista and serve as a backdrop for religious festivals. Revamp or modernisation is not necessary, since the Crucifixion is the most important event for Christians, therefore, as long as the religion exists these sites will stay and the importance of calvaries will remain unchanged.

The link between slow tourism and mitigating overcrowding in urban environment was conceptualised by Pécsek (2018). Calvaries are truly slow attractions compared to churches situated in the urban core. It takes time to get there, and the environment lures people to stay there longer. The green context with the vista is a complete antidote to the busy city centre, so once visitors can be convinced to leave the city, they can easily spend a day in the suburb.

Niche Characteristics

Calvaries are complex attractions as they are green, cultural and spiritual, exuding positive influence on the body, mind and spirit. It can be more like a short outing for some people while others prefer a more meditating experience, and there are those who truly seek a deeply spiritual experience. Whatever their initial motivation is, the calvary will affect visitors in several ways.

The uniqueness of calvaries lies in the fact that no sites have the same cultural, sustainable and spiritual makeup. On one hand, the Crucifixion is at the very core of Christianism; therefore, the simple historical account of events is probably known for most tourists before visiting a European city. On the other hand, a proper on-site interpretation can make each site unique and non-replicable.

It is important to note that the special characteristics of calvaries need a carefully planned marketing strategy. First of all, calvaries should be linked to urban tourism, but always keeping in mind that they are not a simple architectural site but depicting the very essence of Christianity. So, attracting the wrong crowd would be more damming than the current oversight. Therefore, the religious component of calvaries should remain in focus with looking at the list of the most famous Christian perspective, and then a unique selling proposition can be formulated based on the local history and environment.

URBAN TOURISM IN BUDAPEST: RELIGIOUS AND GREEN SIGHTS

Based on the data of the Hungarian Tourism Agency (MTÜ, 2023), 14.2 million guests spent almost 40 million nights in Hungary in 2022, an increase of almost 40 percent compared to 2021, and only five percent below the 2019 level. Almost four million people visited the capital, 82 percent of visitors to Budapest were foreigners. Out of the 23 districts 3 small, inner districts, number 5, 6, and 7 realise about 5 million guest nights each year (figure 3). The results are even more striking when comparing it with the performance of the most popular spa destination, Héviz that welcomes around 1,14 million tourists yearly (Hungarian Central Statistical Office, 2022). It means that extreme tourist concentration with the exception of the 2 Covid-years has not improved.

The capital of Hungary, Budapest boasts a couple of the finest religious buildings in Europe, including St Stephen's Basilica, the most photographed monument and the tallest building in the city, featuring on the lists of top tourism attractions both domestically and internationally. These religious attractions are inundated with tourists all-year-round, causing overcrowding, mobility challenges and conflicts with locals, while a couple of kilometres further some superb religious architecture and numerous spiritual spaces await to be discovered.

Looking at the most famous religious attractions in Budapest, four popular travel sites were examined in May 2023, none of them are sponsored by Hungarian tourism organisations or businesses.

Table 3 shows that in fact, some of the most famous landmarks in Budapest can be classified as religious sites. It can not be disregarded that the Basilica is situated in the most touristic neighbourhood in Pest and the Matthias church is part of the cultural sites on the Buda side. Still, the genuine interest is palpable, so more niche religious sites can capture and build on this momentum. If a cluster of places of interest are available for tourists, then it is easier to lure them to the area. In case of suburban calvaries

Figure 3. The 23 districts in Budapest
Source: https://www.bp18.hu/a-kerulet/a-kerulet-szamokban

Table 3. Top religious sights in Budapest

Sites	*Tripadvisor*	*Touropia*	*Planetware*	*Time out*
St. Stephen's Basilica	*3*	*4*	*6*	*3*
Matthias Church	*7*	*11*	*-*	*6*
Dohány street Synagogue	*-*	*-*	*7*	*5*

Source: own collection, 2023

Table 4. Top green sites in Budapest

Sites	*Tripadvisor*	*Touropia*	*Planetware*	*Time out*
Margaret Island	*12*	*7*	*-*	*10*
Children's railway	*-*	*-*	*-*	*3*

Source: own collection, 2023

the obvious drawcard is the green surroundings. The following table (4) shows the green attractions in Budapest featured on the same four websites.

It can be noted that tourists highly value green spaces (Table 4) as well in Budapest. Margaret island is a green oasis in the middle of the city with promenades and flower gardens and a small Japanese garden for walkers, cycle paths and a jogging track for sports lovers, it is also a great venue for a date or a picnic. Here is where the National Sports Pool is located along with the popular Palatinus lido and thermal bath, a petting zoo and a musical fountain for entertaining visitors. All kinds of bikes, tandems and pedalos can be hired for making attractions more accessible. The ruins of a medieval Dominican convent are historically linked to former resident Margaret, daughter of Béla IV, after whom the island is named. Other parks are also located in the city but with far less attractions and recreational activities.

Looking at this short introduction of the tourism attractions in Budapest, it can be established that city tourists have multitudinous motivations, so the amalgam of leafy areas and architectural heritage can be attractive for urban tourists. In a recent study Reményik et.al. (2021) suggested that Budapest should take measures against overtourism as soon as possible, by diversifying demand as well as new sustainable tourism products need to be developed. In this way, the temporal and spatial spread of tourists can possibly be achieved.

CALVARIES IN THE SUBURBS OF BUDAPEST AND THEIR SYNERGIES

A wide range of historical and contemporary urban calvaries can be accessed from downtown Budapest. This section showcases six calvaries out of the 27 (Csendhegyek, 2023) and the selection was based on the identified and explained general characteristics in the previous chapter. Therefore, it does not deal with the inner city calvaries because they do not fit the purpose of pushing tourists away from the crowded inner core of the city. Those that are out of the way, therefore including them into a city break itinerary would be challenging, were also disregarded. The author aims to prove the point that even though suburban calvaries are similar in many respects, they also have distinctive characteristics worth paying a short visit.

All these calvaries briefly introduced would give an excellent counterpoint to the congested urban core. They are interesting on their own, but with the combination of other activities available close by, these districts offer a unique way of experiencing Budapest by enjoying some quiet, contemplative time. Obviously, a city break has a tight schedule but including one or two suburbs into the itinerary would help to gain a better understanding of the destination proper.

The potential products that can provide a synergy with the calvaries range from wine tourism to active tourism and some more unexpected in between. All the selected complementary products enhance the tourist experience that those calvaries offer and give the opportunity for even deeper immersion into locality and getting a real sense of the place.

- **Calvary in Kiscell – 3rd district**

Óbuda is the most ancient part of Budapest, used to be an independent settlement until 1873 when Buda, Pest and Óbuda merged and became a unified metropolitan. The district is accessible by both buses and trams. One of the oldest calvaries in Budapest is the Kiscell Calvary dating back to over 200 years. The Way of the Cross starts next to the Kiscelli Museum that used to be a monastery, and the blue cross sign leads up to the Chapel of the Holy Blood. There are 14 stations lining on one side leading through an enchanted forest, where detached houses are nestled among the trees. There are three crosses on the hill; Jesus stands in the middle with a thief on either side, surrounded by the statue of Jesus' mother, Mary, and Jesus' disciple, John (Pestbuda, 2023).

The whole site breaths with nature, ideal excursion for nature lovers who prefer hiking, cycling, or forest bathing. However, the real USP of this neighbourhood is the possibility to book a professional or private event at the museum (Figure 4), staging a natural wedding ceremony or a business partner meeting.

- **Calvary in Budafok – 22nd district**

Budafok is easily accessed by busses and offers the most unique experience that is quite rare in metropolitan surroundings. This calvary is situated next to a string of wine cellars that can be explored with the help of a thematic map or on an organised tour. The second biggest wine festival in Budapest is also held there each September. In fact, the calvary itself also owes its existence to a wine maker. Its beginning dates back to 1761 when the cross made of three larch trees was erected. In 1992, a local sculptor and restorer renovated the ruined Calvary and the picture columns and today, the main statue comprises 5 figures. Interestingly, apart from the first and last stations, all the others are lined on one side and residential building are situated on the other side. Between the first and second stations of the Cross, a playground with biblical scenes was created on three levels. In this way, children, or the young at heart can learn about the Holy Scriptures and Christianity in a playful way. On the lower level, the designers

Figure 4. Kiscell Museum
Source: http://kiscellimuzeum.hu/sites/default/files/pictures/magunkrol/kiscell1.jpg

Figure 5. Stations of the Cross, Budafok
Source:https://huszonketto.hu/wp-content/uploads/2021/02/20201003_budafok_pincejarat_0041_k-1024x683.jpg

built - an amphitheatre-like area. During the construction the opinions of the historical churches and the Baptist Church of Budafok were also listened to and integrated in the plans.

The Bible contains numerous references to wine, as a source of joy and celebration. Ephesians 5:18 states, "Go, eat your food with gladness, and drink your wine with a joyful heart, for it is now that God favours what you do." Not surprisingly, monks have played a significant role in the production of beer and wine in Hungary throughout history, often running their own breweries and vineyards. In Hungary the Benedictine Abbey in Pannonhalma runs its own professional winery, so, it is fitting to combine a calvary visit with local wine tasting events.

- **Calvary in Pesthidegkút – 2nd district**

Pesthidegkút used to be an independent settlement and joined Budapest in 1950. From the end of the 1600s to the mid-1700s, German-speaking settlers from the Black Forest, Bavaria, and Austria came to live in this area. Based on estimates the calvary consisting of 14 stations was probably built at the end of the 1700s on a hill (figure 6) and local Germans walked the Way of the Cross from the church to the crosses every Good Friday. However, after World War 2 ended 1,300 out of the 1500 German families were deported, and all the traditions and customs disappeared with them. The site was heavily neglected during the socialist era, and only over the last few decades the sites have been renovated and old customs and traditions have been slowly revived.

Figure 6. Top of the hill, Pesthidegkút
Source: https://kirandulastippek.hu/images/news/article/zoom/p172-19.jpg

This calvary is recommended for those who prefer an active day away from the city. Hikers can choose from several shorter or longer routes to conquer the peak. And those who succeed can enjoy an unparallel panorama of the hilly landscape of Buda.

- **Calvary in Soroksár – 23d district**

Soroksár is the latest addition to the suburbs of Budapest since the settlement used to be an independent town for centuries. It might be a longer trip to access the district but still the tram or bus ride is quite convenient. In the garden of the Szent István Király Church, built in the honour of Our Lady of Fatima in Soroksár-Újtelepi, the Way of the Cross with headstones was made from public donations in 2004. The headstones were prepared by seven wood carvers, but all the bronze station images were made by the same artist. The Way of the Cross begins on the left side of the church, and the stations embrace the church like a mother embraces a child.

Among the many places of interest, the local botanic garden (figure 7) is worth exploring and one of the most exciting organised walks follows a thematic nature trail full of edible wild plants. Similar to wine, herbs and plants are profusely mentioned in the Bible (Dafni & Böck 2019), and even today many monastic herb gardens flourish where monks cultivate fruits and plants and prepare their own healing ointments, creams or cosmetics. So, getting knowledgeable about what nature offers to humankind is a life-long takeaway from this outing.

Figure 7. Botanic Garden, Soroksár
Source: https://storage.googleapis.com/welove-media/d2/soroksari-arboretum-polyak-attila-038.exact1980w.jpg

- **Calvary in Lágymányos – 11th district**

This is the most recent calvary (Figure 8) that was built in 2014 and situated in a newer part of the city. This district is easily accessed by both trams and underground. The area is a mixture of residential blocks of flats and office buildings, and it is the least green and lush among all discussed above. The calvary is a blend of traditional and modern and nestled in the church garden, comprising 14 stations of the Cross. Since it is located in a confined space, no physical effort is necessary to go through the stations. The church caters for a small Korean catholic community as well with Korean masses and programs.

As the underground 4 has the terminal station close by surrounded by a park with the extensive Etele shopping mall in the vicinity, it is a popular gathering place for locals yet completely overlooked by tourists.

CONCLUSION

This chapter produces some new and exciting insights into the field of religious tourism by widening the horizon of pilgrimage tourism. Pilgrimage is associated with long and strenuous walks; however, several mini pilgrimages can be organised within the metropolitan areas, which are close to the centre geographically, yet have similar spiritual benefits to the longer versions. This study challenges the notion of city tourism as well, by taking tourists to the fringe of the city to slow down and breathe. The chapter proves its points through the examples of Budapest. The study disregards calvaries situated in the urban core, because they do not offer the opportunity to get away from the hustle and bustle of the centre. It does not deal with far out suburbs either since their visit cannot be scheduled into a short city break.

Figure 8. Calvary, Lágymányos
Source:https://4.bp.blogspot.com/-cHw3Go5zvqk/VjkP74mNMWI/AAAAAAAAu Iw/WY6qHlwxsHI/s1600/K%25C3%25A9p1.jpg

In order to make a case for calvaries as potential tourism attractions, several characteristics associated to tourism were identified and explored. First of all, they are culturally diverse firmly rooted into the local community, their traditions and customs and reflect the historic era when the given calvary was build.

Furthermore, it was also established that calvaries can be closely linked to sustainable tourism because they are situated in a green environment and can be easily accessed from the centre. So, the involvement of these religious sites into city tourism is highly desirable since they make urban tourism more manageable in terms of the spatial and temporal management of tourists.

They are also sustainable in a sense, that they do not need to be commercialised, they have been around for ages and no necessity has arisen to make any changes on the site in terms of infrastructure or services. They have stood the test of the time because they are complete as they are. Furthermore, all the necessary services are available in the neighbourhood nearby if visitors wish to consume or indulge into other activities. Calvaries were identified as slow attractions because visitors need allocated time to get there and absorb the sight.

No doubt, that calvaries fall into the category of niche tourism attractions even within the context of religious tourism. And as such, special attention should be paid to this characteristic while crafting any kind of promotion involving calvaries. Professionals should be aware that their potential inappropriate actions can easily undermine the symbolism and sacredness of the site. It does not mean that possible synergy effects have to be ignored, but any profane programs should be encouraged outside the territories of the calvaries.

As for recommendations, it would be advisable for tourism professionals, urban developers, and marketing experts to think about religious tourism and city tourism in a more holistic way, otherwise city tourism will remain as unsustainable as it was in the pre-Covid era. Calvaries might never be the landmarks of the city, but as composite attractions perfectly blending nature and culture they might contribute to making sense of the space and to the development of place attachment. As a result, calvaries can put suburbs onto the mental map of tourists, improving the image of the place by showcasing their touristic attributes. City centres will always remain more exciting; however, suburbs can become an antidote to congestions, commotion and noise, so religious sites such as calvaries outside the tourist bubble can provide an alternative enjoyment that can be experienced without substantial development efforts and financial investments.

REFERENCES

Abbasian, S., Onn, G., & Arnautovic, D. (2020). Overtourism in Dubrovnik in the eyes of local tourism employees: A qualitative study. *Cogent Social Sciences*, *6*(1), 6. doi:10.1080/23311886.2020.1775944

Amore, A., Falk, M., & Adie, B. A. (2020). One visitor too many: Assessing the degree of overtourism in established European urban destinations. *International Journal of Tourism Cities*, *1*(6), 117–137. doi:10.1108/IJTC-09-2019-0152

Atout France/ European Travel Commission. (2022). *Restarting tourism for the better: Performance of European tourism before, during & after COVID-19*. ETC. https://etc-corporate.org/uploads/2022/03/2022_Performance-of-European-tourism-before-during-and-beyond-the-Covid-19-pandemic.pdf

Bilska-Wodecka, E. (2000). *Typology of European calvaries*. RUJ. https://ruj.uj.edu.pl/xmlui/bitstream/handle/item/77831/bilska-wodecka_typology_of_european_calvaries_2000.pdf?sequence=1&isAllowed=y

Blau Nautica. (2023). *Underwater Museum*. Blue Nautica. https://www.blue-nautica.com/diving/underwater-museum/

Bouchon, F., & Rauscher, M. (2019). Cities and tourism, a love and hate story; towards a conceptual framework for urban overtourism management. *International Journal of Tourism Cities*, *4*(5), 598–619. doi:10.1108/IJTC-06-2019-0080

Butler, M. (2008). *Turismo de cruceros. Situación actual y tendencias*. Organización Mundial del Turismo.

Calvaries in Budapest (2023). Budapest. *Csendhegyek* blogspot.com/search/label/Budapest

Cohen, E., & Cohen, S. (2014). A mobilities approach to tourism from emerging world regions. *Current Issues in Tourism*, *18*(1), 11–43. doi:10.1080/13683500.2014.898617

Dafni, A., & Böck, B. (2019). Medicinal plants of the Bible—Revisited. *Journal of Ethnobiology and Ethnomedicine*, *15*(1), 57. doi:10.118613002-019-0338-8 PMID:31775790

Grace Fellowship Church. (2020). *Who were on Golgotha during the crucifixion of Jesus?* Grace Fellowship Church. https://www.graceomaha.org/blog/2020/03/19/who-were-on-golgotha-during-the-crucifixion-of-jesus

Hungarian Central Statistical Office. (n.d.). [Data set]. Hungarian Central Statistical Office. https://www.ksh.hu/?lang=en

Hungarian Tourism Agency (MTÜ). (n.d.). *Home.* MTU. https://mtu.gov.hu/

Jover, J., & Díaz-Parra, I. (2022). Who is the city for? Overtourism, lifestyle migration and social sustainability. *Tourism Geographies, 1*(24), 9–32. doi:10.1080/14616688.2020.1713878

Lopez, L. (2013). How long does the pilgrimage tourism experience to Santiago de Compostela last? *International Journal of Religious Tourism and Pilgrimage, 1*(1), 1–14. doi:10.21427/D7C133

Nolan, M. L., & Nolan, S. (1992). Religious sites as tourism attractions in Europe. *Annals of Tourism Research, 19*(1), 68–78. doi:10.1016/0160-7383(92)90107-Z

Park, H. Y. (2013). *Heritage Tourism.* Routledge., doi:10.4324/9781315882093

Pécsek, B. (2015). City Cemeteries as Cultural Attractions: Towards an Understanding of Foreign Visitors. *DETUROPE Cent. Eur. J. Reg. Dev. Tour, 7*(1), 44–61. doi:10.32725/det.2015.004

Pécsek, B. (2018). *A lassú turizmus mint a városi turisztikai desztinációk fejlesztésének fenntartható alternatívája.* Szent Istvan University of Agriculture and Life Sciences.

Pécsek, B. (2021). The transformation of the Budapest Fiumei Road Cemetery into staycation. *Változó világ, változó turizmus*, 323-336.

Benyo, G. (2023). *Calvary above the city.* Rest Buda.: https://pestbuda.hu/cikk/20221105_golgota_a_varos_felett_20_eve_ujult_meg_pesthidegkut_kalvariaja

Sweet, J. (2023). *22 top-rated tourist attractions in Budapest.* Planetware. https://www.planetware.com/tourist-attractions-/budapest-h-ps-bud.htm

Remenyik, B., Barcza, A., Csapó, J., Szabó, B., Fodor, G., & Dávid, L. D. (2021). Overtourism in Budapest: Analysis of Spatial Process and Suggested Solutions. *Reg. Stat., 11*(3), 179–197. doi:10.15196/RS110303

Romanelli, M., Gazzola, P., Grechi, D., & Pollice, F. (2021). Towards a sustainability-oriented religious tourism. *Systems Research and Behavioral Science, 38*(3), 386–396. doi:10.1002res.2791

Settis, S. (2014). *Se Venezia muore.* Giulio Einaudi.

Stations of the Cross. (n.d.) Britannica. online. https://www.britannica.com/topic/Stations-of-the-Cross

Szilágyi, I. (1975). Kálváriák. Budapest: Corvina kiadó.

The Catholic Company. (2023). *Who spread the Stations of the Cross throughout the world?* Get Fed. https://getfed.catholiccompany.com/who-spread-the-stations-of-the-cross-throughout-the-world/

Timeout. (2023). *Things to do.* Timeout. https://www.timeout.com/budapest/things-to-do/best-attractions-in-budapest

Touropia. (2023). *Tourist Attractions in Budapest.* Touropia. https://www.touropia.com/tourist-attractions-in-budapest/

Tripadvisor. (2023). *Budapest, Central Hungary.* Tripadvisor. https://www.tripadvisor.com/Attraction_Products-g274887-a_contentId.40525991899+761525408-Budapest_Central_Hungary.html

UNWTO. (2017). *Annual report*. UNWTO. https://www.unwto.org/global/publication/unwto-annual-report-2017

Vetitnev, A., Chigarev, D., & Afanasiev, O. (2021). Overtourism and Tourismophobia at a Russian resort. *European Journal of Tourism Research*, *28*, 2812. doi:10.54055/ejtr.v28i.2263

ADDITIONAL READINGS

Bräunlein, P. J. (2009). Negotiating Charisma: The Social Dimension of Philippine Crucifixion Rituals. *Asian Journal of Social Science*, *37*(6), 892–917. doi:10.1163/156848409X12526657425262

Holt, R. E. (2023). *Views of Calvary: A Photographer's Perspective of the Cross Through the Kilde, J.H. (2008). Sacred Power, Sacred Space: An Introduction to Christian Architecture and Worship Illustrated Edition*. Oxford University Press.

Lens of Scripture. High Bridge Books ISBN-13: 978-1954943735

Rae, M. A. (2017). *Architecture and Theology: The Art of Place*. Baylor University Press.

KEY TERMS AND DEFINITIONS

Calvary: Religious sites containing a station of the Cross (via dolorosa) and a scene of the passion of the Christ.

Franciscan and Jesuit Orders: These are Christian orders.

Mini pilgrimage: Contrary to the traditional long pilgrimage routes, mini pilgrimages are easily accessible by public transport, close to a residential area. They are ideal one-day trips where quiet and spirituality still can be enjoyed without the strenuous walking.

Overtourism: A higher concentration of tourists causing overcrowding.

Suburban tourism: Urban tourism outside the central area of a city.

Sustainable tourism: Type of tourism that does not make harmful impact on the natural and cultural environment, therefore, beneficial for both the local populace and visitors.

Synergy: The interaction between elements resulting in a greater impact together than individually.

Chapter 4
Contemporary Precedents in Pilgrimage Tourism:
A Case Study of Shri Kashi Vishwanath Temple Corridor (SKVTC)

Sujay Vikram Singh
https://orcid.org/0000-0002-7113-2698
Banaras Hindu University, India

Atul Tripathi
Banaras Hindu University, India

Rashmi Jha
Banaras Hindu University, India

Rajeev Ranjan
Mangalayatan University, India

ABSTRACT

The work purports to present how the Shri Kashi Vishwanath Temple Corridor (SKVTC) Project has propositioned its roadmap of 'Extension and Beautification' of a densely-populated pilgrim hotspot (Kashi Vishwanath Mandir Visistikaran- Saundrikaran Yojana). The scope of this study includes the rediscovered historical temples of the region that were saved from encroachment as a result of this Corridor. The present study fills the gap of a systematic study of the Corridor in terms of relevance in pilgrimage tourism, particularly in the context of the rediscovered historical temples, focusing on heritage sustainability to support the revival of Varanasi's tangible and intangible heritage. The findings of the study, which discussed the rapidly changing landscape of the most important site of the city, the Kashi Vishwanath temple, enabled local communities, tourism stakeholders, and academicians worldwide to encapsulate the diverse complexities of the SKVTC Project and its impact on pilgrimage activities and tourism.

DOI: 10.4018/978-1-6684-4817-5.ch004

INTRODUCTION

Pilgrimage presented 'tourism mobility' in societies centuries before the modern-day concept of tourism (Noga, 2020). These pilgrimages were part of sacred travels performed by ardent religious followers (Pawar, 2019). The city of Varanasi, a popular 'sacred pilgrimage' center, is currently seeing the sharpest rise in pilgrimage tourism since the inception of the Shri Kashi Vishwanath Temple Corridor (SKVTC) in 2019. It also becomes crucial to mark the distinction that 'pilgrimage tourism' also includes tourism done with a strong inclination for experiencing religious or religion-related activities (Morinis, 1992). The 2017 International Congress on Religious Tourism and Pilgrimage, organized jointly by the World Tourism Organization (UNWTO), reflected on the role that pilgrimage tourism plays in the socio-economic and sustainable development of the pilgrimage site. The nature of interaction that pilgrims have during their tour of any city is comprehensive and detail-oriented as there is a need for distinction for specific sites that evoke the sacred (Griffin and Raj, 2017). A key method of achieving this has been through preserving the spirit of the destination, which is further accomplished through tangible and intangible means of displaying heritage. The SKVTC Project has thus far been successful in harnessing these integral factors while being accessible to the majority of the pilgrim population. Therefore, it is imperative to discuss the operational factors that have nurtured the pilgrimage attractiveness of the Shri Kashi Vishwanath Temple at Varanasi.

The present paper focuses on the direct relationship between trans-regional pilgrim mobility on the stakeholders of Varanasi's cultural inheritance. Cultural inheritance in the contemporary urban landscape of Varanasi entails determining maximum efficiency in visitor experience that enables them to perceive heritage as a proponent of collective identity and a source of understanding the present (UNESCO, n.d.). The ambitious SKVTC Project, started off on March 08, 2019, by the Prime Minister of India, Shri Narendra Modi, completed its Phase I on December 13, 2021. This Project gains considerable momentum as a driving force for emboldening aspects of heritage preservation, its connection with the geographical landscape of the city, the related historical mythologies, and the impact of a sustainable approach to religious infrastructure development in a generally perceived 'sacred' city like Varanasi (Divya, 2021).

This study purports to present how the SKVTC Project has propositioned its roadmap of 'Extension and Beautification' of a densely-populated cultural hotspot (Kashi Vishwanath Mandir *Visistikaran-Saundrikaran Yojana*). The scope of this study includes the rediscovered historical temples of the region that were saved from encroachment as a result of this Corridor. These rediscovered temples are placed in the milieu of Varanasi's growing religious heritage. Shri Kashi Vishwanath Special Area Development Board (SKVSADB) has been conserving the rediscovered historical temples, which represent the artistic and cultural legacy of the city in the last three hundred years.

The Corridor's planning has been corroborated with a Detailed Project Report and current status of the Project's Phase. A systematic understanding of new developments at the site, along with its efficiency and impact is required, and the paper attempts at filling this research gap. The heritage conservation process of the Corridor has been evaluated in the context of transforming congested lanes, modern amenities-based architecture, and overall tourism infrastructure (Wan et al., 2018). The strengthening of tourism infrastructure can yield rapid improvement in the quality of pilgrimage tourism to the city as well (Bausch, 2020). Pilgrimage sustained through optimum heritage experience is one of the many effective ways of reaching the sustainable goal of transforming and transferring cultural legacy from one generation to the future (UNESCO, n.d.).

OBJECTIVES OF THE STUDY

1. To examine the development of pilgrimage tourism post-completion of Phase I of SKVTC, Varanasi.
2. To study the pilgrimage attractiveness of the Corridor and its influence on tourism stakeholders.

METHODOLOGY

A review of relevant documents, personal interviews with subject matter experts, interviews with local community stakeholders, and interviews with persons who are either directly or indirectly involved in or affected by the SKVTC initiative are included among these sources. The SKVTC Project's initiatives, such as the preservation of newly found historical temples and the development of tourism infrastructure, are analysed in this document as well. The research was exploratory and conceptual, and it covers secondary data. The secondary data mainly consists of literary source materials collected from books, chapters, newspapers, journals, magazines, and internet sources.

BACKGROUND

Brief History of 'Pilgrimage Tourism' in Varanasi

The origin of asceticism at Varanasi dates back to the creation of Varanasi as a holy place, also known as *Anandavana* or *Anandakanana*- the forest of Bliss (Bhatt, 2004). Since the Vedic and Upanishadic era, Varanasi has been the assembling place for those who renounced the world and came here to practice asceticism (Motichandra, 1962). Shankaracharya received his *jñana* here; Buddha turned the wheel of the law *(dharmacakrapravartana)* at Sarnath. Of the twenty-four *Tirthankaras,* Suparshvanath, Chandraprabhu, Shreyanshnath, and Parshvanath were born here. The holy city sheltered many great luminaries of the Bhakti movement, such as Ramananda, Kabirdas, Guru Nanak, Tulsidas, Vallabhacharya, Chaitanya Mahaprabhu, Dadu, and Raidas.

Situated on the western bank of the Ganga, Varanasi was connected with the Indian state capitals of the time in the Gangetic Valley in the first millennium BCE (Motichandra, 1962). Thus, it became a chief center of art, culture, and education. It also became an outstanding center for textile manufacturing in the early stage of Indian culture. Music traditions of Varanasi carry the legacy of legends like Sitara Devi, Pt. Ravi Shankar, Bismillah Khan, Girija Devi, and Channulal Mishra. By identifying Varanasi as the '*city of music*' under the creative city network, UNESCO has equally acknowledged the contribution of the artists of Banaras Gharana (Kumar, 2003).

Varanasi, one of the world's oldest continually inhabited cities, is also considered one of the Saptapuris, i.e., the 'seven sacred cities' of ancient India (Bhatt, 2004). The place is known for its cultural heritage, of which the river Ganga and Kashi Vishwanath Temple have always played an essential part. They interact with the tangible and intangible heritage of Varanasi in order to experience the historicity of the ancient city. Interestingly, most of the present tangible heritage of the city was primarily developed during the early 18th century CE, post the decline of Mughal empire and during the emergence of East India Company as the governing power (Desai, 2017). Many artist communities, including those of architects, musicians and dancers, were moving across regions in search of patronage, and Benares, being one of the major kingdoms, and later princely state under British Empire, became an epicentre for what many scholars deem, a cultural renaissance (Sandria, 1989). Apart from Kardameshwar temple at Kandwa, stoppages of the most important pilgrimage yatra of the city, the panchkroshi yatra, were rebuilt during this period. The patronage for these temples and structures, which came in the form of ghats, galis, mathas, dharamshalas, and so on, was gained by rulers, zamindars, merchants, and others from various regions of the erstwhile country (Desai, 2017). In addition, the patronage was used to construct various religious buildings. By attesting to the sacred significance of Varanasi, a city that is synonymous with religious antiquity, these benefactors hoped to legitimise their position in the socio-political sphere. Kashi Vishwanath, the most important shrine in Varanasi, was constructed in the year 1777 CE by Ahilyabai Holkar, the queen of Indore (Eck, 1982). Ahilyabai Holkar is also credited with constructing many dharamshalas in the city for the use of pilgrims.

KASHI VISHWANATH CORRIDOR: RESOLUTION TOWARDS HERITAGE PRESERVATION

The transformations of the last three centuries brought out several infrastructure discrepancies hindering accessibility of pilgrimage sites. The Gali (lanes) in the pakka mahal area at the centre of the city attract many tourists. However, accessibility to these lanes remains a significant concern on festive occasions (Sharma, A., 2015). Encroachment, another hindrance to the heritage of the city, shadows the development of pilgrimage tourism to its highest potential. Notwithstanding that Varanasi is itself a densely-populated city, the resulting imagery for pilgrims is a very cramped city with select few spots and inefficient infrastructure. A developmental improvement was required to ease the quests for accessibility, repossessing encroached heritage, and tourism management, all while being environmentally and economically sustainable. To a large extent, these factors have been addressed by the SKVTC Project insofar as, and as the Project progresses, it will be interesting to monitor the reformations in infrastructure development.

The theme of the Corridor Project, "Beautification and Extension of Temple Premises", is to cater to the tangible and intangible cultural heritage associated with the Kashi Vishwanath Temple, Ganga, and the holy city. Shri Kashi Vishwanath Temple Corridor (SKVTC) has taken care of the faith mentioned in the Hindu scriptures. The 6.4 km long riverfront of the Ganga with 84 ghat catches up with the historically developed socio-religious ideas, values, and place consciousness of pilgrims and their faith. They altogether help to form a unique faith scape. The Kashi Vishwanath Temple continued the immortal legacies of Hindu pilgrimage and religious tourism. The essence and importance of the temple are continuously increasing among pilgrims and tourists, requiring more accommodating areas to facilitate the basic amenities and proper space for different spiritual rituals and other necessary customs. The Corridor fulfills the need by extending the temple premises with arranged accommodation and smooth accessibility through Ganga.

Mahatma Gandhi, a luminary in Indian thought and politics, was proud of Hinduism, but it did not prevent him from rejecting and criticising several institutions, ideas, and beliefs that Hindus would ordinarily regard as part of their religion. Gandhiji was deeply pained after visiting Kashi Vishwanath Temple, and he had mentioned his pain in his autobiography 'My Experiments with Truth' (1928). He writes,

"The approach was through a narrow and slippery lane. Quiet, there was none. The swarming flies and the noise made by the shopkeepers and pilgrims were perfectly insufferable where one expected an atmosphere of meditation and communion, and it was conspicuous by its absence".

During the inauguration of Banaras Hindu University on February 4 1916, Gandhiji expressed his deep disappointment with the dirty state of Kashi Vishwanath Temple. He said,

"If a stranger dropped from above on this great temple and he had to consider what we as Hindus were, would he not be justified in condemning us? Is not this great temple a reflection of our character? I speak feeling as a Hindu. Is it right that the lanes of our sacred temple should be as dirty as they are?"

Earlier in 1902, when he was not 'Mahatma' yet, he especially paid a visit to Kashi Vishwanath Temple and had similar feelings. In this context, the SKVTC project, providing comfortable accessibility from the river Ganga, may be regarded as an authentic resolution to the concerns voiced by Mahatma Gandhi. This is because, facilities under the SKVTC, among other, includes-

(i) The 50 feet wide Corridor directly connects Kashi Vishwanath Temple to the River Ganga. Pilgrims can reach the temple from the Lalita and Manikarnika Ghats through the 400-meter-long Corridor, with escalators and ramps for accessibility. The Corridor will culminate into a grand, majestic square before the temple.
(ii) The temple corridor is now spread over five lakh square feet from 3,000 sq.ft., including a civic square in front of the temple, amenities and offices for the trust that runs the temple.
(iii) The temple corridor has the provision of waiting rooms.
(iv) The museum auditorium and *Yagyashalas* will help to safeguard and revive the rich cultural heritage of Varanasi.
(v) The hassle-free lodging facilities in the temple corridor area will help extend the stay of the pilgrims and tourists.
(vi) An authentic food street in the corridor area will help revive the region's gastronomy.
(vii) The proposed *Bhogshala* will help revive the holy city's spiritual rituals and traditions.

Shri Kashi Vishwanath Special Area Development Board (SKVSADB) has been assigned to develop and maintain the cultural, spiritual, mythological, and architectural aesthetics to promote tourism. The board's objectives mainly include conserving, preserving, and protecting the heritage under the administration, beautifying, restoring, and renovating the temple's premises. Shri Kashi Vishwanath Temple Act, 1983 ensures proper and better administration of the temple by adequate arrangement for the preservation and management of the properties, including essential public requirements and facilities for the proper performance of worship to the devotees, including their arrangement for accommodation and other facilities.

Heritage Corridor Relevance

The terms freight corridor, economic Corridor, and cultural Corridor are frequently used to specify a specialty area and promote more accessibility and smoothness. The modern approach of the heritage corridor combines recreation and tourism development with ecological restoration and cultural conservation, which was first introduced in the USA in the 1980s. The minimum Cumulative Resistance (MCR) model and Geographical Information System (GIS) technology are used to develop a corridor. A new approach to the heritage corridor is designed to analyse and identify the potential areas for the heritage corridor (Wei, 2005).

A heritage corridor is expressed as a linear landscape structure that combines the utility of recreational activities, heritage protection, and ecological conservation (Conzen & Wulfestieg, 2001). The importance of cultural and visitor amenities is of significant concern besides cultural activities associated with the tangible heritage, considering the horizontal process of the landscape.

The modern approach and relevance of building a heritage corridor minimise the cumulative resistance and enhance the accessibility of visitors with the best recreational and cultural uses of landscapes (Braslavsky et al., 2002; Byrne, 2016). The specialty of corridor construction is considered a theme-based approach to minimising the flow resistance between the source and destination. The SKVTC, with 50 feet-wide passage accessibility to the temple from Ganga, connects the Gangadhar Shiva with the river Ganga, both having a mythological and cosmological relation. According to the Kashi Khanda of Shiva Purana, the darshan of the jyotirlinga, i.e. Shri Kashi Vishwanath directly after bathing in the holy river Ganga, is considered one of the most pious tirtha for a human (Bhatt, 2004). It renounced them from the cycle of rebirth and ascertaining moksha. This high significance for the pilgrims has been looked after well through the slanting ramp constructed to facilitate direct and easy movement from the river to the temple corridor. For security, free locker rooms are provided to pilgrims which can be accessed through all the gates of the Corridor making it easier for the visitors to perform rituals without the concern for security.

Temple Corridor Proposal, Planning and Detailed Project Report (DPR)

Shri Kashi Vishwanath Temple Corridor (SKVTC) is the dream project of Prime Minister Narendra Modi, who has also represented the Varanasi Lok Sabha constituency since 2014. He laid down the foundation stone of the above ambitious project under the "*Kashi Vishwanath Mandir Visistikaran- Saundrikaran Yojana*" (Kashi Vishwanath Temple- Extension and Beautification Plan, Fig. 1), intending to renovate and revive the soul of the ancient heritage city (Rajeev, 2018). The encroached historical temples are being renovated and are now part of the temple corridor area. Ahmedabad-based HCP Design Planning and Management are carrying out the project, and the estimated project cost is Rs 800 crores.

The Rediscovery of 'Encroached' Heritage

The ancient temples rediscovered in the corridor area are of Shaiva affiliation, now part of the temple corridor project. The temples, in general, enshrine a Shiva *linga* in the sanctum sanctorum (*garbhagṛha*) with images of Surya, Vishnu, Ganesha, and Shakti on the four corners of the principal shrine, showing the rich tradition of *panchadevopasana*. On the ground plan, the temples mainly consist of an *ardha-mandapa*, *mandapa*, and *garbhagrha*. In elevation, the temples generally consist of a *jagati*, *vedibandha*, *jangha*, *varandika,* and *sikhara* followed by *kalasha* and *bijapuraka* with a trident and *dhvaja pataka*. Among the rediscovered temples, Shri Chintamani Mahadeva Temple (latter half of the 19th century) is intricately carved with representations of *apsaras,* animals, and birds (particularly peacocks). The *Torana* on the sanctum doorway of the temple is enriched with minute carvings. The depiction on the outer walls of the temple proper, includes puranic scenes like the churning of the ocean, Vishnu reclining on the multi-headed serpent Ananta, Krishna stealing the clothes of the *gopikas*, *Kalyana-Sundara* (marriage

Figure 1. Shri Kashi Vishwanath Temple Corridor Project map layout
(Source: Times Now, 2019)

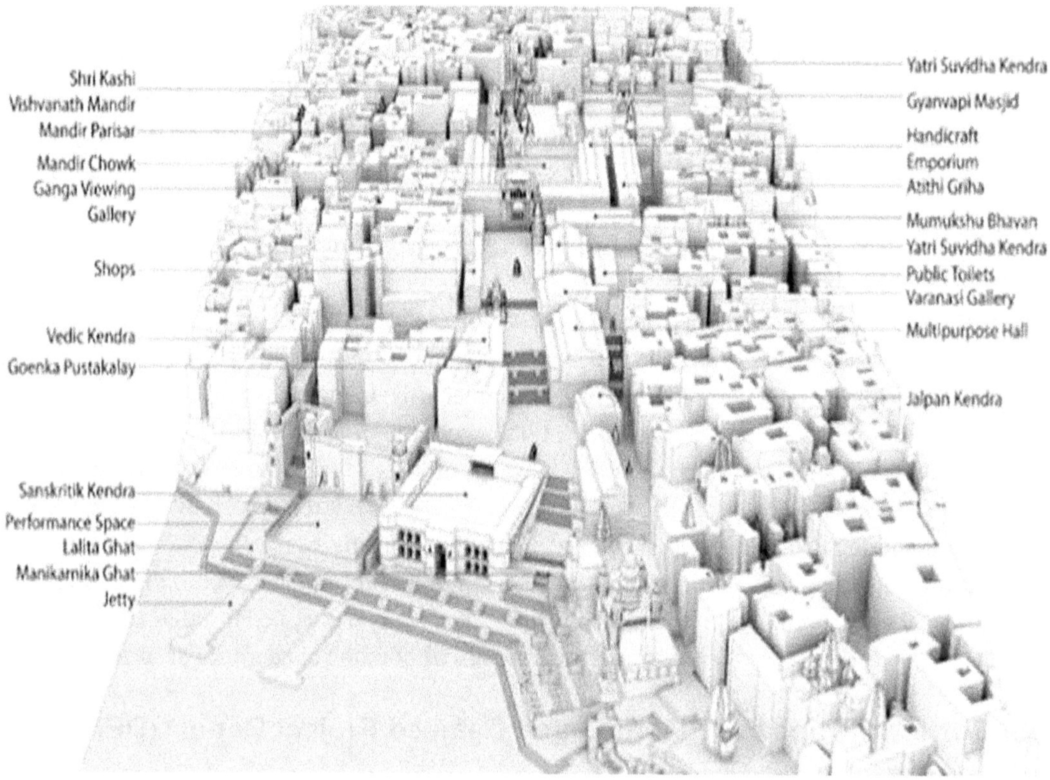

scene of Shiva and Parvati), and different incarnations of Vishnu. The rediscovered temples are mainly of the 18th – 20th century; two are interesting with inscriptions dated V.E. 1877 (1820 CE) and V.E. 1899 (1842 CE), mentioning their names as *Vishveshvara* and *Rudreshvara*, respectively. Interestingly, one of the temples is a replica of the Kashi Vishwanath Temple.

Archaeological sources attest to the religious significance of the corridor area. The presence of Padmeshvara Temple at the gate of Vishwanath Temple is confirmed by a fragment of an inscription dated 1296 CE at the Lal Darwaza Mosque in Jaunpur (Motichandra 1962). It also corroborates the location of the existing Vishwanath Temple. Mukteshvara, another important deity in the vicinity of the Vishwanath temple, was also demolished like the Vishwanath temple. In 1302 CE (1395 CE), a temple dedicated to Manikarneshvara is believed to have been constructed in the *Manikarnika tirtha*. The presence of temples- Vishwanath, Padmeshvara, and Manikarnikeshvara in the corridor area authenticate the site's religious importance. The List of rediscovered temples in the corridor is presented in Table 1.

SKVTC: Heritage Planning, Conservation, and Rejuvenation

The SKVTC project has a holistic approach to transforming the previously congested structures into modern amenities-based architecture with a view of providing accessibility to pilgrims and tourists. In this context, the model Pearce (1997) suggests an overview of the general conservation process. The sequencing of the stages is as follows –

- Identification of the heritage place of the object
- Research and Inventory
- Policy Setting
- Designation and Protection
- Restoration and Development
- Management and Interpretation

From the model, the heritage conservation of the temple corridor has an extricable association with heritage, culture, and tourism. The need for the preservation and conservation of heritage has paved the way for Heritage planning (Ashworth & Tunbridge, 2000). This planning involves tapping on the potentialities of tourism to create widespread awareness among the masses regarding the site's cultural values. This also helps in generating revenue sustainably for the local stakeholders and governing bodies. Therefore, a more contextual approach to heritage preservation for the SKVTC Project could be the conservation policy and practice system exhibited in Figure 2 proposed by The Getty Conservation Institute (2000):

In the relevance of pilgrimage tourism, this latter model of conservation planning appears more inclusive of local stakeholders, religious constructs, and pilgrimage authenticity. This includes generating interest amongst the masses regarding the heritage of a place by defining a set of identities that the site then guards and promotes (Henderson, 2002; Swain, 1989, 1990; Teo & Huang, 1995).

Figure 2. A conservation proposal of policy and planning that integrates stakeholders and social contexts (Source: Erica et al., 2000).

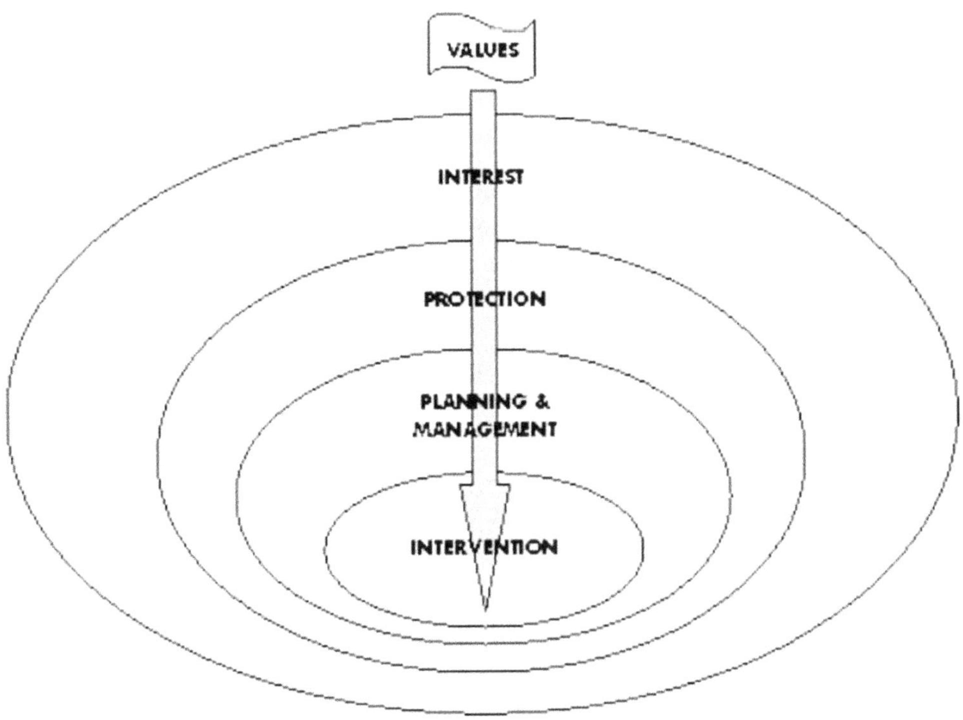

Figure 3. The process of heritage conservation with scope for assessment
(Rani et al., 2018).

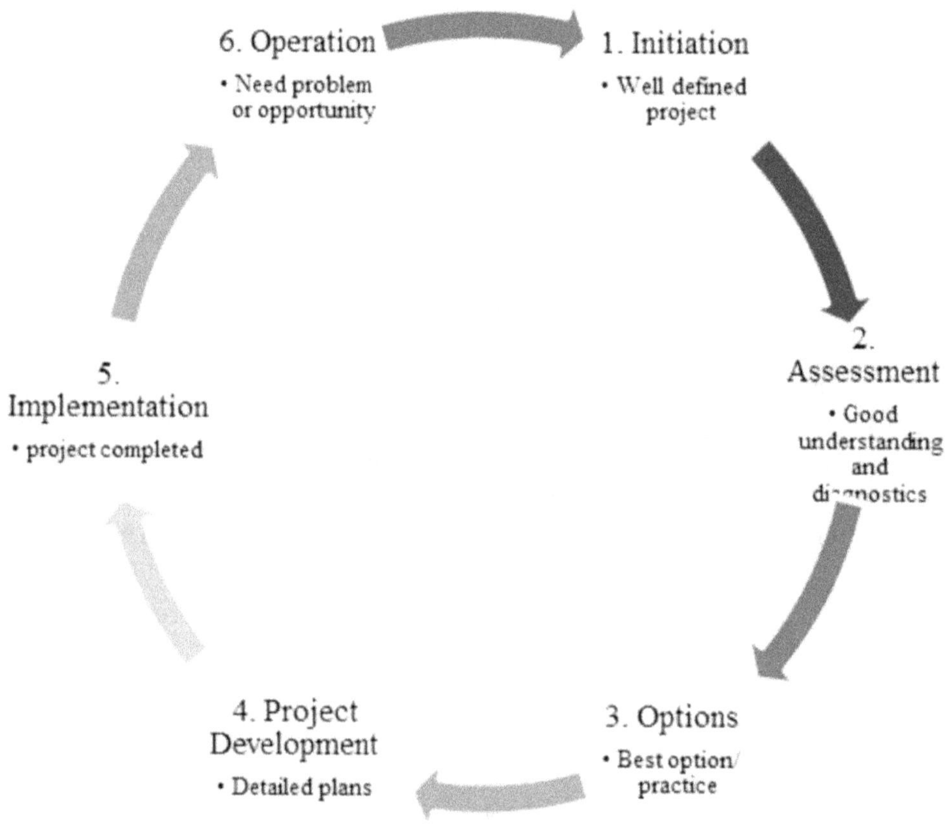

This heritage conservation model (Figure 3) accommodates the religious values of a site by not isolating local stakeholders and analysing the protection of tangible and intangible values. This is done by drafting well-defined guidelines for the project that assimilate a sound understanding of the major and minor strengths and weaknesses of a particular heritage site from the vantage point of conserving its heritage. It curates detail-oriented solutions to safeguard the heritage while maximising tourism potential in order to create an active spot at the heritage site for job opportunities that help gradually impact the economic realities of the locality.

Tourism Infrastructure Development and Revival of Cultural Heritage for Improved Pilgrimage

SKVTC will be a milestone for boosting tourism and employment opportunities in the region; the proposed *Yatri Suvidha Kendra* with a facilitation centre will encourage the duration of pilgrims and tourists and help promote the tangible and intangible heritage of the region. The *Vedic Kendra* will be a corpus of philosophical dialogue, myths, ritual incantations, and spells. The rich content of oral literature and indigenous techniques employed by Brahmin priests will help revive Vedic recitations (UNESCO, 2008). The corridor area may also be an ideal place for curious pilgrims who want to know more about the faith and practices of the *Sanatana Dharma*. Visiting the corridor area can sketch the Hindu pantheon, a sense of Hindu mythology, and metaphysics. The displayed objects in the proposed City Museum will throw more light on the rich artistic heritage of the place.

Varanasi has been a waypoint for traders and travellers and attracts philosophers, writers, and music lovers. It was famous for its weavers, who wove a generous amount of gold and silver zari into their silk brocades, heavily embodied with elaborate patterns and designs. The multipurpose hall under construction can be a center of activities to boost the potential of local emerging artisans and artists of Banaras Gharanas. Moreover, Nilkantha Pavilion, *Bhogshala*, and Spiritual Book Center will also be a center of attraction for pilgrims and tourists. Thus, the corridor development will help revive the rich artistic heritage of Varanasi and may also cater to various segments of tourists for arrival and increasing tourist footfall, as supported by studies by Meyer (2004).

Factoring-In Travel Precedents in Contemporary Uttar Pradesh

The futile gauging of pilgrimage tourism at the SKVTC by excluding other religious sites and upcoming important temple corridors brings a very narrow comprehension of the SKVTC Corridor. Precedents in tourist behaviour are better understood by acknowledging secondary stops and complete itineraries of pilgrims. For instance, many pilgrims who visited the Varanasi Corridor in large groups or with families, tend to have a larger itinerary with Varanasi being one of the many travel destinations. These are mostly South Indian travellers (Karnataka, Kerala, Tamil Nadu, etc.) (Chatterjee, 2022). Their pilgrimage therefore consists of other Hindu holy places in the North India, mainly Haridwar, Rishikesh, Mathura, Vrindavan, Ayodhya, Prayagraj, etc. (Tourism Statistics, 2019). The opening of the Corridor has propagated interest from outstation pilgrims to not just Varanasi, but other north-Indian pilgrimage sites as well. However, with the ongoing construction of Ram Janmbhoomi temple at Ayodhya, pilgrimage at both cities, viz. Ayodhya and Varanasi are seeing a massive boost. It will be interesting to factor if the pilgrims are common, because it will point towards a contemporary precedent in pilgrimage in the state of Uttar Pradesh.

The requirement of acknowledging this direction in tourism is vital, as, according to "The tourist route system – models of travelling patterns" by Thor Flognfeldt Jr., examining tourism as a dynamic function is an integral part of navigating travel experience (2005). This is further explained by the following model, i.e. Figure 4:

This model illustrates diverse travelling options among tourists, from which, one can extract the experience of travelling by an outstation pilgrim discussed above.

The Spotlight on 'Pilgrimage Tourism': Similar Quests for Heritage Corridors

The website of the economic policy division of the Government of India's Ministry of External Affairs reveals that the Government of India has allocated US$185 million in the 2018-19 Budget for the development of tourist circuits under the Swadesh Darshan and the Pilgrimage Rejuvenation and Spiritual Augmentation Drive (PRASAD) schemes. These two schemes are part of the Pilgrimage Rejuvenation and Spiritual Augmentation Drive (PRASAD). According to the National Sample Survey Office (NSSO), the religious economy in India is valued Rs. 3.02 lakh crore, which is equivalent to almost 2.32 percent

Figure 4. Modes of travel according to the modified Campbell (1999)

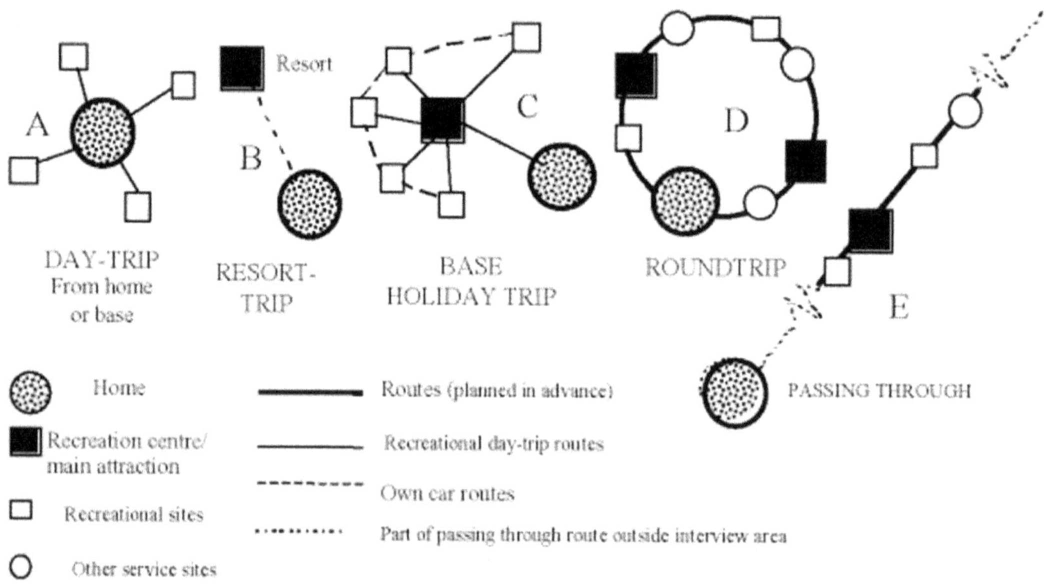

of India's GDP ('Govt. push for temple corridors', 2022). During the second term of the current government, which began in 2019, other historic corridor initiatives have been prepared and planned to address issues that are comparable to those regarding the Kashi Vishwanath Temple. The overarching concerns being congested space, difficulty in access, overcrowding, lack of authenticate information dispensing outlets, security management, and efficient conservation of heritage. A look at some of these corridor plans verifies the start of a new-age tourism infrastructure development, viz. temple corridors.

- **Shri Mahakal Lok Corridor, Ujjain**: In October 2022, Prime Minister Narendra Modi launched the Shri Mahakal Lok Corridor in the city of Ujjain, which is located in the state of Madhya Pradesh. This event was a part of the 856 crore Mahakaleshwar rehabilitation project (Figure 5). The main aim of corridor making is to decongest the entire area and also put special emphasis on restoration and conservation of heritage structure. The Mahakal corridor contains 108 *stambhas* (pillars) and is 900 m in length which depicts Anand Tandav Swaroop (Dance Form) of Lord Shiva. It is built around the old Rudra Sagar Lake around the Plaza area (2.5 hectares) which consists of 200 idols of Lord Shiva.
- **Vindhyachal Corridor, Mirzapur:** This Corridor is situated on Vindhya hills on the bank of the holy Ganga (Figure 6). According to the temple corridor Action Plan prepared for the development of Vidhyavasani Devi Corridor, the construction of the temple park, 50 ft wide *parikrama path* (circumambulatory path), widening and beautification of road and main gate, construction of facade, strengthening and construction of connecting access roads will be provided.
- **Shree Jagannath Temple Corridor Project, Puri:** Began in 2019, it is also named the Shreemandira Parikrama Project (Figure 7), it is perhaps the most extravagantly envisioned project with reportedly an estimated budget of Rs. 3,200 crores ('*Shree Jagannath Temple corridor*

Figure 5. Draft plan for the Shri Mahakal Lok Corridor, Ujjain, Madhya Pradesh
(Source: Times of India, 2022).

Figure 6. Layout plan of Vidhya Corridor
(Source: Hindustan times,2022)

project', 2022). The Week reports, "75m circle around the shrine has been cleared of buildings, and the holy site, with its 65m vimana, is now visible from far away (Sharma,2022). It involves

Figure 7. Proposed Shree Jagannath temple corridor
(Source: Sharma, 2022)

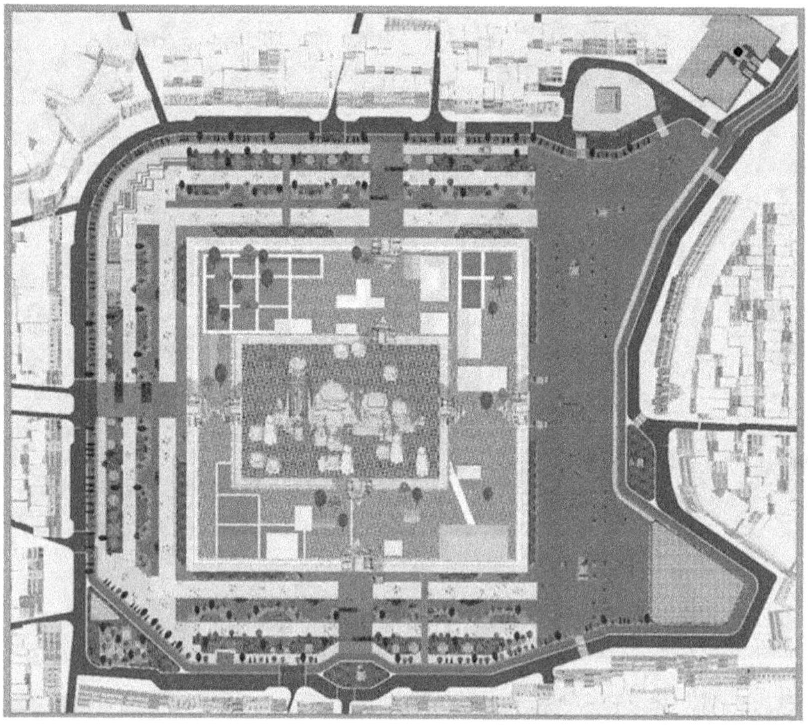

22 aspects including land acquisition, rehabilitation work, road improvement, building of cultural centres, revival of lakes and development of the cremation ground" (Sharma, 2022).

Banke Bihari Corridor, Mathura: In August of 2022, the announcement regarding this Corridor was made for the very first time. Reports from a variety of news organisations estimate that the projected corridor will have a price tag of roughly 248 billion rupees. (Times Now, 2022). It has been stated that five acres of property in the neighbouring area will be purchased in order to accommodate roughly 5,000 worshippers at the same time. On the other hand, the sketch plan and configuration of the corridor are not yet accessible to the general public.

DISCUSSIONS AND IMPLICATIONS

The requirement and relevance of the SKVTC are directly linked with the preservation of heritage and tourism infrastructure development to accommodate the continuously increasing footfalls of pilgrims. The contemporary demand of the visitor's management is to provide them with better amenities and accessible ritual-performing facilities. The project came up with many debates and obstacles due to issues related to encroachment and resettlement, as the demolition of about 291 buildings near the temple triggered some people's resentment and anger against the government (Lazzaretti, 2021). Several temples and idols, including Shiva *linga*, were claimed to be demolished (Lazzaretti, 2021). The firm determination of both central and state governments made possible the inauguration of the first phase on December 13, 2021.

Particular emphasis is given to reviving the lost legacies for maintaining and preserving the cultural heritage of Varanasi and rituals, particularly those associated with the holy Ganga riverfront. The corridor project is dedicated to rejuvenating the religious monuments in the area and restoring the previous essence and relevance of Kashi and Ganga. The most crucial point of this corridor making is the easy accessibility from the Ganga with a smooth 400-meter walk. It will impact the stakeholders and local community as more pilgrims' movements mean more opportunities for business and employment. Traders in the area say that their business has already seen an uptick during the project's construction. Food courts serving diverse cuisines of India is now operational (as of June 2023) and according to the ground staff, receive customers from mostly outside Varanasi which attests that this conservation planning appears to be directly consumed by pilgrims from outside the locality. The entire discussion revolved around making corridors for the preservation, rejuvenation, and socio-economic and cultural upliftment of communities.

RESEARCH FINDINGS AND CONCLUSION

The recently opened Kashi Vishwanath Corridor in Varanasi is being hailed largely as a symbol of the reawakening of India's spiritual awareness. This recognition comes on the heels of the corridor's official inauguration. Shri Kashi Vishwanath Dham was an ambitious project that was created at a cost of roughly Rs 339 crore. On December 13, 2021, Prime Minister Narendra Modi dedicated the first phase of the project to the public. The construction of this phase cost approximately Rs 339 crore. The effort creates a connection between the grounds of the temple and the Ganga River, which provides devotees with a variety of amenities. Following the completion of the spectacular project, the area of the Kashi Vishwanath temple was enlarged from 3,000 square feet to nearly five lakh square feet, making it possible for the temple to accommodate 50,000 to 75,000 pilgrims.

After analysing every aspect of the Shri Kashi Vishwanath Temple Corridor (SKVTC) and all its stakeholders, it is clear that the infrastructure development will undoubtedly benefit both the local community and pilgrims. The temple corridor will also help the initiatives undertaken under the Smart City Project. The state government has successfully resolved the land acquisition and rehabilitation issue of around 1,000 families, many of whom lived in the area for generations. The SKVTC plan set a milestone in the history of Varanasi. The narrow winding lanes (*gali*) and the localities (*mohallas*) are usual characteristics of the holy city as one can find an array of shops dotting the serpentine lanes. Shopkeepers sell traditional Banarasi sarees, dresses, devotional articles, bangles, wooden toys, brass articles, statues, religious books, jewellery, dress materials others. Vishwanath *Gali*, leading to Kashi Vishwanath temple, is famous among these. There were complaints from the pilgrims and tourists regarding the filthy condition of the lanes and particularly the menace caused by stray animals. The safety of visitors was also a significant concern. Throughout the course of the interactions and interviews, the vast majority of the specialists, locals, visitors, and pilgrims expressed happiness with the measures that the government had recently undertaken. They were under the impression that the creation of contemporary infrastructure would attract more tourists, which would in turn open up chances for the local people and any other stakeholders. It is appreciable to see that the historical temples were rediscovered after demolishing the residential buildings are being restored to their original form. Overall, 291 buildings in the acquired area were vacated and purchased by the government. The Corridor is expected to revive the religious significance of the ancient holy city. The corridor project work commenced on January 15, 2020, and will provide pilgrims with easier access from Ganga to Kashi Vishwanath Temple. The expansion and beautification project for the Kashi Vishwanath Temple will ensure visibility of the temple directly from the ghat, which was hidden behind a maze of narrow, congested lanes. The estimated 5 lakh sq. ft. area of the Corridor, will be embellished with Baleshwar stones, Makrana marble, Kota granite, Mandana stones, and red sandstone of Chunar. The Chunar sandstone has been used to make the pillars, whereas Makrana marble for the flooring of the Corridor and Kota granite for designing the Corridor. Mandana stone has been used on the stairs built at the ghat.

FUTURE SCOPE

The 7.35 cr. annual footfall to the Corridor since the inauguration of Phase I is verifying tangible growth in tourism and tourism-related sectors in the city (Dikshit, 2022). It has surpassed the biggest tourist destinations of the country like Goa, which received some three cr. annual tourists. The formation of a concrete scholarly basis of the present status will not only document the functionality of the Project but can also be used as a Case Study for Corridors elsewhere in the country and abroad (Teo, 2000). The related findings will be crucial for local stakeholders in their comprehension of the Project and everything that it entails. Models of tourism ecosystems can be produced by analysing the present study which focuses on sustainability as a driving factor in contemporary religious infrastructure planning and policy frameworks. The cultivation of sustainable cultural policies narrates a unique set of challenges faced by the religious city of Varanasi and how the ambitious Project builds the platform for generating definitive solutions against them.

REFERENCES

Ashworth, G. J., & Tunbridge, J. E. (2000). The tourist city. In *Advances in Tourism Research,* (pp. 53–81). Science Direct. https://www.sciencedirect.com/science/article/abs/pii/B9780080436753500077

Bausch, T., Schröder, T., & Gunya, A. (2020). Pilgrimage as a field of tourism development: The case of Kavala (Northern Greece) and its institutional and cultural challenges. *International Journal of Religious Tourism and Pilgrimage*, *8*(3), 111–129.

Bec, A., Moyle, B., Timms, K., Schaffer, V., Skavronskaya, L., & Little, Ch. (2019). Management of immersive heritage tourism experiences: A conceptual model. *Tourism Management*, *72*, 117–120. doi:10.1016/j.tourman.2018.10.033

Bhatt, G. P. (2004). The Skanda Purana. Motilal Banarasidass Publishing House.

Braslavsky, C., & Fernǹdez, A. L. (n.d.). *A Educaô Secundr̀ia: mudanà ou imutabilidade? - Brasr̀ia: UNESCO, 2002. United Nations Educational Scientific Cultural Organisation (UNESCO).* Brazil Chapter.

British Broadcasting Corporation [BBC] (Director). (2019, October 24). *Kartarpur corridor: India and Pakistan sign deal on Sikh temple project.* BBC. https://www.bbc.com/news/world-asia-india-50138275

Byrne, D. (2016). Heritage Corridors: Transnational flows and the built environment of migration. *Journal of Ethnic and Migration Studies, 42*(14), 2360–2378. https://doi.com/10.1080/1369183X.2016.1205850. doi:10.1080/1369183X.2016.1205805

Campbell, C. K. (1999). *An approach to research in recreational geography.* Department of Geography, University of British Columbia.

Chatterjee, K. (2022). Modern Tourism Development through Sustainable Tourism Approach: A Case study of Pilgrimage Tourism in Varanasi. *The International Journal of Creative Research Thoughts, 10*(5), 361–372.

Conzen, M. P., & Wulfestieg, B. M. (2001). Metropolitan Chicago's Regional Cultural Park: Assessing the Development of the Illinois & Michigan Canal National Heritage Corridor. *The Journal of Geography, 100*(3), 111–117. doi:10.1080/00221340108978426

Desai, M. (2017). *Banaras Reconstructed. Architecture and Sacred Space in a Hindu Holy City.* University of Washington Press.

Di Giovine, M. A., & Elsner, J. (2016). Pilgrimage tourism. In J. Jafari & H. Xiao (Eds.), *Encyclopedia of Tourism.* Springer., doi:10.1007/978-3-319-01384-8_146

Dikshit, R. (2018, May 8). Soon, access Kashi Vishwanath temple directly from ghats. *The times of India.* https://timesofindia.indiatimes.com/city/varanasi/soon-access-kashi-vishwanath-temple-directly-from-ghats/articleshow/64072869.cmsDikshit, R. (2022). 7.35 crore visit Kashi Vishwanath Dham in opening year. *The Times of India.* https://timesofindia.indiatimes.com/city/varanasi/7-35-crore-visit-kashi-vishwanath-dham-in-opening-year/articleshow/96211539.cms

Divya, A. (2021, December 14). Explained: What is changing at the ancient Kashi Vishwanath temple complex? *Indian Express.* https://indianexpress.com/article/explained/kashi-vishwanath-temple-narendra-modi-varanasi-7669709/

Eck, D. L. (1982). *Banaras. The City of Light.* Penguin India.

Enrico, B., Nuccio, M., & Durio, A. (2021). Proximity tourism and cultural amenities: Evidence from a regional museum card. *Tourism Economics, 27*(1), 187–204. doi:10.1177/1354816619890230

Erica, A., Randall, M., & Marta, T. (2000). *Values and Heritage Conservation.* The Getty Conservation Institute.

Freitag, S. (1989). Introduction: The History and Political Economy of Bañaras. In *Cultural and Power in Banaras* (pp. 1–22). University of California Press., doi:10.1525/9780520313392-005

Graham, B. (2000). The past in place: Historical geographies of identity. *Modern historical geographies,* 70-99.

Graham, B. (2002). Heritage as Knowledge: Capital of Culture. *Urban Studies (Edinburgh, Scotland), 39*(56), 1003–1017. doi:10.1080/00420980220128426

Griffin, K., & Raj, R. (2018). The importance of religious tourism and pilgrimage: Reflecting on definitions, motives and data. *The International Journal of Religious Tourism and Pilgrimage, 5*(3), 2–9.

Henderson, J. (2000). Attracting tourists to Singapore's Chinatown: A case study in conservation and promotion. *Tourism Management, 21*(5), 525–534. doi:10.1016/S0261-5177(99)00108-9

Hindustan Times Correspondent. (2022, September 24). Complete Vindhya Corridor in six months: Yogi. *Hindustan Times*. https://www.hindustantimes.com/cities/lucknow-news/complete-vindhya-corridor-in-six-months-yogi-101664044828857.html

Hole, Y., Khedkar, E. B., & Pawar, S. (2019). The significance of pilgrimage tourism to sustainable development with special reference to the Indian context. *African Journal of Hospitality, Tourism and Leisure, 8*(3), 1–12.

India Code [IC]. (1983). The Uttar Pradesh Sri Kashi Vishwanath Temple Act, 1983. In *Digital Repository of State Act*. India Code. https://www.indiacode.nic.in/bitstream/123456789/12272/1/Kashi1983up.pdf

India Tourism Statistics. (2022). *Ministry of Tourism Report 2022*. Market Research Division, Ministry of Tourism, Government of India. https://www.tourism.gov.in/sites/default/files/2022-09/India%20Tourism%20Statistics%202022%20%28English%29.pdf

Indo-Asian news service [IANS]. (2019, March 8). *PM Modi Lays Foundation Stone Of Kashi Vishwanath Temple Corridor*. NDTV. https://www.ndtv.com/india-news/pm-modi-lays-foundation-stone-of-kashi-vishwanath-temple-corridor-2004557

Kang, J. (2005). Suitability analysis of heritage corridor in the rapidly urbanising region: A case study of Taizhou City. Geographical Research, Pecking University, Beijing.

Kirmani, S. (2018, November 29). *Kartarpur corridor: A road to peace between India and Pakistan?* BBC news online. https://www.bbc.com/news/world-asia-46382657

Kumar, R. (Ed.). (2003). *Essays on Indian music*. Discovery Publishing House.

Lazzaretti, V. (2021). The boundary within: Demolitions, dream projects and the negotiation of Hinduness in Banaras. In *Spaces of Religion in Urban South Asia* (pp. 87–99). Routledge.

Meyer, D. (2004). *Tourism routes and gateways: key issues for the development of tourism routes and gateways and their potential for Pro-Poor Tourism*. [Discussion Paper, Overseas Development Institute].

Morinis, A. (1992). *Sacred journeys: The anthropology of pilgrimage*. Greenwood Press.

Motichandra. (1962). *Kashi Ka Itihas*. Mumbai: Hindi Granth Ratnakar Pvt. Ltd.

Noga, C. K. (2020). Pilgrimage tourism-past, present and future rejuvenation: A perspective article. *Tourism Review, 75*(1), 145–148. doi:10.1108/TR-04-2019-0130

Pearce, D. (1997). The roles of the public sector in conservation and tourism planning. *Tourism and heritage management*, 88-100.

Pearce, J. (1997). Commentary by John Pearce. *Bulletin de l'Institut Pasteur*, *95*(3), 182. doi:10.1016/S0020-2452(97)81374-7

Pearson, M., & Sullivan, S. (2013). *Looking after heritage places: The basics of heritage planning for managers, landowners and administrators*. Melbourne Univ. Publishing.

Prasad, N. (2019, June 13). Kashi Vishwanath Temple Corridor's final DPR ready! PM Narendra Modi's dream project is every pilgrim's joy. *Financial Express*. https://financialexpress.com. https://www.financialexpress.com/lifestyle/travel-tourism/kashi-vishwanath-temple-corridors-final-dpr-ready-pm-narendra-modis-dream-project-is-every-pilgrims-joy/1606684

Rangaswami, C. (2017). Ramanathaswamy temple. In C. M. Rangaswami (Ed.), Ramanathaswamy temple (pp. 201–207). Bharti publications.

Rani, W. M., Wan, Tamjes, M.S., & Wahab, M.H. (2018). Governance of Heritage Conservation: Overview on Malaysian Practice. *The Journal of Social Sciences Research*. 847-851. https://ideas.repec.org/a/arp/tjssrr/2018p847-851.html

Richards, G. (1996, ed.). Cultural Tourism in Europe. CAB International, Wallingford.

Richards, G., & Munsters, W. (Eds.). (2010). *Cultural tourism research methods*. Cabi. doi:10.1079/9781845935184.0000

Roche, E. (2019, October 24). India and Pakistan signed a historic agreement to open the Kartarpur corridor. *Livemint*. https://www.livemint.com/politics/policy/india-pakistan-sign-pact-to-enable-indian-pilgrims-visit-kartarpur-11571901789231.html

Sharma, A. (2015). Smart City plan of Varanasi identifies traffic congestion and encroachment as major problems. *The Economics Times*. https://economictimes.indiatimes.com/news/economy/infrastructure/smart-city-plan-of-varanasi-identifies-traffic-congestion-and-encroachment-as-major-problems/articleshow/50226443.cms?from=mdr

Sharma, P. (2022, October 9). Explained: The Rs 3,200 crore Shree Jagannatha Temple corridor project. *The Week*. https://www.theweek.in/theweek/cover/2022/09/30/explained-the-rs-3-200-crore-shree-jagannatha-temple-corridor-project.html

Shri Kashi Viswanath Temple Trust. (n.d.). *Home*. Shri Kashi Viswanath Temple Trust. https://www.shrikashivishwanath.org/

Sigala, M., & Leslie, D. (2016). *International cultural tourism: management, implications, and cases*. Routledge.

Singh, R. P. (2011). Varanasi, India's cultural heritage city: contestation, conservation & planning. *Heritagescapes and cultural landscapes*, 205-254.

Singh, S. V., & Ranjan, R. (2019). Online travel portal and their effect on travel agency: A study on outbound visitors of Varanasi. [IJRAR]. *International Journal of Research and Analytical Reviews*, *6*(2), 387–393.

Sinha, A., Harkness, T., & Harkness, T. (2009). Views of the Taj—Figure in the Landscape. *Landscape Journal*, *28*(2), 198–217. https://www.jstor.org/stable/43323846. doi:10.3368/lj.28.2.198

Swain, M. B. (1989). Developing ethnic tourism in Yunnan, China: Shilin Sani. *Tourism Recreation Research*, *14*(1), 33–39. doi:10.1080/02508281.1989.11014541

Teo, P., & Chang, T. C. (2012). 8 Singapore: tourism development in a planned context. In *Tourism in South and Southeast Asia: Issues and cases* (pp. 117–128). Routledge.

Times Now. (2022, October 19). After Kashi and Mahakal Lok, BJP to build Banke Bihari Corridor: All you need to know. *Times Now*. https://www.timesnownews.com/india/after-kashi-and-mahakal-lok-bjp-to-build-banke-bihari-corridor-all-you-need-to-know-article-94969557

Times of India Report [TOI]. (2022, October 11). All you need to know about Ujjain's Mahakal Lok Corridor. *Times of India*. https://timesofindia.indiatimes.com/city/indore/all-you-need-to-know-about-ujjains-mahakal-lok-corridor/articleshow/94776634.cms

Timothy, D. J. (2000). Sacred Journeys: Religious heritage and tourism. *Tourism Recreation Research*, *27*(2), 3–6. doi:10.1080/02508281.2002.11081214

United Nations Educatiional, Scientific and Cultural Organisation [UNESCO]. (2003). *Tradition of Vedic chanting*. UNESCO. https://ich.unesco.org/en/RL/tradition-of-vedic-chanting-00062

United Nations Educational, Scientific and Cultural Organization report [UNESCO]. (n.d.). Index of development of a multidimensional framework for heritage sustainability. In *CDIS Methodology Manual V7*. United Nations Educational, Scientific and Cultural Organization. https://en.unesco.org/creativity/sites/creativity/files/cdis/heritage_dimension.pdf

United Nations World Tourism Organisation. (n.d.). *Report*: *Tourism and Culture*. UN WTO. https://www.unwto.org/tourism-and-culture

Verma, L. (2021, December 13). PM Modi to inaugurate Kashi Vishwanath corridor in Varanasi today. *Indianexpress.com*. https://indianexpress.com/article/india/pm-modi-to-inaugurate-kashi-vishwanath-corridor-in-varanasi-7669259/

Wei, Y., & Jian, K. (2005). Soundscape and Sound Preferences in Urban Squares: A Case Study in Sheffield. *Journal of Urban Design*, *10*(1), 61–80. doi:10.1080/13574800500062395

KEY TERMS AND DEFINITIONS

Pilgrimage Tourism: Pilgrimage and recreation performed in the milieu of consuming religious cultural heritage (tangible and intangible) outside one's place of residence.

Pilgrimage Attractiveness: A place of interest that pilgrims visit, particularly for its exhibited sacred/religious value.

Heritage Planning: The practice towards preserving, restoring, and managing heritage resources for sustainable purposes.

Cultural Heritage: The set of core values (tangible and intangible) inherited by a community represented through tangible and intangible means.

Rediscovered Temples: These temples from within the Corridor area were hidden from plain sight due to encroachment; although there was active worship in some of these temples, their significance had been largely diminished due to absence from pilgrimage routes.

Heritage Corridor: A designated area enriched with historical/cultural heritage aspects that has been earmarked for the purposes of sustainability.

APPENDIX 1

List of essential temples rediscovered and currently being renovated during the ongoing SKVTC Project:

Table 1. List of rediscovered temples

S. No.	Name of Temples
1	Shri Kumbha Mahadev Mandir
2	Shri Baal Mukundeshwar Mahadev Mandir
3	Shri Neelkanth Mahadev Mandir
4	Shri Brahmeshwar Mahadev Mandir
5	Shri Chandragupt Mahadev Mandir
6	Shri Adivishveshwar Mahadev Temple
7	Shri Gangeshwar Mahadev Temple
8	Shri Amriteshwar Mahadev Temple
9	Shri Jauvinayak Temple
10	Shri Swargadwareshwar Mahadev Temple
11	Shri Ram Temple
12	Shri Chintamani Mahadev Temple
13	Shri Anandakali Temple
14	Shri Mahakal Temple
15	Shri Baba Masan Nath Temple
16	Shri Tarkeshwar Mahadev Temple
17	Shri Ratneshwar Mahadev Temple
18	Shri Panchatan Mahadev Temple
19	Sheetla Mata Mandir
20	Shri Shiv Temple
21	Shri Sankatmochan Temple
22	Shri Karuneshwarai Mahadev Temples
23	Shri Futa Ganesh Vinayak Temple
24	Shri Rudreshwar Mahadev Temple
25	Shri Narmdeshwar Mahadev Temple
26	Shri Shiv Mukaneshwar Mahadev Temple
27	Shri Pramod Vinayak Temple
28	Shri Brahmagupteshwar Mahadev Temple
29	Shri Koneshwar Mahadev Temple
30	Mata Kali Temple
31	Shri Rani Vijayraj Mahoba Temple

Chapter 5
Dark Pilgrimage and Implications in Dissonant Heritage Consumption:
A Short Introduction

Maximiliano Emanuel Korstanje
https://orcid.org/0000-0002-5149-1669
University of Palermo, Argentina

ABSTRACT

In the present book chapter, we have interrogated not only the nature of dark tourism but also the evolution of tourism since the former century to date. We have offered a model based on three stages of tourism, each one conditioned by a turning point or founding event. Classic tourism was originated in the WWII end but finalized as a result of Oil embargo in 1970. A new sustainable form of tourism supported by heritage finally came to stay until the attacks to the US. The surfacing of new morbid form of tourism witnessed the exhaustion of an industry chiefly worried to face unnumberable global risks.

INTRODUCTION

It is not simplistic to say that dark tourism has exponentially grown in recent years consolidating as one of the trending topics and paradigms of the current tourism research (Hooper & Lennon 2016; Light 2017). Leading journals commonly publish one or two papers on dark tourism per issue. This happens simply because dark tourism has captivated not only the attention of journalists globally but also academicians -even some of them coming from social sciences (Shekhar & Valeri 2022; Tarifa-Fernandez, Carmona-Moreno & Sanchez-Fernandez 2022). At a closer look, the abundance of publications has accelerated a dispersion in the produced knowledge as well as the multiplications of definitions and application of the phenomenon (Hooper & Lennon 2016). Dark tourism is often associated with similar terms such as mourning tourism, black tourism, thana tourism or grief tourism -to name some of them-. The debate regarding the nature of dark tourism seems not to be easy to summarize. However, two main families

DOI: 10.4018/978-1-6684-4817-5.ch005

of theories prevail. On one hand, Philip Stone and the thanaptopic tradition focus on dark tourism as a human attempt to contemplate the proper life through the "Other´s death". For this tradition, the secularizing process has relegated religiosity to a private sphere while rising a negation of death. As a rite of passage, dark tourism deals with the complexity of death interrogating furtherly the own finitude through the lens of the "Other´s death" (Seaton 1996; Stone 2012; Stone & Sharpley, 2008). Secondly, the heritage tradition emphasizes dark tourism as a type of pilgrimage based on the need of creating a dark heritage that mediates between the present and a troublesome or bloody past. Dark tourism in essence plays a leading role not only in educating citizens but also reminding the origins of disasters (Cohen 2011; Collins-Kreiner 2016; Olsen & Korstanje 2019). Under some conditions, the formation of this dark heritage is mainly marked by the intervention of authorities that sometimes emulate or lead to ideological discourses (Sather Wagstaff 2016; Tzanelli 2016; Korstanje 2016). In the present book chapter, we go in the opposite direction. We offer a historical sweep on the formation of dark tourism and its conceptual evolution to our days. The chapter is not a summary of the specialized literature but a debate on the evolution of dark tourism and dark pilgrimage as a result of the turn of the century. Though sophistically at a closer look, the argument is not very difficult to understand. Echoing British sociologist J. Urry who claimed that tourism tends to be consumed in decentralized forms, we hold that dark tourism ultimately devotes efforts to resolving the dilemmas of the precautionary platform and risk perception theory to make safer destinations. The turn of the twentieth century witnessed not only more global risks and radicalized violence but also the urgency to impose a new paradigm to mitigate hazards such as terrorism, natural disasters or even the recent COVID-19 pandemic. In this new epoch, the concept of prevention sets the pace invariably to adaptancy. Henceforth, dark tourism consumption emerged from the need to make more resilient destinations in an every-risky -if not uncertain- world (Korstanje 2020).

Dark Tourism and Dissonant Heritage

The term dissonant heritage has originally coined by Turnbridge & Ashworth (1996) to denote any space of massacre to narrate difficult stories or bloody legacies. To some extent, the term dissonant is referenced to the multiple interpretations various groups have along the same event. When the interpretation of the heritage is subject to controversy or attributed to different reasons, there is no consistency in the ways the site is reminded. In the constellations of dark tourism, the site is shaped into a great controversial history associated not only with the macabre but also with mass death and mourning; a point which has been widely investigated in different studies (Hartmann 2013). In other cases, the narratives are systematically constructed to glue long-established asymmetries or privileged relations of power culturally enrooted in the colonial period (Tzanelli 2016; Carrigan 2014). The commoditization of disaster is another point of entry for further discussion. These works stress that dark tourism tends to commercialize disasters as a form of morbid (gothic) spectacles which are widely consumed by an international audience (Korstanje 2016; Potts, 2012). Beyond this discrepancy, dark tourism and dissonant heritage seem to be inextricably intertwined (Bowman & Pezzullo 2009). In her seminal book entitled *Traumascapes: the power and fate of places transformed by tragedy,* Maria Tumarkin (2005) calls attention to the role played by trauma in silencing as well as transforming the remembering process. As she eloquently observes, trauma not only leads to temporal amnesia but also transforms the remembering according to what is tolerable for us. Hence the emerging fabrications, just after a disaster takes a hit, speak to us further of our emotional world than the real facts. Any trauma hides what is far from being self-evident. Dark tourism and dissonant heritage serve as a catalyst that shapes the territory while demarcating those spaces of mourning and grief (Robinson, Heitman & Dieke 2011; White & Frew 2013). Over the recent years, policymakers have delineated the borders of new epistemological guidelines to use dark tourism studies to make more resilient destinations. The advances of academia in this field have shed the light on the new steps to adapt destinations to emerging global threats (Farmaki & Antoniou, 2017). P. Stone (2006) elaborates on a conceptual model to understand the different spectrums of dark tourism consumption which ranges from the darkest to the lightest spheres. While the darkest spectrum rests on spaces of mass suffering dotted with low infrastructure, the lightest products are mainly oriented to entertainment and heritage-centered consumption. Turnbridge & Ashworth (2016) question to what extent darkness is the essence of tourism itself. In the post-9/11 context, the "Other´s suffering" has been marketed as a form of valid spectacle. In a new world where tourist security is far from being granted, the spectacle of dark tourism makes the disaster a criterion of attraction for visitors. At the same time, Fredrich, Stone & Rukesha (2018) applaud the idea of engaging with difficult heritage to make a more connected world. The experiences derived from the visit to spaces of genocide (like in Rwanda) lead visitors to cultivate mutual understanding and reciprocity for the alterity, engaging with values that foster a more peaceful peace. Contrariwise, Bhavna Singh (2020) offers an interesting point in her doctoral dissertation where she mentions some limitations behind dissonant heritage as well as its application to dark tourism. She starts from the premise the Indian model differs notably from the Anglo-Saxon one, which offers a westernized interpretation of death. As Singh highlights, the thanatological model in India does not exclude the presence of an afterlife or religiosity. Anyway, the model contemplates death as an instrument of education and pedagogy. For the memorial sites or those places of reconciliation to be successfully developed, the dissonance should be at least minimized. Having said this, this reconciliation is often frustrated because of political intervention, or in terms of authors, because of "the pervasive politicization of memorialization". In consonance with this, the quintessential characteristic of dissonant heritage is given by a lack of congruence between people, and how they identify with the place. All the involved actors keep a different interpretation of history or the traumatic event. Per her viewpoint, Stone´s model (Stone 2006) has some problems to be applied beyond Western cultures. What is more important, dark tourism as a global concept is alien in Eastern cultures, it was originally embroiled in England and rapidly disseminated to English-speaking countries like the US, Canada, New Zealand and Australia. As Korstanje puts it, the terms dark or thana tourism have been born primarily in England and are widely applied in the Anglophone world. This happens because two main reasons. Firstly, the Anglophone culture has some limitations to understand the concept of predestination. As a result of this, the future as well as death is presented as a closed entity inexpugnable to the human eye. The generated anxiety is sublimated by the articulation of rituals which avoids social disintegration. The "Other´s death" is mainly seen as a criterion of vulnerability that impedes progress and triumph in a Darwinist atmosphere. Social Darwinism, secondly, occupies a central position in the anglophone culture. Life is interpreted as a race of all against all where only a few are the winners, but for good or not, the winner takes all. The "Other´s death" not only feds the proper narcissism but ignites false happiness because I am still in the race (Korstanje 2016). There is a third factor that explains the enthusiasm for dark tourism associated with what S. Lash & J. Urry (1993) dubbed "the economies of signs and spaces". The authors capture the idea that tourism is a direct consequence of capitalist expansion. The technological breakthrough that determined the expansion of the West in the mid of the twentieth century has created faster flights and modes of transport. At the time capitalism grew, a new mobile culture come to stay. Mobilities and tourism are essentially determined by global capitalism. Over the recent decades, a process of reflexibility has created decentralized (abstract) forms of consumption where movements are not necessary. Urry & Lash refer to "virtual tourism" as the triumph of technology over the environment. As they acknowledge, in the global market not only cultures and territories but also people are merchandized as ethnic products consumed by first-world tourists. The policymakers and experts draw the world in two, safe destinations and wild destinations. While the former signals to protected spots where tourist security is preferably the priority of the state, the latter are avoided by international travelers. Lash & Urry helps us understand the evolution of tourism since the inter-war period to date as well as the rise of dark tourism just after the turn of the century. Last but not least, over the years scholars have not only misjudged the nature of tourism (defining it as a mere industry) but also its evolution. An historical approach suggests that tourism has evolved in the time once changed the means of production of western capitalism. As David Harvey (2020) notes, the notable evolution of capitalism has been widely marked by the technological revolution that ultimately optimized the means of production as well as the transformation in travel behavior. At a closer look, our travel behavior seems to be subject to much deep shifts originated in the economic matrix of society. At a closer look, we have identified three stages for tourism evolution: classic tourism, sustainable tourism and the rise of morbid forms of tourism (Korstanje 2022). Our point here lies in the fact that tourism changed according to founding events that substantially re-drawn the conditions of economy. It is noteworthy that the modern tourism industry, as well as the transport system, resulted from some socio-economic factors as the struggle of working unions to reduce the labor hours or the technological revolution in the fields of transport. At the same time economy changed, travel behavior changed as well. In consonance with that, we find a model based on three stages of tourism: classic tourism, sustainable tourism and the rise of morbid forms of tourism. Each one is traversed and conditioned by a turning event that altered the current means of production. The WWII end starts with a new period of peace and the Cold War while classic tourism expands. At the same time, the Arab-Israeli war initiates a new process framed by cultural consumption and the adoption of sustainable practices. The attacks to the US and the rise of global risks accelerated by the climate change affirms in the rise of a new morbid tourism consumption based on the "Other´s pain". This is the plain thesis that guides all the book chapter.

Classic Tourism

Classic tourism goes from the interwar period or since 1910 to 1970. This stage of production in the capitalist system is limited to what specialists termed as "mass-scale production or fordist model". Most certainly, products were fabricated at a great scale with little levels of differentiation. Mass scale production is characterized by manufacturing of large quantities of standardized products often employing assembly lines supported by automation technology. The same can be said about classic tourism. This stage starts through the 30s decade or the WWII end. This facet is mainly marked on a strong fordist model where the tourist product was standardized for a wide audience of consumers. Some experts acknowledge that the period is influenced by "the bubble model" where hosts or local are systematically excluded from the economic benefits of tourism consumption. The foreign visitor physically moves to paradisiacal islands or complexes while avoiding any contact with locals. What seems to be more important, domestic tourism is the only available product affordable for middle class while international flights are reserved for an elite. As D. Harvey puts it, the classic economy was put upside down just after the Oil war and the Embargo of Arab countries to the US. This event resulted from the Arab-Israeli war, produced not only the end of Fordism but also an economic shutdown that led to new decentralized form of tourism consumption. This was the first time Western countries witnessed they lack of the necessary oil (energy) keep their economies working. This decentralization stage characterized by the multiplication of segments, niches and consumers, tailoring each product to the consumer´s mind. This event prompted the segmentation of western economies towards new forms of tourism consumption. The ecological crisis as well as the need of adopting more flexible means of production paved the ways for the rise of a new tourism.

Sustainable Tourism

Sustainable tourism was a stage marked by the urgency to adopt to the economic crisis that whipped the West. This stage begins at 70s decade covering until 2000. In this period, the local "Other" is considered as an object of curiosity while culture and sustainability are the tug of war of tourist research. During this epoch, tourists move with the idea "of experiencing something new or getting an outstanding experience" or simply traveling to "be there" living like the local. As Dean MacCannell (1976) eloquently notes, the concept of heritage derives from a complex process triggered by higher mobilities where subject move towards what he perceives as authentic. The idea of authenticity, echoing Maccannell, mediates between citizens and their institutions. Heritage serves as the glue of society, in a world where religion recedes to the private sphere. Like the totem in the aboriginal culture, from where emanates the political authority of chiefdom, tourism regulates not only the daily frustrations but keeps society united. The apathy for the "Other" -just after the 70s decade- sets the pace to a new period characterized by the rise of heritage and culture as primary forms of consumption. Hence the curiosity of culture as well as the adoption of new sustainable practices, places a leading role during this stage. Host-guest relations are regulated through the introduction of centralized planning. Not surprisingly, it is a time where the quest of authenticity occupies a central position in tourism research. The stage radically finalizes with the terrorist attacks in 2001 and the turn of the century. This event showed the world not only the tourism industry was on jeopardy, but also the impossibility of classic planning to prevent global dangers.

New Morbid Forms of Tourism

the 2001s witnessed a set of radical changes which included the terrorist attacks perpetrated to the US accompanied with an unparalleled economic crack (2008) and an almost irreversible ecological crisis. Unlike the other stages, this period coincides with the importance of building global networks. Having said this, new morbid forms of tourism emerged. Dark, Prison, Thana tourism, Slum tourism or even War-tourism has the same in common: the enjoyment of the "Other´s pain". The notion of culture now sets the pace to death as the main commodity to exchange. The figure of destruction offers a dark sense of beautiness which marks the quest for unique experiences. In this period, the tourism industry does not limit to physical displacement. While travels are virtually organized, digital technologies help tourists to have safer travels and experiences. Last but not least, the local "Other" is designed and consumed according to what the international demand asks for. The recent virus outbreak of COVID19 –far from being a foundational event- affirms the tendency that originated during the 2000s. This point reminds us that With 9/11, the War on terror, and international terrorism, the "Other" was suspected to be a potential enemy who lives like us but can attack anytime and anywhere. Just after 9/11, all we are suspected to be potential terrorists (carriers of a lethal disease). This tendency fed back a culture of fear accelerating the closure of West to the "Non-Western Other". Today, COVID19 far from reversing this tendency aggravates it becoming all in undesired guests! Tourists are suspected to be carriers of a lethal virus and for that; they should be monitored, and isolated. COVID19 continues the same trend originated by terrorism and the so-called War on Terror declared by Bush´s administration guests.

The Next Steps

In a recently-published book entitled *Dark Tourism and Pilgrimage,* Olsen & Korstanje (2019) hold the thesis that dark tourism should be defined as a rite of passage that opens a door between a bloody past and an uncertain future. To wit, heritage and dark tourism are inextricably intertwined. Like heritage, dark tourism serves as a catalyst keeping society working. Most certainly, it does not operate in the constellations of frustrations -as MacCannell adheres- but on death. Dark tourism opens a point of convergence and pilgrimage for those who are not only affected by death, even by their own finitude, but also gives a lesson to community about disasters. All global dangers that potentially threaten the tourist system are packaged and consumed as a cultural product. The intensification of natural disasters, as well as the ecological crisis, without mentioning the radicalization of political violence has led policy makers to develop instruments that made destinations more resilient. These new tourist products are not based on precaution, but adaptancy. The rise of morbid consumption is based on the urgency to adapt -probably leaving the background conditions as they stand- destinations to the new global risks. This is particularly true and has some implications for applied research because of multiple reasons. First and foremost, disasters tend to replicate chronically unless the basis conditions are not tackled down. As a result of this, the spectacle orchestrated around disasters tend to replicate the conditions favoring new similar disasters. Secondly, under some conditions dark tourism crystalizes in a commoditization of disaster. This begs a difficult position: how to correct (resolve) the problem that makes further profits for community? The third point is given by the fact death is inevitable. Also pilgrimage to dark sites involves positive challenges for visitors such as volunteer tourism, the expansion of faith-based organization motivated with accelerating the recovery timeframe, as well as the fund investment to mitigate the negative effects of disaster in poorer groups. Dissonant heritage presents here as the main challenge or problem to overcome, but as an opportunity to bolster a fluid dialogue among stakeholders.

CONCLUSION

In the present book chapter, we have interrogated not only the nature of dark tourism but also the evolution of tourism since the former century to date. We have offered a model based on three stages of tourism, each one conditioned by a turning point or founding event. The classic tourism was originated by the WWII end but finalized as a result of Oil embargo in 1970. A new sustainable form of tourism supported by heritage finally come to stay until the attacks to the US. The surfacing of new morbid form of tourism witnessed the exhaustion of an industry chiefly worried to face unnumberable global risks. The idea of precaution which marked the rational planning over years was substituted by adaptancy or resilience. In this vein, dissonant heritage and dark pilgrimage are key factors that help thinking the future of tourism in a risky world. Here two significant questions arise. How to correct the conditions that led us to a disaster while we earn profit from it? And what is the tourism future in the new Normal or a world without tourists? Quite aside from this, dark pilgrimage has some positive aftermaths that have been above mentioned and discussed.

REFERENCES

Bowman, M. S., & Pezzullo, P. C. (2009). What's so 'dark'about 'dark tourism'?: Death, tours, and performance. *Tourist Studies*, *9*(3), 187–202. doi:10.1177/1468797610382699

Carrigan, A. (2014). Dark tourism and postcolonial studies: Critical intersections. *Postcolonial Studies*, *17*(3), 236–250. doi:10.1080/13688790.2014.993425

Cohen, E. H. (2011). Educational dark tourism at an in populo site: The Holocaust Museum in Jerusalem. *Annals of Tourism Research*, *38*(1), 193–209. doi:10.1016/j.annals.2010.08.003

Collins-Kreiner, N. (2016). Dark tourism as/is pilgrimage. *Current Issues in Tourism*, *19*(12), 1185–1189. doi:10.1080/13683500.2015.1078299

Farmaki, A., & Antoniou, K. (2017). Politicising dark tourism sites: Evidence from Cyprus. *Worldwide Hospitality and Tourism Themes*, *9*(2), 175–186. doi:10.1108/WHATT-08-2016-0041

Friedrich, M., Stone, P. R., & Rukesha, P. (2018). Dark tourism, difficult heritage, and memorialisation: A case of the Rwandan genocide. The Palgrave handbook of dark tourism studies, 261-289. Palgrave.

Hartmann, R. (2014). Dark tourism, thanatourism, and dissonance in heritage tourism management: New directions in contemporary tourism research. *Journal of Heritage Tourism*, *9*(2), 166–182. doi:10.1080/1743873X.2013.807266

Harvey, D. (2020). The condition of postmodernity. In *The New social theory reader* (pp. 235–242). Routledge. doi:10.4324/9781003060963-38

Hooper, G., & Lennon, J. J. (Eds.). (2016). *Dark tourism: Practice and interpretation*. Routledge. doi:10.4324/9781315575865

Korstanje, M. (2016). *The Rise of Thana Capitalism and Tourism*. Routledge. doi:10.4324/9781315457482

Korstanje, M. E. (2020). The dark tourist: Consuming dark spaces in the periphery. In M. Korstanje & H. Seraphin (Eds.), *Tourism, terrorism and security*. Emerald Group. doi:10.1108/978-1-83867-905-720201009

Korstanje, M. E. (2022). COVID-19 and Its Effects in the Global Transportation System: Is This the End of Tourism, at Least as We Know It? In *Challenges and Opportunities for Transportation Services in the Post-COVID-19 Era* (pp. 212–228). IGI Global. doi:10.4018/978-1-7998-8840-6.ch011

Lash, S. M., & Urry, J. (1993). *Economies of signs and space* (Vol. 26). Sage.

Light, D. (2017). Progress in dark tourism and thanatourism research: An uneasy relationship with heritage tourism. *Tourism Management*, *61*, 275–301. doi:10.1016/j.tourman.2017.01.011

Olsen, D. H., & Korstanje, M. E. (Eds.). (2019). *Dark tourism and pilgrimage*. CABI.

Potts, T. J. (2012). 'Dark tourism'and the 'kitschification'of 9/11. *Tourist Studies*, *12*(3), 232–249. doi:10.1177/1468797612461083

Robinson, P., Heitmann, S., & Dieke, P. (Eds.). (2011). *Research themes for tourism*. CABI. doi:10.1079/9781845936846.0000

Sather-Wagstaff, J. (2016). *Heritage that hurts: Tourists in the memoryscapes of September 11*. Routledge. doi:10.4324/9781315427539

Seaton, A. V. (1996). Guided by the dark: From thanatopsis to thanatourism. *International Journal of Heritage Studies*, *2*(4), 234–244. doi:10.1080/13527259608722178

Shekhar, S., & Valeri, M. (2022). Evolving Themes in Dark Tourism Research: A Review Study. *Tourism (Zagreb)*, *70*(4), 624–641. doi:10.37741/t.70.4.6

Singh, B. (2020). *A Critical Analysis of Dissonant Heritage and Dark Tourism in India: The Case of the 'Wall of Truth' Memorial* [Doctoral dissertation, University of Central Lancashire].

Stone, P., & Sharpley, R. (2008). Consuming dark tourism: A thanatological perspective. *Annals of Tourism Research*, *35*(2), 574–595. doi:10.1016/j.annals.2008.02.003

Stone, P. R. (2006). A dark tourism spectrum: Towards a typology of death and macabre related tourist sites, attractions and exhibitions. *Tourism: An International Interdisciplinary Journal*, *54*(2), 145–160.

Stone, P. R. (2012). Dark tourism and significant other death: Towards a model of mortality mediation. *Annals of Tourism Research*, *39*(3), 1565–1587. doi:10.1016/j.annals.2012.04.007

Tarifa-Fernández, J., Carmona-Moreno, E., & Sánchez-Fernández, R. (2022). An attempt to clarify what deserves to remain dark: A long look back. *Tourism and Hospitality Research*, 14673584221110358. doi:10.1177/14673584221110358

Tumarkin, M. M. (2005). *Traumascapes: The power and fate of places transformed by tragedy*. Melbourne Univ. Publishing.

Tunbridge, J. E., & Ashworth, G. J. (1996). Dissonant heritage. *The Management of the Past as a Resource in Conflict.*

Tunbridge, J. E., & Ashworth, G. J. (2016). Is all tourism dark? In *Dark Tourism* (pp. 12–25). Routledge.

Tzanelli, R. (2016). *Thanatourism and cinematic representations of risk: Screening the end of tourism*. Routledge. doi:10.4324/9781315624105

White, L., & Frew, E. (2013). *Dark tourism and place identity: Managing and interpreting dark places*. Routledge. doi:10.4324/9780203134900

ADDITIONAL READING

Dowson, R. International Journal of Religious Tourism an d Pilgrimage. *Journal of Religious Tourism and Pilgrimage, 5*(3), 3-15.

Graham, B., Ashworth, G., & Tunbridge, J. (2016). *A geography of heritage*. Routledge. doi:10.4324/9781315824895

Harrison, R. (2013). *Heritage: critical approaches*. Routledge.

Korstanje, M. (2015). *The anthropology of dark tourism. Exploring the contradictions of capitalism*. CERS.

Korstanje, M., & George, B. (Eds.). (2017). *Virtual traumascapes and exploring the roots of dark tourism*. IGI global.

Korstanje, M. E. (2011). Detaching the elementary forms of dark-tourism. *Anatolia*, *22*(3), 424–427. doi:10.1080/13032917.2011.620800

Korstanje, M. E., & Baker, D. (2018). *Politics of dark tourism: the case of Cromañón and ESMA, Buenos Aires, Argentina. The Palgrave handbook of dark tourism studies, 533-552* (P. Stone, Ed.). Palgrave Macmillan.

Korstanje, M. E., & George, B. (2015). Dark Tourism: Revisiting Some Philosophical Issues. *Ereview of Tourism Research*, *12*(3), 1–15.

Korstanje, M. E., Séraphin, H., & Maingi, S. W. (Eds.). (2022). *Tourism Through Troubled Times: Challenges and Opportunities of the Tourism Industry in 21st Century*. Emerald. doi:10.1108/9781803823119

Liutikas, D. (2021). The expression of identities in pilgrim journeys. In *Pilgrims: values and identities* (pp. 17–34). CABI. doi:10.1079/9781789245653.0017

Martini, A., & Buda, D. M. (2020). Dark tourism and affect: Framing places of death and disaster. *Current Issues in Tourism*, *23*(6), 679–692. doi:10.1080/13683500.2018.1518972

Olsen, D. H. (2021). Fan pilgrimage, religion, and spirituality. In *The Routledge Handbook of Religious and Spiritual Tourism* (pp. 90–110). Routledge. doi:10.4324/9780429201011-8

Olsen, D. H., & Timothy, D. J. (2020). The COVID-19 pandemic and religious travel: Present and future trends. *International Journal of Religious Tourism and Pilgrimage*, *8*(7), 170–188.

Pan, S. Y., Gao, M., Kim, H., Shah, K. J., Pei, S. L., & Chiang, P. C. (2018). Advances and challenges in sustainable tourism toward a green economy. *The Science of the Total Environment*, *635*, 452–469. doi:10.1016/j.scitotenv.2018.04.134 PMID:29677671

Seraphin, H., & Korstanje, M. E. (2021). Dark tourism tribes: social capital as a variable. *Consumer Tribes in Tourism: Contemporary Perspectives on Special-Interest Tourism*, 83-99.

Sharpley, R. (2018). *Tourism, tourists and society*. Routledge. doi:10.4324/9781315210407

Sofield, T. H. (2006). Border tourism and border communities: An overview. *Tourism Geographies*, *8*(2), 102–121. doi:10.1080/14616680600585489

Timothy, D. J., & Olsen, D. H. (2021). Religion, spirituality, and tourism: Emerging and future directions. In *The Routledge Handbook of Religious and Spiritual Tourism* (pp. 461–475). Routledge. doi:10.4324/9780429201011-38

KEY TERMS AND DEFINITIONS

Dark Heritage: A type of heritage conditioned by tragedies or catastrophes that vulnerated human integrity.

Dark Tourism: This should be seen as a new emerging form of tourism marked by the visit to places of mass death, suffering or whipped by disasters.

Disasters: These are mainly disrupting events that for what society has no answers or course of actions.

Dissonant Heritage: A term coined by Turnbridge & Ashworth (1996) to denote any space of massacre to narrate difficult stories or bloody legacies. To some extent, the term dissonant is referenced to the multiple interpretations various groups have along the same event.

Tourism: An activity which consists in visiting places, spaces or cultures while leaving the security of home.

Chapter 6
Hajj Management Chain Feasibility Post COVID-19

Noorliza Karia
https://orcid.org/0000-0002-1449-9979
Universiti Sains Malaysia, Malaysia

Firas Izzat Mahmoud Saleh
https://orcid.org/0000-0002-0562-4264
Universiti Sains Malaysia, Malaysia

ABSTRACT

As a form of spiritual tourism, the Pilgrimage or Hajj is the fifth core pillar of Islam, which involves the global mass gathering and movement of travelers from around the globe to the holy lands in Saudi Arabia. Hajj event is one of the worldwide hubs of health risk in the COVID-19 era, which suspends or limits this annual event for the past two seasons. Towards dealing with this protracted crisis and ensuring the sustainability of this critical event and its resilience to deal with any future concerns, this chapter discusses Hajj-related COVID-19 threats and realities across the chain of Hajj management. Further, it analyses the future outlook of the interlinked domains of health, logistical, and technology, as it provides the feasible, practical strategies and manifestations for the pre-departure processes, management of pilgrims in the holy lands, and post-return processes. The relevant governments and stakeholders can directly consider these strategies to overcome the current crisis and sustain Hajj in the future.

INTRODUCTION

Hajj is the fifth core pillar of Islam of which each Muslim should perform if they are capable of doing so. Despite the significant impact of Hajj on the individual identity of Muslims, the gathering of pilgrims itself is one of the core sublime meanings and aims of Hajj. Muslim, in turn, meeting with other millions of Muslim pilgrims, who came every year from all over the world and met in the holy city of Makkah, to gather and practice the notion of unity with each other beyond the social differences, origins, and cultures under a unified umbrella of justice and solidarity. Moreover, to practice unified activities within limited

DOI: 10.4018/978-1-6684-4817-5.ch006

geographical areas, in unified timeframes, with the same physical appearance, and all in line with the specific significance given by Islam to the public customs. Whereas, among the other forms of Islamic worship, Hajj involves specialised forms of interactions among Muslims. For instance, the gathering of different sexes in the same locations, and the interactions among different nationalities is intensive.

In addition to the spiritual dimension of Hajj in Islam, it includes the performing of kinetic activities. Most of these activities are compulsory and timed activities, in which this worship is not valid without performing them, while some other activities are voluntary activities and where pilgrims seek a reward from God according to the holy Quran. The gathering of pilgrims in Saudi Arabia extends for more than three months, where pilgrims travel to Makkah and stay there for visiting the holy places during their presence in the holy land. However, the most concentrated gathering of more than three million pilgrims takes place for four to five days in certain districts of Makkah.

During the first compulsory day of Hajj, pilgrims travel to Arafat and stay until the sunset and then move to Muzdalifah. On the second day, pilgrims travel to Mina, stone the symbol of Satan, and visit Makkah to perform the Tawaf and the Sa'y. In the third, fourth, and optionally fifth day, they sleep in Mina for two to three nights, stone the symbol of Satan, sacrifice an animal, return to Makkah, and perform the farewell Tawaf.

Further to travel for Hajj, pilgrims can also travel to the holy city of Makkah for Umrah. Unlike Hajj, Umrah can be performed in a couple of hours in the holy mosque in Makkah in any time slot. Therefore, Saudi Arabia receives tens of millions of Muslims to perform Umrah throughout the year in addition to the concentrated gathering of Hajj, which makes this industry one of the most valuable economic industries for Saudi Arabia and all Islamic countries alike. Likewise, for various industries such as tourism agencies, airline companies, operators (Ebrahim et al., 2020). This chapter, however, focuses on Hajj with taking into consideration that what is successful for Hajj is almost applied to Umrah as well.

With the annually increasing number of pilgrims, and the expectations of receiving more than ten million pilgrims from around 180 countries by 2030, the management of Hajj becomes extremely complex and causes various forms of challenges for the Saudi authorities and all stakeholders engaged in Hajj management. Management of this mass gathering entails several complexities that can be in the forms of loss of people in Hajj ritual places (Noor et al., 2020), lack of information and respect of schedules, supply chain management (Makrum, 2020), the multiplicity of ethnics, behaviours, and cultures (Marmot, 2017), and most recently, the complexities and risks of such crowded mass gathering within the current realities of COVID-19 era.

Hajj is considered a major global mass gathering event, and in turn, a risky event on the global epidemic situation. Hence, Saudi Arabia has suspended the Hajj seasons in 2020 and 2021 based on the recommendations of experts since the onset of the pandemic. Despite the acknowledgement of this suspension, more needs to be unpacked on how the mass gatherings and Hajj, in particular, can be managed and sustained during this era and during any similar crises in the future. This, however, requires more understanding of the boundaries and determinants of feasible and sustainable preparedness and response systems (Tambo et al., 2020). As such, to better understand, and therefore, mitigate the challenges faced by the spiritual journeys and spiritual tourism.

Accordingly, the need for the provision of practical guidance on how to deal with this mass gathering and overcome the emerged challenges due to the COVID-19 pandemic and the need for specific strategies under the integrated dimensions of health (Ebrahim & Memish, 2020), technology (Brdesee et al., 2013), and logistical management (Fourati et al., 2017) have emerged. Therefore, to ensure continuity of this important event for Muslims with the protracted nature of the COVID-19 pandemic and the potential mutations of the Virus in the future.

This chapter aims to introduce the research directions concerning the Hajj Management chain and then provide the prospects of future outlooks for the potential strategies in the countries of departure and Saudi Arabia towards the overall feasibility of the Hajj management chain post-COVID-19. Grounded from the discussion of health-related aspects, managerial and logistical issues as well as the relevant technological prospects, this chapter discusses the feasible practical manifestations for ensuring COVID-19 sensitive Hajj management across the whole chain of Hajj management, including the pre-departure processes, management of pilgrims in the holy lands, and post-return processes.

Hajj and COVID-19 Threats

The announcement of COVID-19 as a global pandemic has brought the responsiveness of the whole world to restrict the social interactions and movements in the hope of reducing human losses as much as possible. Yet, it has been recommended to suspend the public events. Therefore, most of the global mass gatherings have been suspended to avoid the effects of mass gatherings on global public health risks (Saleh & Karia, 2020).

Saudi Arabia registered its first COVID-19 case on March 2nd, 2020. This, therefore, and in line with the measures taken by most of the countries, has been followed by a series of measures taken by the Saudi government to control the spreading. Measures included but were not limited to: cancelling all domestic and international flights, suspending entries to the country by e-Visa, banning weddings parties and cancelling public events, and last prohibiting prayers in the mosques along with enforcement of movement restrictions in the holy cities.

The correlation between the Hajj and the spreading of different diseases is a well-known issue. Yet, not new to the experts because of the cumulative knowledge built in the past decades in this regard. Therefore, the observed linkage in Malaysia, South Korea, Iran, India and Singapore between COVID-19 infections and worship places was not coincidental or sudden but confirmed the already existing analogues knowledge. However, this experience did not face by a sudden stopping of life over the globe. Therefore, the consideration of Hajj as a source of threat for the global epidemical situation of COVID-19 has been acknowledged by Saudi Arabia despite the specific essentiality of Hajj in Islam. Therefore, the uncommon decision was taken to suspend Hajj in order to reduce the in-country COVID-19 cases and do not risk the lives of pilgrims who come from 180 countries.

It can be generalised that crowded settings in general increase the risk of infections. As such, the outbreak of infectious diseases is a constant major health risk in Hajj and in any religiously inspired/motivated mass gathering alike. Historically, we can be informed by the consequent outbreaks relevant to Hajj seasons since 2009, such as SARS, meningococcal disease, and Ebola. For instance, Hajj was a key reason for the broad spreading of the 1957's flu pandemic. Likewise, the outbreaks of infectious diseases among the British pilgrims in 1987, 2000, and 2001, and most recently, the outbreak of Middle East Respiratory Syndrome (MERS) which has been confirmed worldwide with 80% of the registered cases in Saudi Arabia and the significant role of returnee pilgrims in the global spreading of the disease. Therefore, the need for sustainable outbreak preparedness and response systems is relevant as a core – even before the COVID-19 era. Hence, to promote and re-enforce precautionary and preventive measures among pilgrims and hosting communities (Tambo et al., 2020).

Hajj as a mass gathering shares most of the characteristics for the rest of mass gatherings, public sport and truism events. However, it can be argued that this specific event has its own specificities that are included, but not limited to: the high average age of pilgrims, the extreme weather, the physical conditions and extreme weather in the holy cities, the level of urbanisation, characteristics of pilgrim, religious and social norms. These specificities, however, make the Hajj a unique event concerning the creation of a risky environmental condition that would enable outbreaks of diseases (Ebrahim & Memish, 2020). Therefore, considering the proper health measures, adequate management of logistical aspects, and technology adoption has been appreciated as potential directions discussed by this chapter towards ensuring quality management of Hajj mass gathering and its related challenges and complexities. Not to mention the cross-cutting issue of the crowd as a mainstreamed factor over these three directions. The next sections of this chapter discuss and contextualise these three directions to better understand the Hajj management chain feasibility framework of post-COVID-19.

Health Measures

Management health of Hajj mass gatherings is correlated with the size of the crowd. The extreme forms of congestions amplify the risks of infectious diseases for pilgrims, which can cause a catastrophic impact on the global spreading of diseases. Therefore, Hajj's lack of health planning can lead to global negative consequences as this global tourism can act as an infection dissemination hub for COVID-19. To that end, Saudi Arabia had extensive investments in the health surveillance and prevention programs, such as but not limited to deployment of more than 25,000 health workers, disseminate travel advice and regulations through international organisations and travel agencies, and dedication of eight well-equipped hospitals for pilgrims in the holy city of Makkah.

COVID-19 does not go beyond mild disease for most people, as most deaths have occurred in the elderly groups with chronic health conditions (Gawthrop, 2020). This, however, most of the pilgrims are mostly elderly and travel from low-level countries concerning the capacities of public health (Memish et al., 2019). Hajj's health risks can vary across non-communicable diseases, communicable diseases, gastrointestinal disorders, and fire and trauma. However, this chapter is specifically focused on the risks of COVID-19 and similar outbreaks that spread very quickly.

Mitigation of health risks in mass gatherings and Hajj can be done through pre-travel preparations for incoming pilgrims. Such as vaccinations, immunisations at entry points, health education campaigns, travel advice, and alerts. It also can be done during Hajj. Such as provision of health awareness and promotion materials, availability of healthcare services, food safety measures, and electronic surveillance of infectious diseases. Learning from the past-outbreaks, several health management strategies have been further acknowledged for risk reduction. Such as vaccination, antibiotic chemoprophylaxis, well-established communicable diseases surveillance system, and compliance with preventative measures. However, the transmission of communicable diseases has been perceived by several scholars as inevitable. Yet, difficult to be prevented in such contexts.

The Ministry of Health (MoH) in Saudi Arabia recommended relying more on individual preventive measures and personal hygiene (Hoang et al., 2020). This, however, was in line with the agreed-upon methods of mitigating infections, such as the use of face masks and alcohol sanitiser. Further, the ministry has recommended pilgrims to follow the travel advice of wash hands with water and soap before taking food, and after coughing, using toilets, or touching animals, use disposable tissues, and avoiding contacting with sick people or sharing with them personal stuffs, and avoid contacting with camels in farms or drink raw milk. However, the pilgrims' knowledge, practices, and attitudes have the most significant impact on responding to any measures, which in turn, increases the risk of infection (Dauda Goni et al., 2019).

The lack of compliance with health advice and recommendations is a well-known issue among Hajj practitioners. In responding to that, scholars have undertaken several studies to explore the implications of religious practices on health and the role of social aspects in risk reduction strategies. Therefore, religion has been recognised as a valuable resource to cope with threats and inform risk reduction plans in such contexts. Likewise, the importance of pilgrims' exposure to health awareness as part of preparing them before their travel in order to reduce infectious disease. Although a multiplicity of factors under the social aspects, the items under this aspect can be categorised into three domains; knowledge, practices, and attitudes (Goni et al., 2020).

In addition to measures and taken in-country, the MoH publishes recommendations and requirements for entry visas of pilgrims for both medical staff and pilgrims in the countries of departure. These measures might be in the context of mass gathering during the global outbreak of COVID-19, which in fact, has to be comprehensively undertaken, as the major risk of Hajj for exposure to COVID-19 is more related to the crowd and huge time-phased influx of people to a mass gathering from one area to another.

To the moment, we do not have agreed-upon guidelines for the number of people gathering in a specific area. But, in May 2020, (WHO, 2020) has published the guidance for "*Key planning recommendations for mass gatherings in the context of COVID-19*". The guidance note stated five key considerations for planning mass gathering events as follows:

1. Assessment of the normative and epidemiological context.
2. Evaluation of the risk factors associated with the mass gathering event. Areas of evaluation include but are not limited to:
    - Physical characteristics of the event
    - Number and characteristics of participants
    - Interaction among participants
    - Duration of event
    - The capacity of the hosting country to detect and manage COVID-19 cases
3. Capacity to apply prevention and control measures. Measures include, but not limited to:
    - Modifying the characteristics of the event, duration of the event, and affecting participants
    - Strengthen preparedness and response to public health issues before, during, or after the event.
    - Risk communication, coordination, and information sharing
    - Surveillance of participants
4. Determination of the overall risk of the mass gathering
5. Evaluating the legacy of the event and the lessons learned

Other recommendations can include but not limited to:

- Enhancing the routine diagnostic at the surveillance ports of entry and healthcare centers.
- Adoption of best practices and treatment standards in spread prevention and control.
- Adoption of practice-oriented and person-centered care approaches.
- Adoption of participatory decision making and evidence-based approaches.

- Relating the Islamic values with illness and healthcare aspects.
- Educate pilgrims on potential medical risks and advise them to seek advice if they experience any symptoms during travel and after returning home.

Even before the COVID-19 era and the invention of its vaccinations, pilgrims were required to take several vaccinations in order to be able to apply for a Hajj visa. Indeed, the same goes for internal pilgrims and workers in Hajj. This attributed to the early attention of Hajj managers and Saudi Arabia to the seriousness of the spread of diseases if happened among pilgrims. Despite the superiority of this mission, it has been described as an overwhelming task, even for a well-resourced country like Saudi Arabia. Here, the need for international cooperation for vaccination campaigns was emerged along with having quotas for pilgrims and rapid returns of pilgrims as integral aspects of Hajj management. The same goes for strategic planning for Hajj public health as it should be undertaken in both custodian countries and countries of pilgrims' origins. As such, it should cover measures or strategies for pre-departure, in the field, and post-return. In fact, all of the stakeholders of Hajj management are invited to adhere to the effective use of mitigation measures, which in turn, might help to overcome the COVID-19 pandemic and unforeseen future outlook. Similarly, the Hajj actors in Saudi Arabia and countries of departures have to collaborate in order to ensure coordinated prevention measures for the diverse pilgrims who have different cultural norms, languages, cultures, and health needs, with no one left behind, particularly those who have health vulnerabilities and want to perform Hajj which has a significant spiritual meaning for them.

Logistical Considerations and Technological Means

The Hajj is costly and an exhausting journey for pilgrims. For instance, the movement of the pilgrim entails movement, and mostly by walking, for around 80 kilometres in the holy sites of Makkah. As such, it is exhausting the Saudi authorities despite the cumulative experience of Hajj management, as it remains challenging because of its multi-countries logistical and cultural considerations and the above-discussed specific constraints as well as the specific nature of this mass gathering.

The Hajj chain and its complexities start from the intention of the pilgrim to travel for Hajj. However, this chain might not end even after returning home, particularly in the COVID-19 era, which enforced a new normal of travelling and movement considerations. To dissect the problem of the concern of this chapter, the Hajj chain passes through three key and interlinked stages as follows: First, the sending country stage, 2) Saudi Arabia stage, and 3) the country of return stage (See Figure 1).

While the multiplicity of sending countries and countries of return is a significant factor that can hinder all of the control measures and the screening processes, the most challenging operation is the movement of more than three million pilgrims within holy places in Makkah. Likewise, the cross-cutting complexity of languages and cultures which overlays the already existing complexities and impedes the control throughout the chain (Elazhary, 2017). In this context, several structured problems confront any logistical aspect in Hajj management, such as the fixed-underplaying infrastructure in the holy places, lack of respect to schedules, lack of guidance and lost pilgrims in the ritual places, and the increased annual number of Muslims who want to perform Hajj.

Crowd management in the second stage of the chain was the biggest concern among the researchers and practitioners of Hajj management (Nasser et al., 2017). Therefore, the focus of the research was on how to minimise the crowd to acceptable levels, and in turn, save the lives of people. However, this research arena has witnessed two main parallel paradigm shifts in the last decade. The first shift was

Figure 1. Hajj management chain

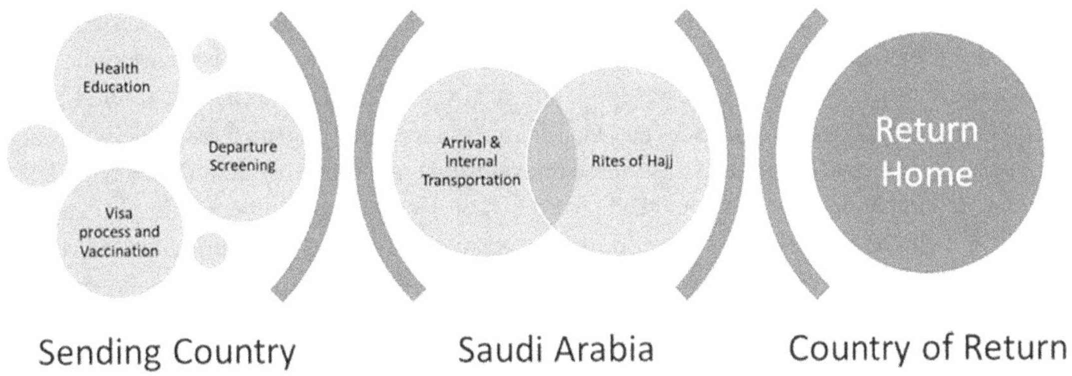

toward more people-centred approaches, in which the reliance on people crowd members themselves as accountable and rational actors instead of solely adopting physical barriers and security measures. In fact, the unified religious beliefs of pilgrims do not reflect the diverse economic and cultural differences among them. However, this diversity does not converge or perhaps impossible to converge with the available resources that can deal with this diversity on the ground, leading to various logistical and management problems for pilgrims and Hajj management authorities. The second shift was related to the adoption of technological approaches for managing crowds and the other forms of Hajj-related crises. Thereby, and despite the increasing recognition of cultural and socio-economic aspects and personal behaviours of pilgrims as key determinants of crowd management, the use of technology systems to monitor and control the crowd during Hajj is not uncommon among scholars. Likewise, the adoption of advanced technologies to measure sensory data of pilgrims for better crowd management and control infection (Yamin, 2015).

The emerging research directions from these two main shifts have provided the opportunity to featured research to go beyond traditional discussions of mass gathering and pursue new research areas such as faith-based health promotion and electronic capacities of disease surveillance. Therefore, to adopt strong systems for reporting and surveillance along with the healthcare measures to control infections in the mass gathering events such as Hajj, which in fact, can promote the efficiency and effectiveness of coordination, cover the lacking of knowledge sharing adoption, and enhance the information technology governance, for the interest of real-time, responsive, and systematic surveillance and control infection systems across the chain of Hajj management.

Generally speaking, it can be said that technology has been either proposed or adopted within a wide range of Hajj-related management areas, such as:

- Using iris scan databases and facial recognition systems to identify missed pilgrims.
- Simulation and analysing movements, and queuing of pilgrims.
- Using wireless and mobile technologies to manage crowds
- Adoption of mobile health applications

However, in light of the increased dependency of social life on technology and the essentiality of internet-enabled and e-government platforms, a key potential form of adoption was related to managing crowd by cloud-based context-responsive data management. This form of technology adoption relies on capturing sensory, spatiotemporal, and contextual data of pilgrims using their smartphones that are equipped with multiple sensors. However, the sources of data include, but are not limited to, social media platforms, available list of services and health information's for pilgrims, maps, vehicle tracking, news, translation services. Therefore, to assist the relevant stakeholders to better document, track data of pilgrims, understand temporal behaviours and crowd and protect the environment (Ahmad et al., 2014).

Therefore, it can be said that the current dominant form of technology adoption for such contextual complexity and during COVID-19 is the big data analysis and mass collecting of the spatiotemporal and geo-tagged data of pilgrims using. This form, in fact, reflects the two shifts of paradigm, and which can be summarised by using the internet-based social platforms in capturing contextual, spatiotemporal, and sensory data of pilgrims for health purposes (Ahmad et al., 2018; Vahdat-Nejad, 2019).

Grounded from this research direction, and in order to put it into practice, the rest of this chapter presents the feasible, practical manifestations and strategies for the countries of departure and in the holy lands focusing on the following aspects:

1. How social media and internet-based platforms are perceived and can be used for the interest of health education?
2. How can Islamic values be harnessed and integrated for mitigating COVID-19 risks?
3. How social media and internet-based platforms can be used for coordinated vaccination programs and departure screening, and after-return processes.?
4. How social media and internet-based platforms can be used for quality reception and transportation of pilgrims in the holy lands?
5. How social media and internet-based platforms can be used for accessing health care services and ensure safe interaction with the inhabitants?
6. How social media and internet-based platforms can be used for ensuring quality and coordinated return screening of pilgrims?

Strategies in Countries of Departure

As one of the COVID-19 consequences, most countries worldwide witnessed a transformation to use internet-based platforms to deal with the movement restrictions and reduce physical interaction between people. In which, most of the governmental services, registration for vaccination, and COVID-19 health support are all provided through these platforms. This "new normal" has supported the trust, acceptance, and familiarity of people with the use of smartphones, social media, and internet-based platforms to access these services. Even elderlies started to positively perceive and use social media and trust it as a reliable source of information. It is almost common among the sending countries that "Facebook" is the most popular and accepted social media platform by the pilgrims' age groups. However, "Instagram" is the preferred platform for youth, in particular for female youth.

Islam prohibits harming others. As such, Islamic teachings are full of verses from the Qur'an and "hadiths" that can be used to raise awareness of pilgrims intending to travel to the Holy Land. Therefore, to be used as an intended content and part of the basic information package for pilgrims about social distancing and COVID-19 prevention measures. Several examples on the content of social media posts or WhatsApp Massaging can be given here, such as "ال ررض الو رارض" which means any harm for you, and harm for others is forbidden.

Usually, pilgrims receive leaflets and brochures about health education upon their reach to the holy lands. Despite the criticality of this kind of information provision or education, there are doubts about this method and its validity during this era of technological advancement or even during COVID-19. We observed that pilgrims were not read the distributed materials at the reception ports, as they were still under travel disorientation. In addition to that, the content still lacks cultural, gender, and age sensitivity. Therefore, the considerations for information provision and health education of pilgrims can be summarised as follows:

- The paper-based materials are no longer suitable or even acceptable by pilgrims themselves in this era, as it perceived as a source of infection. Therefore, a total shift to e-modalities is mandatory.
- While the required health education does almost not vary across sending countries, the experts stated that health education content should be jointly developed between the Saudi authorities in the sending countries in order to ensure cultural sensitivity and ensure lasting impact.
- Both content development and targeting of health education should be customised and inclusive for the different gender and age-sensitive. Therefore, to harness the targeting criteria and capabilities of social media in this regard.

The vaccination programs of COVID-19 in most of sending countries are mostly governmental-led. In March 2021, the Saudi authorities announced that the COVID-19 vaccination is one of the mandatory requests for Hajj, as the vaccination programs in most of the countries around the world have been progressively elaborated and almost have covered all of the at-risk groups. Therefore, the role of non-governmental actors in the sending countries, such as the travel agencies, might be limited to the provision of information on the requirements of Hajj visas to their targeted groups. However, the following strategies can be adopted to ease the preparing of pilgrims before and upon their travel to the holy lands:

- Develop user-friendly and visualised Standard Operating Procedures (SOPs) explaining the requirements and processes of the Hajj visa in collaboration between the Saudi embassies in the sending countries and Hajj agents, and therefore, to be used for information provided on the social media channels.
- Link the governmental data in the sending countries about their vaccinated citizens with the Saudi embassies through electronic means or "web services". Therefore, to facilitate the visa processes.

Basically, Once the pilgrims are ready to travel, they have to be screened when they depart their countries. Hence, the minimum requirements of travelling during COVID-19 should also be applied to them upon departure, such as a negative PCR test for the last three days before departure. Currently, quarantine is not mandatory for those who enter Saudi Arabia. However, this can be changed anytime following any development on the epidemical situation in Saudi Arabia and the sending countries. Not to mention that Saudi Arabia can prohibit any of the countries from sending pilgrims in case the epidemical situation in the country is not safe enough. Therefore, sharing the medical reports and health care data of pilgrims with the health authorities in Saudi Arabia in advance can be helpful. Not only for facilitating the requirements of the entrance to the country, but also, to be able to follow up on the individual health cases of pilgrims based on the available data and to update these medical reports for the use of countries of departure in relation to the after-return health care and follow-ups.

Strategies in The Holy Lands

Based on the developments that happened concerning the reliance and socialisation of mobile phones, the internet of things, and, most importantly huge leap that happened because of COVID-19 in this regard, it seems that the doors are open enough to rely on this advancement and invest in it for the interest of ensuring the continuation of Hajj and save the lives of pilgrims. Therefore, the provision of government-supported and dedicated SIM cards to the pilgrims at the reception ports seems to be a viable strategy. Equipping the mobile phones of pilgrims' and their companions will allow Hajj managers in the holly lands to ensure 24-hours monitoring and tracking of the geo-tagged and spatiotemporal data of pilgrims. The SIM card, however, can be linked with a mobile application that can host the medical data of pilgrims and acts as pilgrim's health passport and formal documentation. Such big data, if well-analysed, will allow Hajj managers to manage the crowd and communicate with the pilgrims about their movements instantly n their mobile phones.

In fact, the movements of pilgrims in the holy sites is a core process of Hajj management. Therefore, it needs specific attention because of the severity of its risks on pilgrims and inhabitants alike. Therefore, the need for a unified platform that links all of the actors responsible for Hajj movements are crucial. Accordingly, Hajj missions, the private institute of Addillah in Madinah Monwarah, Motawifeen institutes in Makkah, and Automobile General Syndicate are all to be connected on this platform under the overall supervision of the Ministry of Hajj and Umrah and other relevant ministries. This approach can facilitate monitoring of the medical conditions of pilgrims and their contacts in case of any infection. The following associated considerations can help to complement the unified platform: 1) tracks of movements to be pre-agreed, 2) directions of hotels to be included in the instant data of movements, and to be updated directly by drivers of buses and trains, and 3) drivers and guiders to be separated from the pilgrims during movement.

Undoubtedly, the collected data through the SIM cards and mobile applications will also inform the management of crowds. In particular, if the movement schedules are already entered in the unified platforms, which can help to track the adherence and send reminders about movements and compare it with the sensory data and maps to ensure social distancing and other measures and pre-established thresholds.

Basically, the dedicated SIM cards and mobile application will provide a map of all health care facilities, and other facilitates in the holy sites. However, accessing these services can be safer if it is based on a booking system whenever possible. The medical staff in the holy sites should be able to check the medical reports of their patients, as they are required to update these reports as well. Therefore, to be able to inform broader dashboards and medical situational maps.

As has been widely done by several countries around the world, Saudi Arabia has developed mobile applications to monitor and track COVID-19 cases, movements, and contacts. The key platform adopted by the Saudi government is called "Tawakkalna" which means "we trust" in English. Tawakkalna is a comprehensive platform enforced by the government to be installed on the mobiles of inhabitants, which allows access to services and facilities in the country. It captures and provides the user with the medical assessment, infections, PCR result, vaccination management, e-learning, and access to the holy lands. Therefore, such a well-established platform can also be extended to be used by pilgrims in order to ensure the safety of all.

The last process of Hajj management in Saudi Arabia is the departure of pilgrims and screening at the departure sites and ports. However, this process is still not automated and requires refugees to be gathered in departure sites or ports in order to finalise the procedures of their return. Therefore, it can be said that the availability of a web-based system as part of the overall electronic platform of Hajj management is a high priority. Consequently, to reduce steps of return procedures and use electronic devices to perform the procedures in the sites of pilgrims, but not in the assembly points or departure ports. This, however, will reduce the crowds and interactions to their minimum levels. The departure procedures should ensure COVID-19 travel requirements and should be ended by ensuring that the medical records of pilgrims are updated in order to be shared with the countries of return accordingly.

CONCLUSION AND FINAL REMARKS

Due to the emerging global situation of the COVID-19 pandemic, Saudi Arabia has suspended or minimised access to Islamic holy lands, which prohibited Muslims from performing one of the core worships perceived by them as a unique and perhaps the most significant occasion in their lives. Given the logistical complexity, social diversity, and specific nature of Hajj management overlaid by COVID-19 restrictions, this chapter aimed to frame the Hajj management aspects within the newly emerged realities and under the dimensions of health, logistics, and technology. To that end, threats of COVD-19 within the Hajj chain context has been firstly discussed, followed by the discussion of health measures, logistical considerations and technological outlooks.

The main identified research stream by this research was the adoption of internet-based platforms and social media to capture the sensory data for pilgrims, track their movements, manage crowds, reduce infections, and ensure the ultimate safety of pilgrims through COVID-19 sensitive Hajj management. Accordingly, the chapter provided feasible, practical manifestations and strategies for the countries of departure and in the holy lands.

The emerged strategies and practical considerations provided by this chapter confirmed the essentiality of international collaboration between Saudi Arabia and sending countries of pilgrims across the whole chain of Hajj, use of internet-based and social media solutions, COVID-19 measures, and travel advice. Further, it shed light on the importance of exploiting and building on the advancements and successes that have already been achieved because of the COVID-19 pandemic in Saudi Arabia and sending countries. Therefore, to allow Muslims to perform Hajj and experience this unique event in their lives. Further, it provided practical considerations that can be directly considered by the relevant governments and stakeholders to allow people to overcome the tragedy of COVID-19.

REFERENCES

Ahmad, A., Rahman, M. A., Rehman, F. U., Lbath, A., Afyouni, I., Khelil, A., & Wahiddin, M. R. (2014). A framework for crowd-sourced data collection and context-aware services in Hajj and Umrah. In *2014 IEEE/ACS 11th International Conference on Computer Systems and Applications (AICCSA)* (pp. 405-412). IEEE. 10.1109/AICCSA.2014.7073227

Ahmad, A., Rahman, M. A., Ridza Wahiddin, M., Ur Rehman, F., Khelil, A., & Lbath, A. (2018). Context-aware services based on spatio-temporal zoning and crowdsourcing. *Behaviour & Information Technology*, *37*(7), 736–760. doi:10.1080/0144929X.2018.1476586

Brdesee, H., Corbitt, B., & Pittayachawan, S. (2013). Barriers and motivations affecting Information Systems usage by Hajj–Umrah religious tourism operators in Saudi Arabia. *AJIS. Australasian Journal of Information Systems*, *18*(1). doi:10.3127/ajis.v18i1.809

Dauda Goni, M., Hasan, H., Naing, N. N., Wan-Arfah, N., Zeiny Deris, Z., Nor Arifin, W., & Abubakar Baaba, A. (2019). Assessment of knowledge, attitude and practice towards prevention of respiratory tract infections among Hajj and Umrah Pilgrims from Malaysia in 2018. *International Journal of Environmental Research and Public Health*, *16*(22), 4569. doi:10.3390/ijerph16224569 PMID:31752224

Ebrahim, S. H., & Memish, Z. A. (2020). Saudi Arabia's drastic measures to curb the COVID-19 outbreak: Temporary suspension of the Umrah pilgrimage. *Journal of Travel Medicine*, *27*(3), taaa029. doi:10.1093/jtm/taaa029 PMID:32109274

Ebrahim, S. H., & Memish, Z. A. (2020). Saudi Arabia's drastic measures to curb the COVID-19 outbreak: Temporary suspension of the Umrah pilgrimage. *Journal of Travel Medicine*, *27*(3), taaa029. doi:10.1093/jtm/taaa029 PMID:32109274

Elazhary, H. (2017). Cloud-based context-aware mobile applications and framework for Hajj and umrah management. *International Journal of Computer Trends and Technology*, *47*(2), 106–115. doi:10.14445/22312803/IJCTT-V47P115

Fourati, J., Issaoui, B., & Zidi, K. (2017). Literature Review of Crowd Management: A Hajj Case Study. In ICINCO, (1) (pp. 346-351). doi:10.5220/0006472103460351

Gawthrop, M. (2020). Advising Hajj and Umrah travellers in general practice. *Practice Nursing*, *31*(4), 156–161. doi:10.12968/pnur.2020.31.4.156

Goni, M. D., Naing, N. N., Hasan, H., Wan-Arfah, N., Deris, Z. Z., Arifin, W. N., Hussin, T. M. A. R., Abdulrahman, A. S., Baaba, A. A., & Arshad, M. R. (2020). Development and validation of knowledge, attitude and practice questionnaire for prevention of respiratory tract infections among Malaysian Hajj pilgrims. *BMC Public Health*, *20*(1), 1–10. doi:10.118612889-020-8269-9 PMID:32114986

Hoang, V. T., Gautret, P., Memish, Z. A., & Al-Tawfiq, J. A. (2020). Hajj and Umrah Mass Gatherings and COVID-19 Infection. *Current Tropical Medicine Reports*, *7*(4), 1–8. doi:10.100740475-020-00218-x PMID:33169095

Makrum, M. (2020). Hajj, its islamic perspective, management and role of supply chain in Indonesian perspectives. [IJSCM]. *International Journal of Supply Chain Management*, *9*(1), 794–804.

Marmot, M. (2017). Inclusion health: Addressing the causes of the causes. *Lancet*, *391*(10117), 186–188. doi:10.1016/S0140-6736(17)32848-9 PMID:29137870

Memish, Z. A., Steffen, R., White, P., Dar, O., Azhar, E. I., Sharma, A., & Zumla, A. (2019). Mass gatherings medicine: Public health issues arising from mass gathering religious and sporting events. *Lancet*, *393*(10185), 2073–2084. doi:10.1016/S0140-6736(19)30501-X PMID:31106753

Nasser, N., Anan, M., Awad, M. F. C., Bin-Abbas, H., & Karim, L. (2017). An expert crowd monitoring and management framework for Hajj. In *2017 International Conference on Wireless Networks and Mobile Communications (WINCOM)* (pp. 1-8). IEEE. 10.1109/WINCOM.2017.8238202

Noor, T. H., Atlam, E. S., Malki, Z. S., Alzighaibi, A. R., Hashim, H., & El-Agamy, R. (2020). Android/iPhone Mobile Application for Quick Response Pilgrims Campaign Locator. *Journal of Computer and Communications*, *8*(05), 1–13. doi:10.4236/jcc.2020.85001

Saleh, F. I. M., & Karia, N. (2020). Benchmarks for INGOs' effective responses during COVID-19 pandemic. *Benchmarking*, *27*(10), 2863–2886. doi:10.1108/BIJ-04-2020-0157

Tambo, E., El-Dessouky, A. G., Khater, E. I., & Xianonng, Z. (2020). Enhanced surveillance and response approaches for pilgrims and local Saudi populations against emerging Nipah, Zika and Ebola viral diseases outbreaks threats. *Journal of Infection and Public Health*, *13*(5), 674–678. doi:10.1016/j.jiph.2020.01.313 PMID:32265162

Tambo, E., El-Dessouky, A. G., Khater, E. I., & Xianonng, Z. (2020). Enhanced surveillance and response approaches for pilgrims and local Saudi populations against emerging Nipah, Zika and Ebola viral diseases outbreaks threats. *Journal of Infection and Public Health*, *13*(5), 674–678. doi:10.1016/j.jiph.2020.01.313 PMID:32265162

Vahdat-Nejad, H., Asani, E., Mahmoodian, Z., & Mohseni, M. H. (2019). Context-aware computing for mobile crowd sensing: A survey. *Future Generation Computer Systems*, *99*, 321–332. doi:10.1016/j.future.2019.04.052

World Health Organization (WHO). (2020). *Key planning recommendations for mass gatherings in the context of COVID-19: interim guidance, May 29th 2020 (No. WHO/2019-nCoV/POE_mass gathering/2020.3)*. World Health Organization.

Yamin, M. (2015). Secure and healthy Hajj management: a technological overview. *American Academic & Scholarly Research Journal, 7*(3).

Chapter 7
Pilgrimage Tourism in Hong Kong?
Case Study of Tourists' Visitation to Chinese Folk-Religion Temples

Po Man Tse
The Hong Kong Polytechnic University, Hong Kong

Pamela Sau Ying Ho
https://orcid.org/0000-0001-6033-0683
The Hong Kong Polytechnic University, Hong Kong

ABSTRACT

Pilgrimage tourism is gaining more attention in tourism literature in the past decade. Most of the studies focused on exploring the motivation of tourists visiting destinations where their religion originated. This study aims to explore the motivation of tourists visiting local religious sites in lower devotion cities which is not promoted for tourists' visitation. Adopting qualitative research methods, participation observation and semi-structured interviews were conducted for data collection. The findings of this study contribute to the current literature in pilgrimage tourism and provide insights for tourism bodies in determining elements of religious sites for the development of new tourism attractions in non-secular destinations.

INTRODUCTION

Pilgrimage tourism was considered one of the oldest forms of leisure travel (3G E-learning LLC, 2021), the early evolution of the tourism industry was based on religious activities (Senbeto, 2023). Pilgrimage tourism is a type of tourism that entirely or powerfully motivates tourists for the achievement of religious attitudes and practices (Saidovna, 2022). There are approximately 600 million religious trips recorded every year worldwide (UNWTO 2014, Ruttanavisanon and Agmapisarn, 2022). Due to the popularization of spiritually motivated travel, there is growing interest to understand the of pilgrimage tourism. Researchers found tourists travelling to sacred places in Thailand, India or Israel etc. are extremely loyal

DOI: 10.4018/978-1-6684-4817-5.ch007

to such tourists' attractions and destinations (Kim, Kim & King, 2020). Unlike these destinations, Hong Kong was not one among the sacred places, tough there are a huge variety of religious groups in Hong Kong, such as Christianity, Buddhism, Confucianism, Taoism, Hinduism, Islam, Judaism and Sikhism (GovHK, 2016). Including Taoism, none of them are originated from Hong Kong, but significant number of tourists are spotted visiting temples in Hong Kong, practising the behavior like any other believer. This study aims to explore the motivation of tourists in visiting local religious sites in destinations where the religion is not originated nor highly promoted by the tourism bodies.

BACKGROUND

Pilgrimage

Pilgrimage is "a journey resulting from religious causes, externally to a holy site, and internally for spiritual purposes and internal understanding" (Barber, 1993, cited from Collins-Kreiner & Kliot, 2000), it is "a quest for a mystical or magical religious experience – a moment when they experience something out of the ordinary marks a transition from the mundane secular world of their everyday existence to a special and sacred site" (Collins-Kreiner, 2016). But the word "pilgrimage" has been applied in a broader context in recent years, such as site visits to famous graveyards (Collins-Krenier & Kliot, 2000).

In tourism literature, the term pilgrimage tourism was not well-defined and still lacks synthesis and holistic conceptualization (Collins-Kreiner, 2018). Literature on related context has just started to grow in the 1990s and marked a rapid increase in the last decade (Progana & Santos, 2023). In recent years, scholars have attempted to better define the term and have invented other exprcssions to study tourists' behavior. Collins-Kreiner (2020) explained that rather than religious motivations, researchers in pilgrimage tourism have also begun to discuss about modern ideas in the spiritual context. Therefore, different terms such as religious tourism, spiritual tourism were also adopted in different studies.

Although concepts in pilgrimage tourism overlaps with religious tourism and spiritual tourism, researchers have attempted to classify and qwertyuo identify differences between them. Recently Zheng et al. (2023) suggested "Pilgrimage involves religious experiences combined with religious travel, including two integrated components, namely a traditional religious journey and a modern secular journey." The researchers (Zheng et al, 2023) claimed that the purpose of pilgrimage tourism is originated from religious beliefs and involves religious behavior. Other researchers such as Shou, Ryan & Liu (2009), claimed pilgrimage is a multi-facet tourism product and can attract tourists with different motivations. Pilgrim tourist may travel to the religious sites for religious purposes but with no-religious motives (Fladmark, 1998, cited from Shou et al., 2009). Shou et al. (2009) found tourists' motivations in visiting religious sites reveal intertwining results of religious motives coupled with other travel motives, such as heritage and culture. According to Shou et al. (2009), destinations can be classified into four different dimensions, which two of them represents the more conventional places for pilgrimage, whereas the other two dimensions may attract 'non-believers" who may visit due to pure curiosity as the places possess heritage and cultural values.

A tourists can be identified as pilgrims as long as they are "engaged in a quest for mystical or magical religious experience" (Collins-Kreiner, 2020), which they have experienced something extraordinary that is different from their everyday existence in the mundane secular world. Few recent articles have a broader scope in exploiting the term "pilgrimage tourism", Tian et al. (2023) has applied "pilgrimage tourism" in researching about young fans who travel for idol-related purposes; Hao et al. (2020) explored anime tourists' behavioral patterns and identified it as "anime pilgrimage tourism"; O'Regan et al. (2019) has reconceptualised gambling tourism to Macau as "Chinese Pilgrimage".

Religion and Tourism in Hong Kong

Protected by the Basic Law, religious freedom is enjoyed by the residents in Hong Kong (GovHK, 2016), there are different religious groups in Hong Kong such as Christianity, Buddhism, Confucianism, Taoism, Hinduism, Islam, Judaism and Sikhism. All of these groups have a considerable population of followers and religious establishments built for local followers (GovHK, 2022). The Christian community population which includes Protestants and Roman Catholics is estimated to be approximately 1.3 million (GovHK, 2021). With approximately 1 million followers for both Buddhism and Taoism, there are over 400 Buddhist temples, 300 Taoist abbeys and temples in Hong Kong (GovHK, 2021). There are also an estimated number of 300,000 Muslims, 100,000 Hindus and 12,000 Sikhs in Hong Kong (GovHK, 2021).

In terms of tourism, only a few of the religious establishments and local festivals are being promoted officially on the Hong Kong Tourism Board website. The Big Buddha and Po Lin Monastery located in Lantau Island is one of the most iconic religious sites in Hong Kong, it is considered as "10 things every visitor must experience in Hong Kong" in the website. However, none of the Christian churches can be found easily on the website, only 2 relevant results are shown through searching for the keyword "church". Besides, there are no information about St John's Cathedral, which is the cathedral of the Anglican Diocese of Hong Kong Island, and mother church to the Province of Hong Kong and Macao. For the section "Muslim travel" of the website, it was focused on Halal food and Muslim friendly places instead of introducing religious spots for tourists in general.

Taoism in Hong Kong

Taoism, in a native Chinese religion, advocating harmony with the nature and simple living, was first established formally in the late Donghan dynasty, with a history of more than 1800 years (GovHK, 2022). Taoism have spread very widely, from the coastal areas in the east to the north-western inland areas and to Hong Kong (GovHK, 2022). There are currently over one million Taoist followers in Hong Kong with more than 300 Taoist abbeys and temples established, in which most of them are open to the public (GovHK, 2022).

Taoist in Hong Kong practice polytheism. The followers believe in many different deities residing in the supernatural world, and the gods are classified into different ranks (Liu, 2003). As followers believe different deities have different areas of expertise, Jade Emperor is the supreme deity, other gods in the middle rank includes the Empress of Heaven, Tian Hou (Tin Hau); the god of war Guandi; the god of the South Sea, Hongsheng and the god of the North, Beidi. And the Earth Gods are on the lowest level of the hierarchy they will seek help them in the area of needs (Liu, 2003). Temples are important religious sites within Chinese societies (Ly and Tan, 2023). Hong Kong was a fish village in the early 1900s and highly depend on the sea for trade in the decades after, the most commonly prayed deities are those related to the sea or weather for fishing and trading. Amongst all, the Emperess of Heaven (Tin Hau), which is the protector of seafarers is very significant for worship. The most popular temples in Hong Kong is Tin Hau Temple, other notable Taoist temples include the Man Mo Temple and the Che Kung Temple (GovHK, 2016).

Formed by representatives of major Taoist temples and Taoists in 1961, The Hong Kong Taoist Association have been organising a wide range of religious and cultural activities to promote Taoism in Hong Kong, also the well-being of society through educational, social and charitable services (GovHK, 2022). Taoist will pay regular temple festivals to celebrate special dates and the birthdays of their patron deities (Liu, 2003), for example, to honor the Supreme Patriarch, Taishang Laojun and the Taoist Day on the second Sunday of March (GovHK, 2022). However, the official holiday of Hong Kong has neither mentioned about the deities' birthdays nor the Taoist Day as an official holiday.

It is an interesting yet significant observation made by Formichi & Connor (2015) who found that some of the people in Hong Kong who visit local temples doesn't claim to practice "Taoist" as their religion, and rather regard themselves as having no religious beliefs. These people may refer to their religious behavior as "worshipping deities" (baishen) (Liu, 2003), and they just ask for the blessings of the god in the area of needs, but they do not worship all Taoist deities and may not understand Taoism thoroughly.

Wong Tai Sin Temple

Wong Tai Sin Temple is located at Chuk Yuen Village, Tai Sin, North Kowloon. Administered by Sik Sik Yuen it is a major tourist attraction. The temple pays tribute to the Taoist monk, Wong Tai Sin (Hong Kong Tourism Board, 2022). Master Wong Tai Sin refers to Wong Cho Ping, who was known as Immortal Huang (Liu, 2003) and also as Red Pine Immortal as he was known to live in hermitage (Sik Sik Yuen, 2022). Wong Tai sin is a deity introduced to Hong Kong from Guangzhou, the capital of Guangdong province. In the beginning, only a small shrine was built to worship Wong Tai Sin in 1921 (Lui, 2033). Due to the growth of number of immigrants from Mainland China, especially in the Guangzhou area, the number of worshippers of Wong Tai Sin has gradually increased. A few establishments in the vicinity are also named after Wong Tai Sin. In 1958, the large government housing project nearby was named Wong Tai Sin. In 1980s, the administrative district and the subway station was also named as Wong Tai Sin (Liu, 2003). Tourists who wish to visit the temple has to be careful especially when searching the name on Google map or asking a taxi driver to drive them to the location. To be exact, the name Sik Sik Yuen Wong Tai Sin Temple should be used, or at least Wong Tai Sin Temple. According to the Hong Kong Tourism Board (n.d.), amongst all Taoist temples, Sik Sik Yuen Wong Tai Sin Temple, which is dedicated to the god of healing Wong Tai Sin, is suggested to be the busiest and luckiest temple in Hong Kong.

Research Gap

The Taoist religious tourist sites are located mainly in the less populated and beautiful scenic areas (Mu et al., 2007), and the most famous Taoist mountains and temples are all located in Mainland China such as Mount Qiyun in Anhui province, Mount Qingcheng in Sichuan province, Mount Wudang in Hubei province and Mount Longhu in Jiangxi province (GovHK, 2022). Interestingly, Hong Kong is not the best destination to experience Taoist heritage or culture as Hong Kong is not known as a Taoist city. In addition, the Taoist deities such as Wong Tai Sin were born in Zhejiang, originated in Guangzhou, not Hong Kong. It is rather not understandable for pilgrims or tourists to pay visits to local temples which are not the birthplace or of the god or the place where the deity was originated. Currently, most literature in pilgrimage tourism focused on the famous pilgrimage destination, this study aims to contribute

Figure 1. Photo of Sik Sik Yuen Wong Tai Sin Temple
Source: Hong Kong Tourism Board (n.d.)

to the current research gap in exploring the motivations of tourists' visitation to local religious sites to shed light on the development and promotion of religious attractions in non-pilgrimage destinations.

METHODOLOGY

This research has adapted a non-participant observation component and was followed up with a semi-structured interview. The observation was conducted by two researchers in early May 2023. The results of the observation were served to crystalize the data collected from the semi-structured interview. Research on pilgrimage stresses into the importance of exploring the pilgrim's thoughts and opinion about their experience (Collins-Kreiner and Klior, 2000). Twelve semi-structured interviews was conducted to collect in-depth information for this study. Interviews were recorded and transcribed for data analysis. Content analysis was applied for data analysis, coding system was used by the researchers. The codes were discussed and compared to identify different themes and dimensions.

FINDINGS AND DISCUSSIONS

Type of Tourists

All of the interviewees in this study belong to the same type of tourists, they claimed there are leisure tourists in Hong Kong, none of them purely visit Hong Kong just to experience the heritage and culture, and none of them are Taoist. According to 3G E-learning LLC (2021), tourists visiting Sik Sik Yuen Wong Tai Sin Temple can be categorized as "Ethnic tourist", who are driven by the desire to experience something different purely due to curiosity. As tourists are experiencing something different in a religious context, they may also be identified as pilgrims according to Collins-Kreiner's 2(020) definition.

"We just to include a bit of everything to experience this city, from shopping, entertainment, scenery to cultural sites." (Interviewee 1)

"I have not been to Chinese temples before, so I have added Wong Tai Sin Temple and Po Lin Monastery into my itinerary." (Interviewee 8)

In contrast to the findings of similar studies in pilgrimage tourism to famous monasteries in China (Wong et al., 2013) found that there is a combination of different types of tourists' visitation, namely "Pilgrims", whom are the believers; "Leisure tourists", whom visit the religious site for sightseeing and relaxation; to "Cultural tourists", whom are interested in historical, heritage and cultural themes.

Authenticity

Similar to most studies, tourists traveling to Hong Kong are seeking authentic experiences. According to the official website of Sik Sik Yuen (2022), it also stated that authentic experience of the neighbourhoods is one of the common purposes for visitation as the Wong Tai Sin Temple and as it is classified as one of the national intangible cultural heritage of China in Hong Kong since 2014. Local religious sites are essential elements to learn, practice, and experience local culture and religion (Ly and Tan, 2023), therefore Chinese temples are also popular attractions to tourists apart from the traditional pilgrims. Besides, temples are normally located within the neighbouring areas in Hong Kong, it may provide "a quieter and closer cultural connection to the local lifestyle, and are attractive to tourists looking for more genuine, back-of-house cultural experiences" (Ly and Tan, 2023).

"Some tour websites have recommended Wong Tai Sin as a cultural attraction spot and I have also read about it in other bloggers' sites. One of the suggested tours on TripAdvisor have included this attraction in the Discovery tour to the Gems of Hong Kong" (Interviewee 7)

"We are not Taoism believers; we just want to experience something local in the neighbourhood." (Interviewee 9)

The majority of the interviewees claimed that they would like to explore the cultural practices of the local Hong Kong people at Taoist temples. All of them have experienced the fortune telling practices Kau Chim (Lottery poetry). The practice involves few steps begins with the querent (person asking the question) holding a long cylinder bamboo cup of flat bamboo incense sticks while asking their question (Psyminds, 2020). The querent will then tilt the cylinder slightly downwards and shake it till a stick falls out. Each stick has a number on it, and the querent can then get a corresponding written oracles for the answer to their question. One of the interviewees explained that even if she does not really believe in the oracles, she finds it interesting to try out what she has seen in movies. Another interviewee even suggested that such activities are tourists must do activities, just like what she would do at the shrines when she travels to Japan. Similar findings are found in study of Wong et al. (2013), the researchers suggested that tourist's not practicing Buddhist belief, wished to experience the religion and the ambience it entails and will also follow the acts of traditional religious obeisance such as lighting incense sticks in the Buddhism Monasteries and placing food on the tables before the statues of Buddha.

"I have always wanted to try shaking the cylinder, this is always a scene in the Hong Kong movies." (Interviewee 10)

"To understand the local culture, the best thing to do is to follow the customs and tradition. I have always followed all the local practices when I visit religious sites, for example, I have tried to throw stones into a hoop placed at the cliff when I visit a shrine in Miyazaki, Japan last month." (Interviewee 12)

Architecture

In most studies, popular tourists' religious attractions have great architectural appreciation value as well, tourists also visit places of important religious events or places with sacral architecture (Raja et al., 2017). Although Sik Sik Yuen Wong Tai Sin Temple may not compare to the famous religious architectures such as Notre Dame Cathedral in France, Sagrada Familia in Spain nor Taj Mahal in India, Sik Sik Yuen Wong Tai Sin Temple exhibits some architectural value in demonstrating how a traditional Taoist temple looks like with its signature red pillars, golden rood, memorial arches and yellow latticework (Conference lodge, 2022).

"A temple located at the heart of Kowloon Peninsular is like a serenity in the midst of the busy districts. It is interesting to visit such an architecture located among the skyscrapers around the area." (Interviewee 6)

"Since I am not visiting Mainland China in this trip, visiting Wong Tai Sin Temple might as well can give me an idea of what a Chinese Taoist temple is like, how is it different from other Chinese Buddhism Monasteries." (Interviewee 8)

Tourists' Religious Practices

Apart from the common themes identified in religious tourism literature, there is also an underexplored motivation for pilgrims. Few tourists with different religious or cultural background are committed to worship their gods even when they are away from their hometown. Two interviewees from Thailand have expressed that they are used to visit temples when they are travelling to pray for safety. As they see Taoism is quite similar to Buddhism due to the fact that both religions have similar gods, share similar teachings and beliefs, so the Thai tourists have decided to visit Wong Tai Sin Temple for blessings. Similar results were also found in another interviewee's explanation, the interviewee who visited Wong Tai Sin Temple for cultural exploration claimed that he is a Christian and has planned to visit St. Andrew's Church, a church that is near his hotel with English section that he has found online, for Sunday services during his stay in Hong Kong.

Blessings

Unlike most religions, each Taoist temple is established mainly to worship a particular Taoism deity and Taoist deities are classified into ranks and have specific powers in answering specific needs (Liu, 2003). Sik Sik Yuen Wong Tai Sin Temple is known as the god of healing. Word of mouth and the success of other people may have motivated tourists to visit this particular Taoist temple. Few tourists mentioned that they have heard of the story or research on the "successful stories" of people worshiping at the temple and have their prayers being answered. Three interviewees even stated that they believe in Wong Tai Sin's power of healing and have purposely visit this temple and hope their prayers will be granted as well. One of the interviewees stated that he will also visit another particular Taoist temple for another blessing.

"I somehow believe the powers of Wong Tai Shin as my friend who has visited before told me it works for her, so I followed the customs for the blessing." (Interviewee 2)

"I heard there are different powers of the Taoist deities, I will also visit Che Kung Temple in Shatin for the blessing of luck in this trip to Hong Kong. My friend told me that I have to turn the fan-bladed wheel of fortune three times and buy a small windmill home." (Interviewee 5)

Suggestion for Tourism Bodies to Develop Pilgrimage Attractions

There are a numerous of benefits in developing pilgrimage tourism, first of all, it may diversify the tourism industry in the host city, promoting an additional type of tourism (Ugli, 2023). Its significant deliverables also includes economic benefits such as national integration, increasing job opportunities and foreign exchange earnings (Hole et al., 2019). There are also significant cultural and social benefits to the host city (Ugli, 2023), tourists' experiences in engaging religious behavior in local religious sites may lead to cultural exchange and promote social harmony. Last but not least, researchers have proved that pilgrimage tourism may leverage green innovation by promoting environmental conservation, encouraging pilgrims to comply with environmental codes of conduct (Senbeto, 2023).

Marketing and Promotion

Sufficient amount of tourism resources should be allocated to develop religious sites and attract international visitors. Currently, there is a lack of information of the local religious sites for tourists in general. For example, there is just a small paragraph in Hong Kong Tourism Board's website introducing Sik Sik Yuen Wong Tai Sin Temple, and the official website of the temple is only available in Traditional Chinese. More information may be provided in the tourism board websites while support may be offered to assist official websites to be translated into different languages and accommodate different tourists. Apart from the generic information of address and transportation method, information of tourists' interest may also be provided on all of these platforms.

Participatory Experience

Visitor interpretation centers may set up at the attraction as some tourists have relatively little knowledge of the religion that they are experiencing (Wong et al., 2013). Telling a story may assist tourists in better understanding the history and culture behind and it may also arouse tourists' interest. "It can also instill the destination with new meaning and signalize the tourists how to interpret the cultural tourism products." (3G E-learning LLC, 2021). Wong (2013) also suggested multi-media and high-end technology may be applied like Walk With Buddha, to tell the story of the legends and gods. Other than the more serious and traditional storytelling, it may also extend to theatrical performances and more creative entertainment.

Organization and promotion of special religious events may also encourage visitors' engagement and therefore lead to tourists' satisfaction (3G E-learning LLC, 2021). Local temples or other religious establishments may also announce information of events on their websites and welcome tourists' participation.

Address Tourists' Religious Needs

Different religious followers commit to different religious practices, even when they are travelling, for example Christians would like to go for Sunday services. Special menus are also required for few religions followers, like Halal food for Muslim and Kosher menu for Jews. A tourist destination should be able to address tourists' religious need as "defining the extent to which the ability and inability to serve these needs affects the tourist decision making process" (Raja et al., 2017). The tourism board of Hong Kong might have neglected the importance of religious needs to attract different tourists from all over the world. Tourism board websites should not only highlight the iconic religious sites as tourists' attractions, they should also include a section on local religious establishments that are suitable for tourists to perform their duty as a follower of a certain religion. For special local religious sites like Wong Tai Sin Temple, tourism boards may also include information on the beliefs and description associated with the deities and temple for most locals, it will also create an ease for tourists' when they plan travel itinerary and thus searching for the specific requirements and blessings they are seeking.

LIMITATIONS AND SUGGESTIONS FOR FUTURE RESEARCH

There are few limitations in this study, such as although the non-participation and semi-structured interviews are adopted for the study, the data collected are based on qualitative study on a limited number of interviewees. To generalize the results, future studies should duplicate the research in other Taoism temples and integrate with quantitative elements. Besides, the study was based in Hong Kong, hence, its viability for other destinations and non-pilgrimage tourism areas may not be entirely applicable. The demographic variables could be also integrated and accessed in future research to collect more accurate results based on the study area.

CONCLUSION

This study contributes to the current literature on pilgrimage in exploring the motivations of tourists in visiting religious sites in lower devotion cities. Research findings supports current findings on pilgrimage tourism an provide further insights on tourist motivations and behavior at local religious sites. The study also explores the opportunities and provide recommendations on developing pilgrimage tourism and visitor attractions in non-secular destinations.

REFERENCES

China Discovery. (2022). *Wong Tai Sin Temple – An active Taoist Temple.* China Discovery. https://www.chinadiscovery.com/hong-kong/wong-tai-sin-temple.html

Collins-Kreiner, N. (2016). The lifecycle of concepts: The case of 'Pilgrimage Tourism'. *Tourism Geographies: An International Journal of Tourism Space. Place and Environment., 18*(3), 322–334.

Collins-Kreiner, N. (2020). Pilgrimage tourism-past, present and future rejuvenation: A perspective article. *Tourism Review, 75*(1), 145–148. doi:10.1108/TR-04-2019-0130

Collins-Kreiner, N., & Kliot, N. (2000). Pilgrimage tourism in the Holy Land: The behavioural characteristics of Christian pilgrims. *GeoJournal, 50*(1), 55–67. doi:10.1023/A:1007154929681

Conference lodge. (2022). *Culture recreation.* Conference Lodge. http://www.conferencelodge.hk/en/discovery/culture-recreation.html

Formichi, C., & O'Connor, P. (2015). Introduction: Overlooked religions in Hong Kong. *Asian Anthropology, 14*(1), 3–7. doi:10.1080/1683478X.2015.1025594

Government of Hong Kong. (2016). *Hong Kong: The Facts Religion and Custom.* Government of Hong Kong. https://www.gov.hk/en/about/abouthk/factsheets/docs/religion.pdf

Government of Hong Kong. (2021). Religion and Custom. Government of Hong Kong. https://www.yearbook.gov.hk/2021/en/pdf/E21.pdf

Government of Hong Kong. (2022). *Hong Kong: The Facts Religion and Custom.* Government of Hong Kong. https://www.gov.hk/en/about/abouthk/factsheets/docs/religion.pdf

Griffin, K. (2018). The Importance of Religious Tourism and Pilgrimage: Reflecting on definitions, motives and data. *The international journal of religious tourism and pilgrimage,* 5(3), 2-9.

Hao, X., Zhang, X., & Mai, P. (2020). Anime pilgrimage tourists' behavioral patterns and their homologous affection: Taking anime film your name. As an example. *Luyou Xuekan, 35*(1), 95–108.

Hole, Y., Khedkar, E. B., & Pawar, S. (2019). The significance of pilgrimage tourism to sustainable development with special reference to the Indian context. *African Journal of Hospitality, Tourism and Leisure,* 8(3), 1–11.

Hong Kong Tourism Board. (n.d.) *Wong Tai Sin: zen in the city.* Hong Kong Tourism Board. https://www.discoverhongkong.com/in/explore/neighbourhoods/wong-tai-sin/zen-in-the-city.html

Kim, B., Kim, S., & King, B. (2020). Religious tourism studies: Evolution, progress, and future prospects. *Tourism Recreation Research, 45*(2), 185–203. doi:10.1080/02508281.2019.1664084

Lin, L. H., Ho, Y. L., & Lin, W. H. E. (2013). Confucian and Taoist work values: An exploratory study of the Chinese transformational leadership behavior. *Journal of Business Ethics, 113*(1), 91–103. doi:10.100710551-012-1284-8

Liu, T. (2003). A Nameless but Active Religion: An Anthropologist's View of Local Religion in Hong Kong and Macau. *The China Quarterly, 174*, 373–394. doi:10.1017/S0009443903000238

Ly, T. P., & Tan, X. (2023). Temple keepers in religious tourism development: A case in Macao. *Journal of Heritage Tourism, 18*(1), 67–83. doi:10.1080/1743873X.2022.2134019

Ma, L., & Tsui, A. S. (2015). Traditional Chinese philosophies and contemporary leadership. *The Leadership Quarterly, 26*(1), 13–24. doi:10.1016/j.leaqua.2014.11.008

Mu, Z., Li, H., Jian, W., Ji, L., Yan, J., & Xiting, L. (2007). Religious tourism and cultural pilgrimage. *Religious Tourism and Pilgrimage Festivals Management*, *14*, 98.

O'Regan, M., Choe, J., & Di Giovine, M. (2019). Reframing and reconceptualising gambling tourism in Macau as a Chinese pilgrimage. *Tourism Geographies*, *21*(3), 508–528. doi:10.1080/14616688.2018.1545248

Progano, R. N., & Santos, X. M. (2023). Introduction to Host Communities and Pilgrimage Tourism: Asia and Beyond. In *Host Communities and Pilgrimage Tourism: Asia and Beyond* (pp. 3–14). Springer Nature Singapore. doi:10.1007/978-981-19-9677-1_1

Psyminds. (2020). Kau Chim- A Chinese Fortune-Telling Practice. *Psy Minds*. https://psy-minds.com/kau-chim-fortune-telling/

Raja, J. (2017). Religious Tourism and Pilgrimage in Non-religious Country: The Czech Republic. In Raj, R., Griffin, K., & ProQuest. Conflicts, religion and culture in tourism (pp. 57-81). CABI Religious tourism and Pilgrimage series.

Ruttanavisanon, W., & Agmapisarn, C. (2022). Religious tourism in Thailand-A review of Thai literature from 2012-2021. *Social Science Asia*, *8*(1), 12–25.

Saidovna, V. F. (2022). The Usage of Pilgrimage Tourism Terms in the English Language. *International journal of inclusive and sustainable education, 1*(4), 66-70.

Senbeto, D. L. (2023). The Greener, the Better? Probing Green Innovation in Pilgrimage Tourism Destinations. *Journal of Hospitality & Tourism Research (Washington, D.C.)*. doi:10.1177/10963480231151674

Shuo, Y., Ryan, C., & Liu, G. (2009). Taoism, temples and tourists: The case of Mazu pilgrimage tourism. *Tourism Management*, *30*(4), 581–588. doi:10.1016/j.tourman.2008.08.008

Sik Sik Yuen. (2022). *Cultural Affairs. Wong Tai Sin Beleif.* Sik Sik Yuen. https://www2.siksikyuen.org.hk/en/index

Tian, J., Li, T., Chen, R., Yang, K., Li, P., & Wen, S. (2023). Converting idol worship into destination loyalty: A study of "idol pilgrimage tour" in China. *Journal of Research in Interactive Marketing*. doi:10.1108/JRIM-01-2023-0020

Ugli, V. A. A. (2023). The importance and advantages of the development of pilgrimage tourism in Uzbekistan. *Galaxy International Interdisciplinary Research Journal*, *11*(5), 304–307.

Wong, C. U. I., Ryan, C., & McIntosh, A. (2013). The Monasteries of Putuoshan, China: Sites of secular or religious tourism? *Journal of Travel & Tourism Marketing*, *30*(6), 577–594. doi:10.1080/10548408.2013.810996

Chapter 8
Comparative Study of Marian Pilgrim Sites of Velankanni and Poondi in Tamil Nadu, India and Lourdes, France

Shaheed Khan
https://orcid.org/0000-0002-5920-6703
Dharthi NGO, India

Freeda Maria Swarna M
https://orcid.org/0000-0002-9649-090X
Dharthi NGO, India

ABSTRACT

Pilgrim tourism across the world has attracted visitors from far and wide, irrespective of the orientation of religion. Be it the Sanctuary of Our Lady of Lourdes in France or the Sanctuary of Our Lady of the Rosary of Fatima in Portugal, Mecca, and Medina for the Believers in Islam, the Char Dham (four abodes) Badrinath, Dwarka, Puri, and Rameswaram in India or even the Chota (smaller) Char Dham of Yamunotri, Gangotri, Kedarnath, and Badrinath in the Indian State of Uttarakhand, or Jerusalem, which the Semitic Religions of Judaism, Christianity, and Islam consider to be Holy and are often visited by travellers from afar. The current research is an engagement of the authors having visited the Marian Destinations of the Sanctuary of Our Lady of Lourdes in France, which attracts more than five million visitors annually. The chapter compares the way the two Marian Pilgrim destinations in the state of Tamil Nadu, India; Basilica of Our Lady of Good Health, Velankani, and Madha Basilica, Poondi, can learn and unlearn from the Lourdes construct and become world renowned destinations.

INTRODUCTION: CHRISTIANITY IN INDIA

It is believed that Christianity in India dates back to the earliest followers of Jesus in the first century C.E. Today, there are millions of Christians in the country, although they make up just 2.4% of India's massive

DOI: 10.4018/978-1-6684-4817-5.ch008

population. South India is home to about half of the Christians in the country, and Christians make up a relatively large share of people in India's more sparsely populated Northeast, where most Christians belong to tribal communities (Salazar, 2021). When one does a deep-dive into the History of Christianity in India, the name of Saint Thomas, one of the apostles of Lord Jesus comes into perspective. St. Thomas travelled to India as early as 52 C.E. making India among the earliest places in the world where Christianity spread (Storytrails, 2021). The apostle is believed to have landed in Kerala (Payaloor) and the religion took root there. He then walked across to Mylapore in present-day Chennai, spoke of a new God in a new land, and left his footprints on the sands of time. The population of Christians of all affiliations in India is about 32 million as per the census (Findeasy, 2023). From the affiliation's perspective, the Protestants are the largest in number, followed by Roman Catholics, Orthodox and Christians of other denominations. Table (1) clearly enumerates the affiliations and the population estimates of all segments:

Keeping in the estimated population in a context, the paper has focussed on the Marian/Catholic destinations of Velankanni and Poondi Churches in Tamil Nadu and tried to understand the construct of Lourdes, France as a Pilgrimage site, which attracts visitors from far and wide.

THE TOURISM PRODUCTS AT THE MARIAN DESTINATIONS, THE FEASTS

Velankanni

Located on the East Coast of India, Velankanni is well connected by all modes of transport, from Bangalore (444 Km), Chennai (325 Km), Coimbatore (360 Km), Madurai (245 Km) Tiruchirapalli (156 Km) and other cities in India. Pilgrims travel, by rail, road, and air (Tiruchirapalli being the nearest airport). Though tucked away in a far corner of Tamil Nadu, the easy accessibility ensures that pilgrim travel to Velankanni for pilgrimages throughout the year.

Ramakrishnan (2019) speaks of the eleven-day event from August 29 to September 8 which attracts pilgrims in the million from far and wide. Sehgal (2018) mentions, the feast that commemorates the Lady of Good Health, to whom the shrine is dedicated. The origin of the festival dates to the sixteenth century and since then, it is organised each year with great aplomb. The feast is organised on the 29th of August every year, kick starts with flag hoisting. The flag is hoisted continuously for the next 9 days as well, till the closing of the feast. The feast plays host to more than 3 to 3.5 million pilgrims who come from all over the country. Speak to Puthota Leon (Leon, 2023), a Manager, Hotel Dravidar (www.hoteldravidar.in), which overlooks the Church complex, he says, "it is the fervour that is exhibited by

Table 1. The church affiliations in India and the population

	Affiliations	Estimated Population
1	Protestants (Including the Church of North India and Church of South India and other denominations)	18,860,000
2	Roman Catholics	10,570,000
3	Orthodox (Malankara Orthodox Syrian Church, Jacobite Syrian Orthodox Church, and Malabar Independent Syrian Church)	2,370,000
4	Other Christian denominations (Mormons: Community of Christ and The Church of Jesus Christ of Latter-day Saints)	50,000

(Findeasy, 2023)

Figure 1. Velankanni, India
(Google Maps, n.d.)

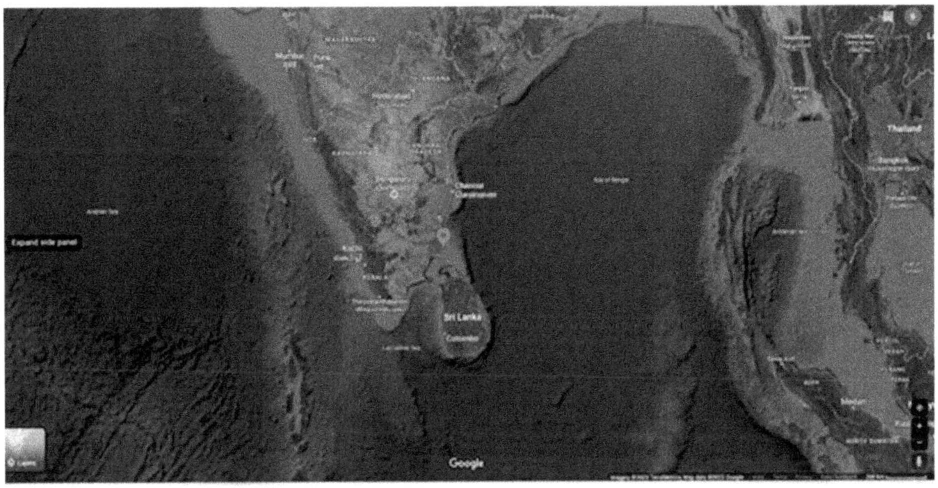

the pilgrims from Goa (a former Portuguese Colony in India on the West Coast, Arabian Sea) where one can understand the belief in Virgin Mary. Pilgrims travel 1,267 Km to the coastal town on the now famous Vasco-Vailankanni Express, to the 'alter of faith' to seek blessings of the Holy Mother." Leo echoes the fact that, "the pilgrims from Goa, plan well in advance and ensure the comforts are taken care, in fact, during the feast, a good number of them, stay for all the days, savouring the best of the holy site." Valsan (2012) considers the trip to Velankanni during the feast as one of the longest of 'Pilgrim Journeys' that the Catholic Christians from within India and even from Sri Lanka, besides other nations. Besides the common thread of faith and devotion linking Velankanni with Goa and Vasai near Mumbai (earlier Bombay), regions which had a strong Portuguese connect. Valsan (2012) speaks of the fact that, the original edifice of the shrine was built by the Portuguese sailors who according to a story and the subsequent belief were rescued from the stormy seas by the Holy Mother.

The Basilica of Our Lady of Good Health, also sanctified as the Sanctuary of Our Lady of Velankanni, is a Catholic Basilica that attracts pilgrims from far and wide, and is active all around the year. The small town, which is a Town Panchayat (the lowest unit of Administration in India); transforms into a religious centre in sixteen-century and it is believed that the Franciscan Missionaries with Portuguese patronage were sent to Nagapattinam (13 Km north of Velankanni) and Velankanni to spread the faith. Nandagopal a tour operator (Vijaya Travel & Related Services) (https://www.vtrs.co.in) based in Pondicherry, mentions that, "I get requests from people across India, and on a few occasions even from other nations to help organize trips to Velankanni." Nandagopal mentions, that, "it is a matter of faith and I have seen in my discussions, the link and relationship of Velankanni and Lourdes in France has ensured that, those who cannot afford a life-time trip to Lourdes, venture out to Velankanni and pray to the Holy Mother; similar to what the Hindus and Muslims believe in their travels to Kashi (Benaras/Varanasi) and Mecca respectively." Sehgal (2018) aptly brings in the fact of Velankanni known as the 'Lourdes of the East,' the Velankanni Church is one of the most revered pilgrimages for Catholics in India.

The sixteenth-century has its own set of miracles which pilgrims' belief happened and there are ample testaments in the presence of churches and artefacts that will prove a point. One of the earliest of miracles is when Mother Mary with Child Jesus saves Portuguese sailors and soldiers from a deadly sea-storm

Figure 2. Exterior pictures of the Basilica of Our Lady of Good Health
(Pictures by authors)

in the area and brought them to safety. The oral lore and popular beliefs have the visitors and pilgrim to lay their trust in the hands of the Mother. Thomas (2009) mentions that a Portuguese ship sailing from Macao to Ceylon (Sri Lanka) was caught in extreme weather in the Bay of Bengal. The terrified sailors invoke the aid of the Virgin Mary under her title of 'Star of the Sea' and the raging storm subsides and the entire crew of about 150 is saved from drowning. The event is celebrated as the 'feast day' of the Nativity of Mary on 8th September every year. The sailors built the shrine and continued to visit the place as a mark of thanksgiving whenever their voyages brought them to the area.

The other two Miracles are known as the 'Milk Miracle' and 'Crippled boy's Healing' are engrained in the minds of the pilgrims, who discuss the same with elan and place their requests to the Mother. Hopler (2018) elucidates the two miracles and mentions that according to believers, in the 1500s there were several miracles that happened during apparitions of the Virgin Mary in Vailankanni, India. These events are now known as "Our Lady of Vailankanni" or "Our Lady of Good Health."

Hopler (2018) provides for one of the finest of rendition of the 'Milk Miracle' story, where a shepherd boy whose name was Tamil Krishnannesti Sankaranarayanan was feeling tired on a hot summer day as he walked from Velankanni to Nagapattinam to deliver milk to his master. So, Tamil lay down under a banyan tree beside a pond to nap. After a strong gust of wind began to blow, he woke up and was startled to see an apparition of Mary standing before him, holding a child in her arms. Both Mary and the child (Child Jesus) had a glow of white light emanating from them. Mary asked Tamil if he would give her son some milk to drink, and Tamil agreed. The child then drank all the milk that Tamil offered from his pot, and Mary thanked Tamil before they disappeared. Tamil apologized to his angry master for arriving late in Nagapattinam and explained why he had been delayed. After hearing the story, his master became angry that Tamil had given away some of the milk he was supposed to deliver. But when Tamil and his master looked inside the pot to see how much milk was left over, they saw it miraculously filling up. It kept filling up until the milk spilled out over top of the brim. Astonished, Tamil and his master ran back to the pond, and Mary and the child appeared to them there as an encouraging vision before disappearing. Thereafter, the pond became known as *Matha Kulam* ("Our Lady's Pond").

The 'Crippled Boy healing' puts the belief mechanism into the pilgrims and generates lot of fervour and hope. Hopler (2018) speaks of how a crippled boy who supported himself and his widowed mother selling buttermilk to thirsty travellers were at work at a place called Nadu Thittu (which means "central mound") in the Velankanni area, when he saw a bright light appear suddenly. Then he saw two figures inside the light: Mary and a child she was holding. Both Mary and the child wore white clothes that emanated a dazzling light. Mary asked the boy for a cup of buttermilk for her son, and he handed her a cup and watched as the child drank it all. Then Mary thanked the boy for being generous. Afterward, Mary asked the boy to go to Nagapattinam to tell a certain Catholic man living there that she and her son had appeared in the area, and that he should build a church there to commemorate the apparition. The boy told Mary that he'd like to go, but he couldn't walk on his own, so he had to wait for his mother to arrive at the end of the day and carry him there. But Mary urged him to try, and when he did, he realized that he had been miraculously healed of his disability. The boy then ran to Nagapattinam and discovered after talking with the man Mary had told him to visit that Mary had appeared to him in a dream the night before and told him to expect the boy's visit. The man built a chapel where the apparition had appeared and placed a statue on the altar of Mary holding Jesus Christ as Child Jesus. The statue became known as Our Lady of Good Health.

Velankanni as a pilgrimage destination has an incredible movement of people from far and wide who come to savour the benefits of the Holy destination. Many come to thank for having received what they had asked for, some come on their annual journey to see Our Lady of Good Health, some come with requests for the future. Pilgrims of hues comes with an intent to get blessed. Some just up in Velankanni, and use the public utilities, sleep in front of the Church, and get back. A good number who can afford, stay in the hotels, Church accommodation, and home stays for a day or two, or even a week and get back. For those who live very close to Velankanni, it is like a day trip to invoke the blessings of Mother Mary. Not surprising, pilgrims of all religions, come, besides, Christians, Hindus, Muslims, and others do come pay their obeisance and concretise their love for the Mother of God. As, Charles a contractual employee of the Church, put it, "people come, they want to stand in front of the Mother, they may not even know the prayers to be recited, but in their own words, communicate with fervour and faith, contribute something to the Church, carry, the holy oils, the souvenirs, candles, flowers back home with a promise that they will come back to be blessed by Mother." When the authors spoke to a visitor in the hotel lounge and asked, is it your first time in Velankanni, the prompt reply was; "I visit Velankanni at least four-times in a year. I do not feel tired coming here, I feel I am blessed. My family too comes with me and they too get a feeling of goodness. What more does one need in life."

Poondi

Located in the heart of Tamil Nadu, Poondi is well connected by all modes of transport, from Bangalore (389 Km), Chennai (336 Km), Coimbatore (259 Km), Madurai (167 Km) Tiruchirapalli (36 Km) and other cities in India. Pilgrims travel, by rail, road, and air (Tiruchirapalli being the nearest airport). Though the location appears off-beat, the pilgrims do visit Poondi on a regular basis from far and wide.

Consecrated as a Minor Basilica in 1999 by Pope John Paul II, Our Lady of Lourdes Basilica, locally known as *Poondi Madha* Basilica is a Catholic pilgrimage centre which has attracted visitors and pilgrims alike. Located between the rivers, Cauvery and Colleroon, in the Alamelupuram Panchayat

Figure 3. Poondi, India
(Google Maps, n.d.)

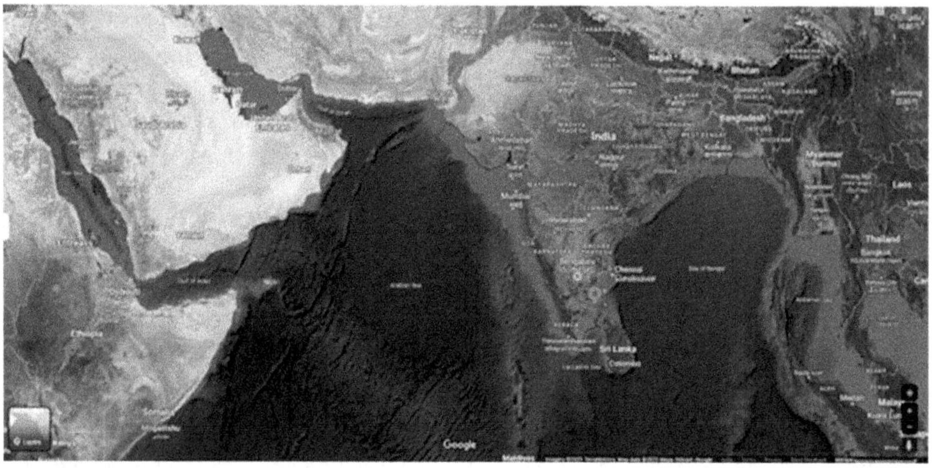

in Thiruvaiyaru Taluk of Thanjavaur District, Poondi is three kilometers from Thirukattupalli (Poondi Madha Basilica, 2023).

It was in the eighteenth century (1714-1718) an Italian Jesuit missionary, Rev. Father Constantine Joseph Beschi SJ who was locally known *Veeramamuniyer* built a church called, 'Mary Queen of Immaculate Conception.' It is in this dome of the church the statue of Miraculous Lady of Poondi is placed (Poondi Madha Basilica, 2023). Padmapriya (2011) speaks of how three statues came to India and one was brought to the Poondi Shrine. The Holy Mother had appeared to Bernadette (Later Saint Bernadette) at Lourdes in 1858. Statues of Our Lady of Lourdes had been sent to different parts of the world. The Statue was brought by Rev. Fr. Darres, of the French Missionary Fathers (MEP Fathers). Joseph an ardent devotee of The Mother from Bangalore, mentions that, "the image of The Mother is so very bright, it is believed that, the brightness has been there from the time of installation. One gets to see something special in the image of the Virgin." Joseph, further states, "the image of the Mother ensures a focus and brings about a devotion in oneself. Considering that the image of the Mother has come from Lourdes, I always visualize myself in the Holy Town of Lourdes and would like to visit the Holy place at least once in my life." The linkage to Lourdes truly evokes not only a spirit of devotion, but also brings one in connect with the realities of life. The one that Saint Bernadette had, one of devotion and belief.

A visit to the Church environs that one realizes the fact that, the posting of Rev. Fr. Loudes Xavier on 1st September 1955 as the Parish Present that efforts were made to improve the condition of the Church. Considering that the Church was already more than two hundred years, the central roof was on the verge of collapse, so the same had to demolished without the front elevation, the altar and dome being disturbed. With no money to get to the repairs, Fr. Xavier had to wait, and Prayed to Our Lady of Poondi, to help solve the issue. An inner voice predicted to Fr. Xavier that the central roof of the Church would come down on a particular day and time, and it did. Fr. Xavier who administered the Church till August, 1972 has been hailed as one of the finest who made efforts to ensure the devotees visit and seek blessings of the Virgin (Poondi Madha Basilica, 2023).

Rev. Fr. Rayappa who was the Parish Priest from 1972 to 1991 (Poondi Madha Basilica, 2023) was instrumental in getting a small relic of the Original Cross of Jesus to the Church in 1976 and the same can be seen below the Statue of the Holy Mother (Padmapriya (2011). A priest of the Church who wanted to remain anonymous confirmed that on the 5th Friday of Lent, the feast of the Holy Cross, the Relic, is taken out in procession from Michaelpatty to Poondi Madha Basilica, which is followed by a Mass Celebration. Throughout the year pilgrims proceed to Poondi; but the place is agog with activity on 8th September every year, which is the Birthday of Holy Mother.

The rural environment of Poondi may not have many hotels and restaurants to invite the visitors and pilgrims. But the few that are to be found are used by those who come from a distance. But most happen to be day trippers or the ones who come late at night, sleep on the pavements without any fear, use the public utilities and get back to their homes. A local flower seller mentioned, "on a routine day, at least four to five hundred pilgrims come to the Church, and if about ten percent were to buy the flowers, we can manage our families. Mine is the second generation that is living out by selling flowers in the precincts of the Church." Clearly it is a relationship that the local community enjoys with the Church that has ensured lot of harmony.

Figure 4. Pictures of the Poondi Madha Basilica
(Pictures by authors)

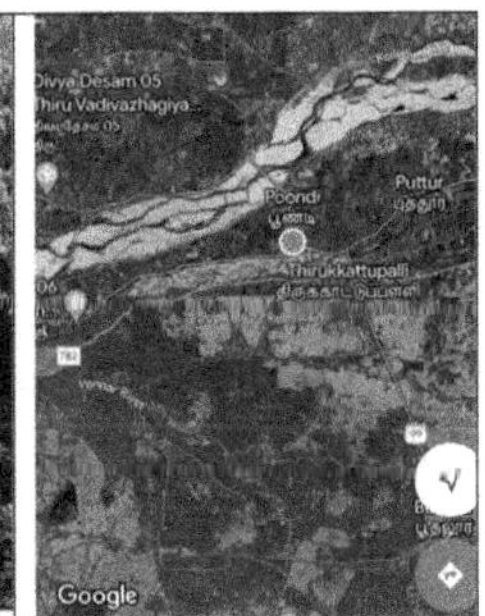

Picture of the Relic (Poondi Madha Basilica, 2023).	The Highway Department, has placed a hoarding near the Church (Picture by the authors	The location of GPS location of Poondi from authors mobile.

Lourdes

For Indians a trip to Lourdes is like a once in a life opportunity. Both the authors got an opportunity to visit Lourdes and combine it with few other Marian destinations in Europe. Having visited a few of the Shrines dedicated to events of Lourdes, viz., Our Lady of Lourdes Shrine, Perambur, Chennai (1897); Our Lady of Lourdes Shrine, Villianur town, Pondicherry (Puducherri) (1886); Our Lady of Lourdes Church in Ulsoor (Halasuru), Bangalore (Bengaluru) (1938), and yes, the Shrines in Velankanni and Poondi; the authors found the visit to Lourdes, France, as something more than being blessed. The visit was a dream come true and far from the watching the live feed from the Grotto daily sitting in our homes.

Like in case of the stories of Velankanni and Poondi, the simple story and the apparitions at Lourdes, which encompassed within itself the story of the peasant girl Bernadette Soubirous (later Saint Bernadette) and where the Immaculate Conception appeared eighteen times to Bernadette (Dichoso, 2015). Adetunji (2022) mentions that, Bernadette said that the figure identified herself as "the Immaculate Conception" and instructed the her to dig into the earth and drink the water she found there. In other messages, the lady asked for a church to be built there so priests could come in procession. Church officials were quickly alerted to Bernadette experiences and were initially concerned about the truth of her account. After investigating, the local bishop became convinced that Mary had indeed appeared to the young woman. Popes later encouraged veneration at Lourdes, and in 1933, Bernadette herself was canonized as St. Bernadette.

Britannica (n.d.) speaks of Lourdes as a pilgrimage town, Hautes-Pyrenees department, Occitanie region, located in south western France and south west of Toulouse, with a population (Klimiuk & Moriarty, 2021) of about 15,000 inhabitants. The town is situated at the foot of the Pyrenees on both banks of a torrent, the Gave de Pau. Though the town had its own history in the Medieval times, the contemporary importance of Lourdes dates from 1858, wherein, from February 11 to July 16, Bernadette Soubirous a fourteen-year-old girl had numerous visions of the Virgin Mary in the Massabielle grotto on the left bank of the stream.

Figure 5. Lourdes
(Google Maps, n.d.)

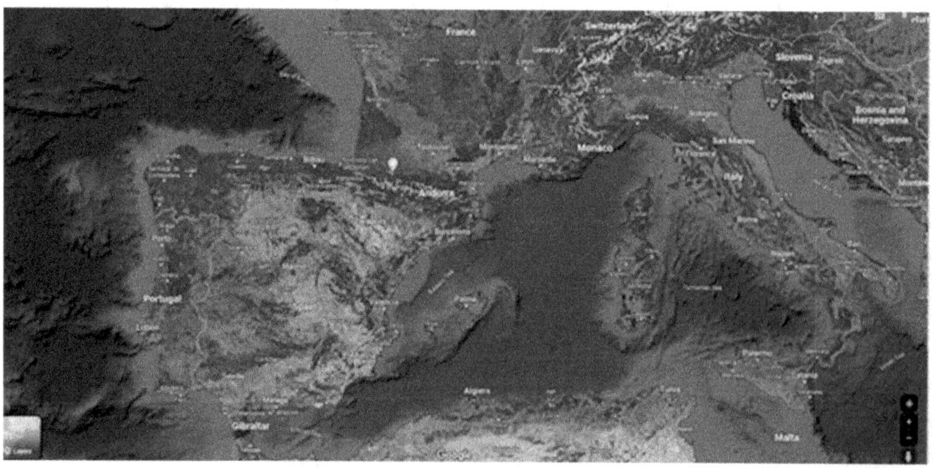

Turner and Turner (1978) speaking about Roman Catholic shrines and the pilgrimages in general and Lourdes in particular, mention, "in Lourdes there is a sense of living communitas, whether in the great singing processions by torchlight or in the agreeable little cafes of the back streets, where tourists and pilgrims gaily sip their wine and coffee. Something of Bernadette has tinctured the entire social milieu, a cheerful simplicity, a great depth of communion." Klimiuk & Moriarty (2021) speak about the pilgrimage experience in Lourdes, mention that, the level of experiences is multiple. These include prayer and devotion at the Grotto site, where the visions are believed to have taken place and actively participating in daily collective worship, including Holy Mass and processions. Many pilgrims visit the baths and water faucets, which provide water from the spring believed to have been discovered by Saint Bernadette. Francois et. al (2012) mention that, since 1858, seventy cures at Lourdes have been recognised by the Roman Catholic Church as being medically inexplicable and "miraculous". However, many thousands of unexplained cures have been reported to the Bureau des Constatations Médicales or Lourdes Medical Bureau. True every pilgrim comes to Lourdes with a prayer and to imbibe the best that the place has to offer. Gesler (1996) writes, Lourdes' is an attraction for the millions as a place of healing. No doubt, it may not be about one's health, but the mental agony one is facing, or the issues on a day-to-day basis, that pilgrims come and move around the Holy town imbibing what is is best for them.

Velankanni, Poondi, Lourdes, the Similarity, the Synergy for Tourism Promotion

Apparitions, stories, individuals, common people, are part of the construct in the three pilgrim sites of Velankanni, Poondi and Lourdes. Though each destination has marched ahead and ensured an attraction for itself under the tourism, travel, and hospitality (TTH) syndrome; places like Velankanni and Poondi have a lot to learn from the experiences of Lourdes, which today is a Global Destination. Though we have seen the number of visitors is large in case of Velankanni, a lot must be done when it comes to Poondi.

Velankanni

Though pilgrim tourism has evolved over a period. The opportunities to better the tourism construct is always available for Velankanni. As Nandagopal put it, "the management of people, the vehicular traffic, the cleanliness facet, waste management, involvement of the local community, the community-visitor interaction, helping with free accommodation for those who cannot afford, food for the needy, the management of alms seekers, should be considered as important for Velankanni."

Poondi

Though one does not see the rush of Velankanni at Poondi; it can and should be developed as a good pilgrimage destination. As Nandagopal, ascribed, "converging Poondi with Velankanni and other nearby destinations should be considered; this will surely ensure a large number of visitors coming to Poondi."

Lourdes

Laurent Duval (Duval, 2022) a resident of Lourdes and a person who ferries pilgrims to and from the airport and also takes them in and around Lourdes, mentions, "when one compares Lourdes to a Paris

Figure 6. A grab from the live feed from the Grotto at Lourdes
(TV Lourdes, 2023)

or a London, oh….those are major destinations, with all the wherewithal; Lourdes in a way appears sedentary; but the visitors are many. Come summer, the Lourdes is agog with activity; the winter is no doubt lean, but the visitors are a constant; it may be a trickle, but the pilgrims are always present, at no time we feel that the place is bereft of visitors." The town provides for a variety of housing from affordable to the expensive, the restaurants too cater to the needs of one and all. One can travel with ease from any part of the world.

When the authors were informally speaking with the pilgrims in Velankanni and Poondi and asked about Lourdes, there was an immediate sense of religious fervour and wanting to go to the place, but then each of the pilgrims realized the distance and difficulty of visiting Lourdes, but are happy that they are part of the Velankanni and Poondi milieu and are in a position to seek the blessings that they need.

From a Tourism, Travel and Hospitality (TTH) perspective, it is best suggested that the Church along with the Leadership and the Government Tourism mechanism, and the private sector collaborate to ensure the best of Lourdes comes into Velankanni and Poondi and one can create pilgrimage destinations that cater to one and all, the visitors and the local community.

REFERENCES

Adetunji, J. (2022, February 9). How Lourdes became a byword for hope. *The Conversation.* https://theconversation.com/how-lourdes-became-a-byword-for-hope-174574

Poondi Madha Basilica. (2023) *Our Basilica.* Poondi Madha Basilica. https://www.poondimadhabasilica.org/about/

Charles. (2023). Personal Communication. [Discussion].

Dichoso, T. J. (2015). Lourdes: A uniquely Catholic Approach to Medicine. *The Linacre Quarterly, 82(*1). Catholic Medical Association. https://www.doi.org/10.1179/2050854914Y.0000000034

Duval, L. (2022). Personal Communication [Interview].

Findeasy. (2023). Population and more. https://www.findeasy.in/christian-population-in-india/

Francois, B., Sternberg, E., & Fee, E. (2012). The Lourdes medical cures revisited. *Journal of the History of Medicine and Allied Sciences*, *69*(1), 135–162. doi:10.1093/jhmas/jrs041 PMID:22843835

Gesler, W. (1996). Lourdes: Healing in a place of pilgrimage. *Health & Place*, *2*(2), 95–105. https://www.doi.org/10.1016/1353-8292(96)00004-4. doi:10.1016/1353-8292(96)00004-4

Hopler, W. (2020, August 27). The Virgin Mary's Apparitions and Miracles in Vailankanni, India. *Learn Religions*. https://www.learnreligions.com/virgin-marys-miracles-in-vailankanni-india-124597

Joseph. (2023). Personal Communication. [Interview].

Klimiuk, J., & Moriarty, K. J. (2021). The Lourdes Pilgrimage and the impact on Pilgrim quality of life. *Journal of Religion and Health*, *60*(6), 3775–3787. doi:10.100710943-021-01398-0 PMID:34505260

Leon, P. (2023). Personal Communication. [Discussion].

Lourdes, T. V. (2023). *TV Lourdes – Le Sanctuaire de Lourdes en direct.* [Video]. YouTube. https://www.youtube.com/watch?v=Zz2lqOMiBeo

GoogleMaps. (n.d.) [Lourdes]. https://www.google.com/maps/place/65100+Lourdes,+France/@41.4860998,2.2709957,1414979m/data=!3m1!1e3!4m6!3m5!1s0xd57d3f0e00a9549:0x406f69c2f3d7c20!8m2!3d43.091463!4d-0.045726!16zL20vMDE4cjN3?entry=ttu

Google Maps. (n.d.) [Poondi]. https://www.google.com/maps/place/Poondi,+Tamil+Nadu+613105/@16.4673358,68.613474,3755947m/data=!3m1!1e3!4m6!3m5!1s0x3baaebe35ef12321:0x81de0812de77a633!8m2!3d10.8655294!4d78.9411821!16s%2Fg%2F1vf99t15?entry=ttu

Google Maps. (n.d.) [Velankanni]. https://www.google.com/maps/place/Velankanni,+Tamil+Nadu+611111/@10.9260104,80.9059162,2319868m/data=!3m1!1e3!4m6!3m5!1s0x3a556ed46f2b695d:0xb2fb9d52ccf6410e!8m2!3d10.6819262!4d79.8436796!16zL20vMDR2bjUz?entry=ttu

Nandagopal. (2023). Personal Communication. [Discussion].

Padmapriya, K. S. (2011, December 20). *Poondi Madha Basilica*. Aalayam Kanden (Temples I saw). www.aalayamkanden.blogspot.com/2011/12/poondi-madha-basilica.html

Ramakrishnan, S. (2019, August 30). Annai Velankanni Festival: All you need to know about the 11-day event. *India Express*. https://indianexpress.com/article/cities/chennai/annai-velankanni-festival-all-you-need-to-know-about-the-event-5951112

Salazar, A. M. (2021, July 12) *8 key findings about Christians in India*. Pew Research. https://www.pewresearch.org/short-reads/2021/07/12/8-key-findings-about-christians-in-india/#:~:text=Some%20say%20Christianity%20in%20India,2.4%25%20of%20India's%20massive%20population

Sehgal, S. (2018, February 6) *Velankanni Church*. Sky Scanner. https://www.skyscanner.co.in/news/velankanni-church

Storytrails. (2021, April 13) *When did Christianity come to India? The story of St. Thomas, St. Thomas Basilica, Chennai*. [Video]. YouTube. https://www.youtube.com/watch?v=I9M0wx5N24k&t=2s

Thomas, W. (2009, August 2) Our Lady of Health, Velankanni, India. *Catholic Voice*. https://web.archive.org/web/20110213110628/http://www.catholicvoice.ie/features/2009/08/02/our_lady_of_health_velankanni_india

Turner, V., & Turner, E. (1978). *Image and Pilgrimage in Christian Culture: Anthropological perspectives*. Basil Blackwell.

Valsan, B. (2012, September 9) Goa to Vailankanni: A journey of faith. *Times of India*. https://timesofindia.indiatimes.com/city/goa-to-vailankanni-a-journey-of-faith/articleshow/16317746.cms

Chapter 9
Role of Indian Temple Architecture in Tourism Development

Sachin Yadav
Mizoram University, India

Dwijomala Hanjabam
Mizoram University, India

Anjali Verma
WASH-I, India

ABSTRACT

Temple architecture can be a significant draw for tourism, as temples are often considered to be important culture and historical landmarks. Many people are drawn to temples for religious and spiritual reasons, while others are interested in the art, architecture, and history of these structures. Various styles of temple architecture can be seen in different parts of the world. Tourism related to temple architecture can take many forms, ranging from guided tours to individual temples to multi-day trip exploring entire temple complexes. During the visit, tourists may also join in the ceremonies and cultural activities to understand local culture and tradition. In addition to being cultural landmarks, temples can also have economic and social impacts on the local communities. In this chapter, the authors are exploring the potential of tourism in various temple architecture sites through different case studies.

INTRODUCTION

While travelling through different cities, tourists will find different heritage structures. These structures may be different types of structure such as palaces, monuments, and other types of architectural marvels. These buildings are highlighting the glory of history, tradition, and culture of different communities. This is the reason, why art and architecture is integral part of the Indian society. The different detail of

DOI: 10.4018/978-1-6684-4817-5.ch009

building defines the evolution of different communities in India. The best example of Indian Architecture can be found in Harappan civilization that's have a distinctive type of town planning. After the Harappan period, the different architecture styles have been observed through various communities i.e Hindu, Jain, and Buddhist. The middle age period certified the amalgamation of Persian and regional style of architecture (Berger and Luckmam, 1967). Thereafter the foreign rulers period brought the influence of Western architectural style to India. Perhaps Indian architecture is a combination of regional styles and foreign influences which have provide it a unique architecture. Therefore, it will provide an immense opportunity to explore as a scholar and will provide more authentic information to the tourist.

Indian architecture progressed in various duration in different region of the country. On one side, we are witnessed the archaic evolution of Indian architecture from ancient time to historic period, transformation in Indian architecture create a great impact on tourism development (Curtain and Gaither, 2007). Organically, the inception and putrefy of many dynasties in the country derived the style of Indian architecture. Because of the different communities from outside brought their tradition and a different way of living have contributed in the evolution in Indian architecture. This differentiation can be observed comfortably by the tourist while visiting the different places in India.

Architecture is not a recent occurrence. It started a long time ago, when humans used to carve out space in the mountain in the form of shelter. Because of inherent curiosity in the human being, they develop and evolved such kind cave to live safe and secure life. So, their evolved architecture was an amalgamation of their needs and imagination and their capacity to develop these kind of shelters. These structures are also an incorporation of the local and regional context inspired from tradition, economic prosperity, social requirement, and their belief system in the religious building of different eras. It also accommodated the local and regional cultural traditions and social requirements, economic prosperity, and religious practices of different times. So, the study of Indian architecture helps us in understanding the cultural diversities and rich heritage of India.

Indian economy depends on the various sector, i.e., manufacturing, agriculture, infrastructure, etc. But now, the economy of India has a huge impact because of the Tourism industry. It has grown exponentially over a decade. Undergoing changes in environment, economy and society as well as natural and manmade disaster, tourism industry has made its way on the basis of situations arising and ultimately tourism industry widely recognised as a fastest growing business around the world. Additionally, tourism industry proved his potential to provide employment to the skilled and unskilled workers. So this suggests that tourism can play a vital role in enhancing the economy of developing and developed countries. As indicated by the tourism industry, it will not only improve the source of income; it will also improve the livelihood of various stakeholders.

Pilgrimage tourism evolved on the basis of differences in people belief and traditions. This kind of tourism gives opportunities to study these historical structures (Bayjh, 2018). During the visit, the tourist may find some interesting story about these structures and can relate with the different historical events. Pilgrimage tourism provides an opportunities to extend the faith and religious teaching values. Presently, Pilgrimage tourism is not only for the specific devotee, this will also provide a platform for interaction in different communities. The success of pilgrimage tourism is depends on the various supporting infrastructure provided in the neighbourhood (Bayjh, 2018).

India's heritage built form, we can understand tangibly as well as intangibly. It is creating an impact on global tourism because of the diversity in culture and tradition. Many researchers have emphasis on the relation between natural diversity and traveling identity to encourage various form of sustainable tourism. Pilgrimage tourism is the most accepted way of tourism, globally.

This study main objective to highlight the role of cultural identity in the development of temple architecture. In this respect, the chapter elaborates how temple architecture in India evolved while carrying the culture and tradition is on the most priority. In present context, it is serving in the national economy by providing employment and service to the different stakeholders.

TEMPLE ARCHITECTURE STYLE IN INDIA

India is a diverse country and, has huge potential of tourism development through different style of temple. Jammu as old town has been flourishing as a pilgrimage town because of several temples located in the city.

Earlier Jammu town is famous for the existence of cluster of temple in and around the city. In existing scenario, Jammu administrative area have various types of new and old shrine. Due to multiple in-migration and out migration happened in the city, The people still have their association with these sacred buildings. There are two architecture styles are prevalent in ancient temples i.e Pahari style and Shikara style. The Pahari style of architecture consists of various characteristics i.e Slanting roof covered with wooden slits. Vertical construction of the Pahari style shows the reflection of religious developmental growth. This style of construction have a central space where people can comfortably gather and do interaction. While using the space, we can observe the different ornamentation and carving done on wood. The Bhillawar and Kirmachi temple were raised in the Pahari style of construction.

The different architecture styles of temples in Jammu shows the amalgamation of Kashmiri and Himachali culture and tradition. Some peculiar details of both cultures can easily seen in plant and elevation of the structures. The ancient temples of Jammu show the sangam of Kashmiri and Himachali Building style. Massive stone construction and sun dried bricks made these structure unique in term of construction (Karar, 2010).

In every 50-100 meter, the traces of temples is recognised. The most famous temple is Ragunath temple and it is located in the heart of the city. Maharaja Gulab singh was founder of the Ragunath temple in 1835 AD, later it was completed during his son Maharaja Ranbir Singh ruling as king in 1860 AD. The internal walls of the temple was cladded with golden sheets on the three sides. The Ragunath temple have seven shrines in its temple complex, each shrines have its own tower that is highlighting the religious developmental growth. Due to its structural importance, it is recognised as a largest temple complex in the northern india. After 130 years of built heritage, the reflection of mughal architecture will found in the form of arches, niches and surface parallelly we can found the interior orientation done in gold sheet.

In the near proximity of Jammu, the Bahu fort is not famous as religious structure, but it also acts as an architectural marvels for the city. It is located 5 km far from the city and it is lied on the bank of the river tavi. Bahu Fort, which also serves as a religious temple, is situated about 5 km from Jammu city on a rock face on the left bank of the river Tawi. This fort and building may be the oldest ones in Jammu city. The fort was upgraded and restored by Dogra Kingdom more than 300 years after it was first built by Raja Bahulchan. The ruling diety of Jammu, the goddess Kali, also known as Bave Wali Mata, has a temple inside the fort. This temple is composed of white marble and is built on a platform that is 4 feet high. According to some researcher, the plinth above it, shows the reflection of modernism.

Kirnachi is also a famous temple, located in Jammu region. It is located 9km far from Udhampur and 64 km from Jammu town. Although seldom recognised outside of its own country, this community boasts a special architectural heritage. In the past, Kirmachi held a significant position because of

Figure 1. View of Ragunath temple

Figure 2. Top view of Bahu Fort in Jammu

their cultural heritage. This may be determined from the four magnificent and commanding temples. The convergence of Pagoda and Gandhara building styles is an intriguing aspect of the design of these temples. The Shikara temples of Himachal are very similar to Kirmachi temple. Numerous sculptured artefacts adorn the temples's exterior walls. The larger temple's main enterance contains columns that are adorned with flowers, foliage, and tiny figurines in the form of Varah Avtra and the three headed shiva. The details and quality of stone sculptures is prominent in the whole temple complex.

Ladan Kotli temple is situated in Udhampur town. At Ladan Kotli, there is only one ancient structure left temple i.e Jalandharia temple. This tmple is made out of Shikara style of construction common in Jammu region. In front of the temple is a stone lined spring with clear water. The temple is particularly interesting since it is the only temple in Jammu region with sensual themes depicted on the outside sculptured walls (Page and Hall, 2014).

It is evident to the mind that houses of worship, particularly temples, were the focal point of educational and political activity in ancient times. The essential aspect that everyone should remember is that

Figure 3. View of Kirmachi temple highlighting the construction detail.

it captivates everyone who chances to look at it. Though. I haven't even gotten to the Kashmir section of temple and mosque architecture, which is even more spectacular. My suggestion for accentuating the region's temple design is to encourage officials to revitalise and conserve these magnificent temples of Jammu region by including them on the major tourist circuit, such as Raghunath Temple (owing to its location). By doing so, this tourism-friendly state may add another feather to its cap, as well as a significant boost to its economy. Now, Kirmachi temples are under the control of archaeological survey of india to restore or maintain the whole structure.

After showcasing the various architectural styles of temples in Jammu, further temples from various parts of India must be examined in this chapter. The sites of the Indus Valley civilisation, some of which are situated in India, are a highly intriguing area for travellers to visit and study. Excavations at Harappa, Mohenjodaro, and other Indus Valley Civilisation sites showed the existence of a relatively advanced urban civilisation with superior town planning and engineering abilities. The highly developed drainage system, as well as the well-planned roads and dwellings, demonstrate that India formerly had a complex and highly evolved civilisation. The Harappans appear to have primarily built three sorts of structures: public baths, pillared halls, and residential houses (Page and Hall, 2014)..

The most notable aspects of Harrapan architecture are their exceptional town planning abilities and towns based on a clear geometric pattern or grid layout. Roads intersected at right angles and were meticulously built out. Because the Indus Valley villages were built on the river's banks, they were frequently devastated by large floods. Despite this disaster, the Indus Valley people rebuilt new towns on the same places (Rao and Suresh, 2013). During excavations, layer upon layer of towns and buildings were discovered. The fall and eventual annihilation of the Indus Valley Civilization, which occurred somewhere around the second millennium BCE, is still a mystery. Lothal, in Gujarat, also features the ruins of a dockyard, demonstrating the importance of commerce.

Figure 4. View of Ladan Kotli temple

India entered an important period in her history around the sixth century BCE, when two new faiths evolved, Jainism and Buddhism, which aided in the development of early architectural style. Buddhist stupas were created in locations where Buddha's bones were kept, as well as prominent sites where important events in Buddha's life occurred. Stupas were constructed from massive mounds of mud surrounded by neatly burned little standard bricks. One was erected in his birthplace of Lumbini, the second in Gaya, where he obtained enlightenment under the Bodhi Tree, the third in Sarnath, where he gave his first sermon, and the fourth in Kushinagar, where he died at the age of eighty, obtaining Mahaparinirvana (Rao and Suresh, 2013).

Because they were prominent architectural structures in the country, Buddha's burial mounds and locations of pivotal events in his life became essential sights for tourists. These became major venues for the sangha, Buddha's order of monks and nuns. Monasteries (viharas) and centres for preaching, teaching, and learning sprung up in such locations. Congregational halls (chaitya) were also established for instruction and interaction between the ordinary people and the monks.

THE STUPA OF SANCHI AND SARNATH

The stupas of Sanchi and Sarnath are icons of Mauryan architectural excellence. The Sanchi Stupa's entrances, which have magnificent sculpture representing scenes from Jataka legends, are examples of the artists' talent and artistic sensibility.

Gandhara art arose later as a result of the mixing of Greek and Indian art. Buddha and Bodhisattavas were sculpted in the form of Greek gods in stone, terracotta, cement-like substance, and clay.

Another area known for Buddhist architecture is Nagarjunkonda. The Gupta period represents the commencement of free-standing Hindu temple architecture. The temple in Deogarh (Jhansi district) has a central shrine or garbhagriha where the deity's picture was put as an example. Another outstanding example of this time may be seen at Bhitargaon (Kanpur district).

CAVE ARCHITECTURE STYLE

Another distinguishing aspect is the evolution of cave building, which represents an important period in the history of Indian architecture. Between the second and eleventh centuries C.E., more than a thousand caverns were dug. Among these were the Maharashtra caverns of Ajanta and Ellora, as well as the

Figure 5. Buddhist Stupa at Sanchi, Madhya Pradesh

Figure 6. Buddhist temple at Nagarjunkonda

Orissa cave of Udaygiri (Rao and Suresh, 2013). These caverns house Buddhist viharas, chaityas, and mandapas, as well as pillared temples dedicated to Hindu gods and goddesses.

Basalt collected in the form of volcanic rocks was used to construct the Ajanta Caves. These rocks collected as a result of long-ago volcanic eruptions. The craftsmen carefully chiselled the granite, since fissures appeared during the procedure.

Figure 7. View of Ajanta Caves

CONCLUSION

The changing planet has had a significant impact on all aspects of human civilization. This expanding world of capitalist business has also had an impact on religious organisations. A big religious organisation, a temple, is undergoing self-adjustment to match the changing societal trend (Gunn,1994). A temple is converting itself from a centre of soul and a main venue of sacred practise to a prominent tourist attraction. This study examines how temples in India establish cultural identity in order to increase cultural capital.

The first step is to increase the financial worth of cultural capital. Temples entice tourists by commercialising their existing attractive architecture. Temples with cultural-religious heritages can draw large numbers of tourists, particularly adherents of diverse religions, on the idea that worshipping a sacred item or place brings wealth and merit. They may not only enjoy the travel to the temples, but they can also pay respect to the pilgrimage and absorb the grandeur of architecture inside the destination temple (Hall,1992). However, it is important to note that this method of building cultural identity cannot be used by all temples. It might happen to temples with a lengthy history and cultural-religious legacy in particular.

The faith that variety in various temples is the second approach to build cultural identity. As one of the most fundamental religious teaching principles is to respect monks, visiting and seeking for blessings from gods and goddesses is popular in various religions. People hold the utmost regard for monks who perform great deeds or have the capacity to regulate magical chanting to benefit people. They keep going back to the temple where he used to live. The sacred emblem is something they normally carry back with them. As a result, the various temples may enchant a vast number of tourists indefinitely (Gladstone, 2013).

Following that, some temples choose to forge their cultural identity by fusing religious conviction with communal faith. This mix results in a unique tradition that might pique the interest of travellers.

The fourth method temples might develop their cultural identity is to create a distinctive landmark. This landmark was built specifically for tourism purposes. It is the most popular of the five cultural identity development methods since it is an excellent option for temples that do not have existing cultural-religious heritages. However, if this method of cultural identity formation is chosen, the temple must be confident that its landmark is religious in nature. Furthermore, it must be distinctive, massive, and gorgeous enough to draw people's attention.

Another strategy to develop a cultural identity is to build a tradition that both the temple and the community can share. Those manufactured traditions are typically religious in origin, therefore they constitute a cultural commodity with a selling signature in the tourism industry.

In conclusion, the cultural identity creation of many temples in India may be understood as shifting non-financial value goods to financial value ones. This is consistent with Karl Polanyi's idea of fictional commodity formation. Like Polanyi's (1944, 2001) concept of labour, land, and money, a holy statue, holy spot, great monk, religious belief, and local tradition are not actual commodities. They were never intended to be sold. They can, however, be converted into fake goods and sold in the tourism sector. Although tourism is not completely unrelated to religion and culture, the capitalist economy fosters a close tie between them. Due to the need to make more money, temples in India are attempting to turn themselves into renowned tourist destinations. This phenomenon diminishes their former significance as a spiritual centre for religious practise and the heart of the community.

As previously said, the cultural identity building of temples in India is for tourism purposes. This is generated from the current world's economic system, which influences not just people's way of life, but every aspect of human civilization. Temple, as a significant religious organisation, cannot remain in the world unless it changes. As a result, the temple must make every effort to transform itself into a tourist attraction.

REFERENCES

Bayih, B. E. (2018). Potentials and Challenges of Religious Tourism Development in Lalibela, Ethiopia. *African Journal of Hospitality, Tourism and Leisure*, *7*(4).

Berger, P., & Luckmam, T. (1967). *The Social construction of reality: A treatise in the sociology of knowledge*.

Curtin, P. A., & Gaither, T. K. (2007). Global public relations and the circuit of culture. Thousand Oaks, CA: sage. doi:10.4135/9781452224817.n3

Du Cros, H., & McKercher, B. (2014). *Cultural tourism*. Routledge. doi:10.4324/9780203790601

Gladstone, D. L. (2013). *From pilgrimage to package tour: Travel and tourism in the third world*. Routledge. doi:10.4324/9780203689363

Gunn, C. A. (1994) Tourism Planning, New York, Taylor and Francis

Hall, C. M. (1992). *Tourism Planning: Policies, Processes and Relationships*. Prentice Hall.

Karar, A. (2010). Impact of pilgrim tourism at Haridwar. *Anthropologist*, *12*(2), 99–105. doi:10.1080/09720073.2010.11891138

Page, S. J., & Hall, C. M. (2014). *The geography of tourism and recreation: Environment, place and space*. Routledge.

Rao, N., & Suresh, K. T. (2013). *Domestic tourism in India, The Native Tourist*. Routledge.

https://www.nios.ac.in/media/documents/tourism_337_courseE/337_Tourism_Eng/337_Tourism_Eng_L8.pdf

www.google.com/images

Chapter 10
A Perspective on Animal Cruelty:
Conservation in Religious and Mythological Tourism in South Asia

Debasish Batabyal
https://orcid.org/0000-0003-0191-9804
Amity University, Kolkata, India

Pratim Chatterjee
https://orcid.org/0000-0002-4136-8157
Amity University, Kolkata, India

Somnath Chatterjee
Aliah University, India

Namrata Chakrabortty
JIS University, India

ABSTRACT

Given the growing importance of conservation and preservation of animals and other species, the purpose of the chapter is to present a review of literature around the subject. The search engine Google Scholar has been used for searching the most cited paid and unpaid articles based on chosen keywords. A total of 250 documents represented the population size, in which only 80 were found cited and relevant for the study, though only sixty-six were finally selected due to certain constraints. Key themes from the literature focus on varying cruelty-conservation paradox as a religious belief and consequences for pilgrimage, fairs, festivals, and special events have been highlighted. This chapter is a collection and compilation of fact-finding evidence with paradoxical features. Understanding these features has twofold aspects, viz. identifying new and allied segments for future pilgrimages and seeking a background for research on sustainable pilgrimage linking ancient religious inscriptions recognizing animal welfare with the most modern SDGs.

DOI: 10.4018/978-1-6684-4817-5.ch010

INTRODUCTION

It has been common in almost all communities and cultures for companion animals, or pets, to share living quarters with people around the world (Savishinsky, 1983; Serpell, 1987, 1996; Gray &Young, 2011). While attachment theory exhibits emotional relationships between the infant and caregiver (Bowlby, 1958; Ainsworth & Bowlby, 1991), social support theory exhibits evidence that humans with good social connections are healthier and that there are significant medical outcomes of loneliness (Lynch, J.J., 1977; 2000; South, et al., 2011). The concept of animal sacrifice in any religion is believed to have been inhibited in this pain of sacrificing the most near and dear ones. On the other hand, social support theory suggests and strengthens a human-animal bonding of coexistence since the primitive age.

In Islam, the sacrifice of animals is accepted as a complete obedience and demonstration of submission from the morning of the 10th to the sunset of the 13th *Dhu al-Hijjah*, the 12th lunar month of the Islamic calendar. In Hinduism, animal killing on specific occasions in most of the Shakti temples is based on faith and belief that it stimulates her violent vengeance against the buffalo demon. Similarly, different types of animals or birds are sacrificed, the slaying of goats, chickens, pigeons, ducks, fishes, male water buffalos and such other animals are sacrificed in different religions (Fuller, C.J,2004). In Bhutan and remote Sikkim Himalaya, animal sacrifice was in vogue in Bonism and shamanism during the pre-Buddhist period (Erschbamer,2019; M.,Palni, L., 2019; Roy, M., 2020). As winter spreads across the country, people move from their residences to mountains, caves, ruined houses, streams and giant trees to worship their local deities, mostly associated with animal sacrifice. With the acceptance of Buddhism, the practice of killing animals is now replaced by offerings of joints and cuts of meat (ShaNga-not a fresh or sacrificed meat but naturally dead animal's meat).

The 23rd Tirthankara Parswanath accelerated *Jainism* and *Ahimsa* in the 9th century BC which raised voices in favor of animal rights in religious practices on the Indian subcontinent. Also, the *Dravidian* philosopher *Valluvar* in his *Tirukkural* emphasized and invoked *Ahimsa*(tolerance) and moral vegetarianism at individual level. Interestingly, the Parsees or *Parsis*, one of the oldest extant ethnoreligious groups of the Asian subcontinent practicing Zoroastrianism, perform their funerals dedicated mostly to birds and animals. A dog is essential in the funeral process because it can see death and the body is taken to the tower of death where the vultures feed on it.

Another group of scholars such as Kuznets (1955) and Morris (2013) worked on an animal protection index, reducing cruelty, and encouraging vegetarianism from a socio-economic perspective, not necessarily focusing absolutely on community cultural issues incorporating animal stress, cruelty, entertainment and sacrifice directly. However, Holst &Martens (2016) found a strong voice against animal consumption and killings in greater democracy and Kuznets et. al (1955) evidenced that vegetarianism prevailed over the very poor and very rich. Fennell &Sheppard (2020) focused directly on utilitarianism in tourism and developed a scales of justice framework on two different normative ethics models, varying from deep, intermediate and shallow to no justice (Shahi, A., et al., 2008; Fennel, D. A. (2014). Their study dealt with more contemporary issues but the core area of granting ritualistic cruelty inhibited in culture and heritage. Some renowned scholars' studies are animal specific, discussing elephants, pandas etc. and identifying and explaining animal consumption, killings and stress and subsequent effects on economic and socio-political issues in tourism industry (Fennell & Sheppard, 2020; Cohen, 2010; Bansiddhi, 2020; Duffy et.al., 2011). This school of scholars found how the common tourists are critically varying and compromising in nature.

Therefore, a systematic understanding and body of knowledge of the human-animal interrelation and interdependence should be studied in religious tourism, fairs, and events. This study of literature review explores such uses and abuses of animals in religious tourism, fairs, festivals, and events along the continuum of primitive cruelty to modern conservation with a broader timeframe of changing values, cultural rigidities, modern perception, tolerance and awareness towards responsibility and sustainability (Duffy & Moore, 2011; Shaheer et.al., 2021).

OBJECTIVES

Any place of attraction showing cruelty and pain towards living creatures goes against sustainable values and principles around the world. Many scholars advocated for vast and complicated animal ethics theories, with regional variation, into the tourism lexicon (Fennell, et al., 2012; Duffy et.al., 2011). Responsible tourism is defined as making better places for people to live in and better places for people to visit by ensuring that operators, hoteliers, governments, local people and tourists take responsibility and take action to make tourism more sustainable (Cape Town declaration, 2002; Kerala declaration, 2008). Fennell, D. (2014) advocated responsible tourism not only for the welfare of the tourist in wildlife tourism but also for the welfare of the animals in other destinations. As an approach, the diversity of the world's cultures, habitats and species and the wealth of our cultural and natural heritage, as the very basis of tourism, should be protected in different ways across all generations world-wide (Goodwin, H., 2016). Therefore, it is an urgent need to identify and promote all such means and ways for a better future in tourism. In doing so, this study focuses and reviews animal cruelty-conservation paradoxes while identifying new and allied religious, mythological tourism in onc hand and promoting ways and means of sustainable and responsible tourism principles on the other hand in religious centers of intensive mass tourism.

METHODOLOGY

This study analyzes and reviews a range of literature which provides an insight into the unethical use and/or abuse of animals in religious tourism in a wider perspective. Till April 2021 reviews were carried out and all earlier cited papers meeting the criteria were also included. This was the last time the search results were accessed (sorted by citation) to decide whether they should be part of the review. Purposively, the search engine Google Scholar has been considered for searching the most cited paid and unpaid articles based on the selected keywords such as 'animal sacrifice (and) tourism', 'sustainable and responsible pilgrimages' 'tourism animal relation', 'animals (in) religious tourism' and 'abuse (and) utilitarianisms (of) animal (in) tourism' respectively. In each case, a maximum of five pages of results were considered and only cited articles have been examined. A total of 250 documents represents the population size in which only 80 are found cited as well as relevant for the study. Due to financial constraints and partial availability of some articles, a final total of sixty-six articles were used in this study. More than fifty percent relevant and cited articles are found to be on or from 2010, representing contemporary literature for more than a decade.

FINDINGS AND DISCUSSIONS

Is there such a thing as a cruelty-conservation paradox in cultural, religious, and heritage tourism?

Yajur Veda mentions that no person should kill animals helpful to all (Dwivedi & Tiwari, 1987). Again, animal sacrifice was allowed in the Vedic rituals, mostly for special occasions. Later, frequent animal sacrifice superseded the Vedic philosophy of love and care for them. The same is evident in Islam and other religions. It may be that other important geo-physical and socio-political, natural reasons are involved however, animal sacrifice, killing etc. has become a vehicle of exchange where it is offered to God in return for something. In South and Southeast Asian regions, the gender, age, number of animals to be sacrificed and even how to kill that animal are prescribed and practiced in culture and religion. Culture, community and religious affiliations have direct association with the ritual slaughter of animals in many countries and, as such, the concerned communities have been successful in getting their views enacted into law (Lerner & Rabello, 2007).In India, tourists also visit and pay respect to or view shrines and temples where animal sacrifice is offered to gods or goddesses, mainly the goddess Kali (Tul, 2015). This is how places or religious centers of animal sacrifice are sources of attraction. The cults of Saint Anthony of Padua at Puliyampatti, a famous Tamil Catholic pilgrimage area, follows the Hindu culture of animal sacrifice as offerings to the gods. Generally, goats are used as offerings and are beheaded outside the church, after which the pilgrims offer the goat skin to the god as part of their Tamil Catholic tradition (Sébastia, 2004). On the South Asian subcontinent, Muslims and low caste Hindus are mostly found performing rituals through animal sacrifice. Kali worship is based on tantric tradition where sacrifice is necessary to satisfy the god. Sometimes animals, their presence and their sacrifice are an extensive part of black magic and dark tourism. Even if the courts stop the sacrifice, it still secretly takes place in many states in India (Moodie, D., 2019). Animal sacrifice is mentioned as the barbaric and tribal forms of sacred power. Domestic animals like buffalo, sheep or goat are used mainly as offerings and as part of the main rituals to gods in different parts of South and Southeast Asia. Moreover, additional animals are also often sacrificed by others to fulfill their personal vows (Hitlerite, 1980). The modern transformation of tantric tradition as part of a tourist destination is observed on the Indian subcontinent, also incorporating Bhutan and Nepal. Countless goats, pigeons, fish, buffalo and other animals are brought for slaughtering by pilgrims and tourists. Different non-Vedic offerings are made as a part of rituals on different puja like Durga puja and Kali puja (Urban, 2019). The goddess Kali or Durga and their different consorts (*Bhairava*)are worshiped throughout Asia, including some places in Tibet in China as seen in figure 1 below. According to Hindu mythology, these are the main places that have had specific rituals invoking Shakti through animal sacrifice for several thousand years. The rituals of animal sacrifice are even popularized in many other places on this sub-continent deriving mostly from these places.

Animal offerings to gods as a means of sacrifice also induce tourism which is observed in different religions. At the *Kamakhya* temple in Assam, India and *Gadhimai* temple in Nepal, buffalos are beheaded and offered to the goddess. The ritual attracts many locals in addition to national and international tourists, who come to behold the tradition (Urban, 2016). Animal sacrifice is part of the old shamanic tradition in Sikkim. As part of the ritual, these animal sacrifices are made to worship or show loyalty to the gods and to calm or tame the supernatural. Though at present animal sacrifices can create some problems, these rituals are still practiced as a part of heritage from forefathers, and it also depicts one's own tradition and identity (Marlene, 2019). Religious and pilgrimage treks are one of the most popular tourist activities on the Indian subcontinent but killing animals for rituals, catching birds, fishing and cutting trees are a matter of concern (Gupta, 1999). In ritual slaughter, animal suffering is very likely so this kind of act can never be morally justified on the grounds of tradition, involvement of culture, religiously required practice or a source of income from tourism. Domestic animals which make the most noise while dying are mostly offered as a part of tradition (Behrens, 2009). Snake charming was a tradition of nomads and tribes and snake festivals are stills observed in many places where hundreds of different species of snakes are taken together for invoking the mother goddess of snake. Though this type of religious practice and entertainment is justified, may be as a tradition, yet not a sustainable practice at all. Rather, exchanging snakes, removing poisonous teeth of these species cause scarring or even death to snakes (Mohanty et al., 2004).

Animal fairs also cause animal suffering in many ways, often commemorating religious occasions in Indian subcontinent. Foot and mouth disease and other animal illness is only treated locally in those fairs and lack of proper treatment claims the lives of hundreds of cattle in traditional cattle fairs in India (Galav et al., 2010). Several festivals are also considered as a tourist commodity and attract a lot of tourists and locals at the expense of these animals. Camel wrestling in Turkey is one such source of income for the locals through tourism though the camels are very likely to be injured or killed in this violent sport (Donlon et al, 2010).

As a part of controversial cultural tourism, abuse of animals is observed on almost all continents of the world (Malchrowicz-M., et al.,2020). Animals in contest against each other, animal in contest against men, animals used as religious sacrifices, animals being hunted, animal used as means of religious processions and transportation, animals used as temple or church attractions, animals used as tourist picture props in temple complex, flesh or meat of animals as ritualistic special food etc. are surely a threat to the safety and security of these affected animals (Shastri, J.L.; 1970; Shastri, B.,1994; Ranaweerage, E., et al., 2015).

Till the first decade of the present century, there were lot of evidences of strong commercial and cultural interests like poaching, mining and animal sacrificing as rituals in the pilgrimage centers of the Himalayan region of the South Asia (Suzanne, L. C., 2002; Singh, S., 2006). Harsh training is also very common for animals that are used in different temples, religious festivals, animal shows etc. Camel festivals, desert tours, camel fairs and camel tourism attracts millions of tourists each year in Jaisalmer, Rajasthan. Camels which are generally owned by the locals are extensively used in tourism throughout the year, especially between November to March. The rest of the time they are used for work or as carriers (Shackley, 1996). Animal slaughters are performed in different cultures for different purposes. Like in Islam and Judaism, ritual slaughters are performed per guidelines and to make the meat fit for human consumption. In Hinduism, the blood of the animal is offered to the deities per the rituals which are performed in temples and festivals. Individual devotees also offer animal sacrifices to show gratitude for fulfilling certain prayers. On the country, there are countries like Sri Lanka observes bans of animal sacrifice and promotes advocacy against animal cruelty (Boaz, 2019). Some scholars, mentioned how improper safeguarding of animals adversely affects conservation measures in restricted areas (Lubbe et al., 2017).

CONCLUSION AND LIMITATION

The purpose of this paper is to provide an overview of the animal cruelty conservation paradox in religious tourism, religious fairs, festivals, events to conceptualize the new niche tourism segments supplementing highly intensive mass tourism at religious sites in one hand, and to highlight sustainable pilgrimages through adoptions of sustainable and responsible tourism principles and values specifically in those sites. It is crucial that academics and practitioners realize the dire need to expand the type of research being conducted in this area of religious tourism, religious events, fairs and festivals. It is surprising that, despite a huge socio-political relevance, cultural rigidity, religious belief and, above all, this paradoxical behavior is ignored while studying animals in religious tourism, religious fairs, festivals, and such other events in Indian subcontinent. There are many instances of cultural and religious faith, rituals and amusement through animal killings, trafficking, fighting as a mean of promotion of a temple, festival, or places of attraction. The world we live in today should not only require but rather demand that we implement policies which dilute religious tourist traffic in densely populated pilgrim centers through new types and forms of tourism, and carefully adopt responsible and sustainable tourism principles in many popular religious sites for a peaceful coexistence.

Business houses and entrepreneurs can identify and utilize many profitable opportunities in different forms of religious tourism, religious fairs, festivals and events with an understanding of this body of knowledge, e.g. non-superstitious tourism, promotion of animal welfare with hymns, slokas and sacred words, concerted and coordinated planning for rehabilitation of animals and pets in destinations etc. In addition, policy makers may initiate a mandatory accreditation of responsible religious tourism destinations keeping in view this paradox. Animal cruelty, sacrifice, trafficking, consumption etc. may be incorporated while promoting and accrediting any destination, reducing the possibility of boycotting destinations (Shaheer et al., 2019). Moreover, this article advocates responsible religious practices, sustainable, sacred, and secular pilgrimages (Singh, S., 2006; Hyde, et al., 2011; Ambrosio, V., 2015; Marine-Roig, E., 2015) mythological tourism, religious fairs, festival and event tourism.

Scholars may conduct further research on the reasons for the concentration of the Hindu pilgrim centers sacrificing animals in Indian subcontinent. Academics can contribute by proposing and studying both the theoretical and practical impacts of this relatively hidden and much avoided area of study for an immediate sustainable and green future (Patange,et al., 2013; Hole, et al., 2019; Gairola, S. U.,2020).

The limitations of this study are in line with the limitations of practicing sustainable management. Levels of acceptance, tolerance, awareness, and implications of utilitarianism are varying from society to society, community to community and country to country in the region (Winter, C., 2020). Also, this area lacks even the preliminary records of animal killing, sacrifice, diseases of animals due to abuse etc. in the study region.

REFERENCES

Ainsworth, M. D. S., & Bowlby, J. (1991). An ethological approach to personality development. *The American Psychologist*, *46*(4), 331–341. doi:10.1037/0003-066X.46.4.333

Ambrosio, V. (2015). *Sacred pilgrimage and tourism as secular pilgrimage. Religious Tourism and Pilgrimage Management–An International Perspective*. CABI.

Bansiddhi, P., Brown, J. L., Thitaram, C., Punyapornwithaya, V., & Nganvongpanit, K. (2020). Elephant Tourism in Thailand: A Review of Animal Welfare Practices and Needs. *Journal of Applied Animal Welfare Science*, *23*(2), 164–177. doi:10.1080/10888705.2019.1569522 PMID:30689409

Barja, I., Silvan, G., Rosellini, S., Pinerio, A., Gonzalez, A., Camacho, L., & Illera, J. C. (2007). Stress physiological responses to tourist pressure in a wild population of European pine marten. *The Journal of Steroid Biochemistry and Molecular Biology*, *104*(3-5), 136–142. doi:10.1016/j.jsbmb.2007.03.008 PMID:17467268

Behrens, K. (2009). TonyYengeni's Ritual Slaughter: Animal Anti-Cruelty vs. Culture. *South African Journal of Philosophy*, *28*(3), 271–289. doi:10.4314ajpem.v28i3.47859

Bertella, G. (2014). The Co-creation of Animal-based Tourism Experience. *Tourism Recreation Research*, *39*(1), 115–125. https://www.trrworld.org/. doi:10.1080/02508281.2014.11081330

Boaz, D. N. (2019). The "Abhorrent" Practice of Animal Sacrifice and Religious Discrimination in the Global South. *Religions*, *10*(3), 160. Advance online publication. doi:10.3390/rel10030160

Bowlby, J. (1958). The nature of the child's tie to his mother. *The International Journal of Psycho-Analysis*, *39*, 350–373. PMID:13610508

Cape Town Declaration on Responsible Tourism. (2002). *International Conference on Responsible Tourism in Destinations*, Cape Town, South Africa.

Donlon, J. G., Donlon, J. H., & Agrusa, J. (2010). Cultural tourism, camel wrestling, and the tourism 'bubble' in Turkey. Anatolia. *International Journal of Tourism and Hospitality Research*, *21*(1), 29–39.

Duffy, R., & Moore, L. (2011). Global regulations and local practices: The politics and governance of animal welfare in elephant tourism. *Journal of Sustainable Tourism*, *19*(4), 589–604. doi:10.1080/09669582.2011.566927

Dwivedi, O. P., & Tiwari, B. H. (1987). *Environmental Crisis and Hindu Religion*. Gitanjali Publishing House.

Erschbamer, M. (2019). Taming of supernatural entities and animal sacrifice, The synthesis of Tibetan Buddhism and local shamanistic traditions in Northern Sikkim (India). *Études mongoles et sibériennes, centrasiatiques et tibétaines, 50*. . doi:10.4000/emscat.3915

Fennel, D. A. (2014). Exploring the boundaries of a new moral order for tourism's global code of ethics: An opinion piece on the position of animals in the tourism industry. *Journal of Sustainable Tourism*, *22*(7), 983–996. doi:10.1080/09669582.2014.918137

Fennell, D. A. (2012). Tourism, animals, and utilitarianism. *Tourism Recreation Research*, *37*(3), 239–249. doi:10.1080/02508281.2012.11081712

Fennell, D. A. (2013). Tourism and Animal Welfare. *Tourism Recreation Research*, *38*(3), 325–340. doi:10.1080/02508281.2013.11081757

Fennell, D. A., & Sheppard, V. (2020). Tourism, animals and the scales of justice. *Journal of Sustainable Tourism*, *29*(3), 314–335. doi:10.1080/09669582.2020.1768263

Fuller, C. J. (2004). The camphor flame: popular Hinduism and society in India (Revised and Expanded ed.), Princeton University Press.

Gairola, S. U. (2020). Protection of environment through green pilgrimage. *International Journal of Ecology and Environmental Sciences*, *2*(3), 199–203.

Galav, P., Jain, A., Katewa, S. S., & Nag, A. (2010). Animal healthcare practices by livestock owners at Puskar animal fairs, Rajasthan. *Indian Journal of Traditional Knowledge*, *9*(3), 581–584.

Goodwin, H. (2016). *Responsible tourism: using tourism for sustainable development*. Goodfellow. doi:10.23912/978-1-910158-84-5-3101

Goodwin, H. (2021). *Responsible Tourism in India. Government of India Institute of Secretariat Training & Management*. ISTM.

Gray, P. B., & Young, S. M. (2011). Human–pet dynamics in cross-cultural perspective. *Anthropos*, *24*(1), 17–30.

Gupta, V. (1999). Sustainable tourism: learning from Indian religious traditions. *International Journal of Contemporary Hospitality Management, 11,* 91–95. //. doi:10.1108/09596119910250751

Hole, Y., & Snehal, P. (2019). The significance of pilgrimage tourism to sustainable development with special reference to the Indian context. *African Journal of Hospitality, Tourism and Leisure*, *8*(3), 1–11.

Holst, A., & Martens, P. (2016). *Determinants of animal protection policy: a cross country empirical study*. Politics and Animals.

Hyde, K. F., & Harman, S. (2011). Motives for a secular pilgrimage to the Gallipoli battlefields. *Tourism Management*, *32*(6), 1343–1351. doi:10.1016/j.tourman.2011.01.008

Kuznets, S. (1955). Economic growth and income inequality. *The American Economic Review*, *45*, 1–28.

Lerner, P., & Rabello, A. M. (2007). The Prohibition of Ritual Slaughtering (Kosher Shechita and Halal) and Freedom of Religion of Minorities. *The Journal of Law and Religion*, *22*(1), 1–62. doi:10.1017/S0748081400003210

Lubbe, B. A., Preez, E. A., Douglas, A., & Wessels, F. F. (2017). The impact of rhino poaching on tourist experiences and future visitation to National Parks in South Africa. *Current Issues in Tourism*, *22*(1), 8–15. doi:10.1080/13683500.2017.1343807

Lynch, J. J. (1977). *The broken heart: The medical consequences of loneliness*. Basic Books.

Lynch, J. J. (2000). *A cry unheard: New insights into the medical consequences of loneliness*. Bancroft Press.

Malchrowicz-M, E., Munsters, W., Korzeniewska-Nowakowska, P., & Gravelle, F. (2020). Controversial Animal Tourism Considered from a Cultural Perspective. *Tourism/Turyzm, 30,* 21-30. . doi:10.18778/0867-5856.30.1.14

Marine-Roig, E. (2015). Religious tourism versus secular pilgrimage: The basilica of La Sagrada Família. *The international journal of religious tourism and pilgrimage, 3*(1), 25-37.

Mohanty, G. N., & Mohanty, S. N. (2004). Living with Snakes: The Lifestyle of the Snake Charmers (SapuaKela. *Adivasi*, *44*(12), 24–40.

Moodie, D. (2019). On blood, power, and public interest: The concealment of Hindu sacrificial rites under Indian law. *The Journal of Law and Religion*, *34*(2), 1–18. doi:10.1017/jlr.2019.24

Morris, M. C. (2013). Improved animal welfare is more related to income equality than it is to income. *Journal of Applied Animal Science*, *16*, 272–293. doi:10.1080/10888705.2013.768921 PMID:23795689

Palni, L. (2019). Tibet: The Sacred Landscape in Transition. The State of Ecology of the Tibetan Plateau.

Patange, P., Srinithivihahshini, N. D., & Mahajan, D. M. (2013). Pilgrimage and the environment: Challenges in a pilgrimage centre in Maharashtra, India. *International Journal of Environmental Sciences*, *3*(6), 2269–2277.

Ranaweerage, E., Ranjeewa, A. D. G., & Sugimoto, K. (2015). Tourism-induced disturbance of wildlife in protected areas: A case study of free ranging elephants in Sri Lanka. *Global Ecology and Conservation*, *4*, 625–631. doi:10.1016/j.gecco.2015.10.013

Roy, M. (2020). *Tourism and Its Environmental Implication: A Case Study OfKaziranga National Park.* JISU Journal of Multidisciplinary Research. https://www.researchgate.net/publication/344225631

Savishinsky, J. S. (1983). Pet ideas: The domestication of animals, human behavior, and human emotions. In A. H. Katcher & A. M. Beck (Eds.), *New perspectives on our lives with companion animals* (pp. 112–131). University of Pennsylvania Press.

Sébastia, B. (2004). Like an Indian God, Hinduisation of the cult of Saint Anthony of Padua in Tamil Nadu. *Auxlieuxd'Etre.*

Serpell, J. (1987). Pet keeping in non-western societies: Some popular misconceptions. *Anthrozoos*, *1*(3), 166–174. doi:10.2752/089279388787058443

Shackley, M. (1996). Community impact of the camel safari industry in Jaisalmer, Rajasthan. *Tourism Management*, *17*(3), 213–218. doi:10.1016/0261-5177(96)00008-8

Shaheer, I., Carr, N., & Insch, A. (2021). *Spatial distribution of participation in boycott calls: a study of tourism destination boycotts associated with animal abuse.* Anatolia, An International Journal of Tourism and Hospitality Research.

Shahi, A., & Pizam, A. (2008). Towards an ethical framework for animal-based attractions. *International Journal of Contemporary Hospitality Management*, *20*(6), 679–693. doi:10.1108/09596110810892236

Shastri, B. (1994). *KalikapuraneMurtivinirdesah.* Motilal Banarsidass Publications.

Shastri, J. L. (1970). *The Siva Puraṇa.* Motilal Banarsidass.

Singh, S. (2006). Secular pilgrimages and sacred tourism in the Indian Himalayas. *GeoJournal*, *64*(3), 215–223. doi:10.100710708-005-5649-8

South, N., & Wyatt, T. (2011). Comparing illicit trades in wildlife and drugs: An exploratory study. *Deviant Behavior*, *32*(6), 538–561. doi:10.1080/01639625.2010.483162

Suzanne, L. C. (2002), Animals and the Concept of Dignity: Critical Reflections on a Circus Performance. *Ethics and the Environment, 7*(2), 104-126. https://www.jstor.org/stable/40339038

Tul, H. (2015). Kali's Tongue: Shame, Disgust, and the Rejection of Blood and Violence in Vedic and Hindu Thought, *International Journal of Hindu Studies, 19*, 301–332. . doi:10.1007/s11407-015-9181-2

Urban, H. B. (2016). Desire, Blood and Power– Georges Bataille and the Study of Hindu Tantra in Northeast India. *RevistaCientíficaArbitrada de la Fundación MenteClara, 1*(3), 26–48. doi:10.32351/rca.v1.3.22

Urban, H. B. (2019). 'The Cradle of Tantra': Modern transformations of a Tantric Center in Northeast India from Nationalist Symbol to Tourist Destination. *South Asia, 42*(2), 256–277. doi:10.1080/00856401.2019.1570609

Venu, V., & Goodwin, H. (2008, March). *The Kerala Declaration on responsible tourism*. In Incredible India, Second International Conference on Responsible Tourism in Destinations, Kerala.

Winter, C. (2020). A review of research into animal ethics in tourism: Launching the annals of tourism research curated collection on animal ethics in tourism. *Annals of Tourism Research, 84*(10), 29–89. doi:10.1016/j.annals.2020.102989

Chapter 11
A Study of Issues and Challenges of Global Pilgrimage Tourism

Abhimanyu Awasthi
Amity University, Haryana, India

Akshay Nain
Amity University, Haryana, India

ABSTRACT

Pilgrimage is one of the oldest and most fundamental forms of population mobility known to human civilization, and its political, social, cultural, and economic ramifications have always been and continue to be significant. Pilgrimages and tourism are inextricably linked. The tourism business attracts a considerable number of "local visitors," who mostly visit pilgrimage sites. Pilgrimage tourism contributes significantly to travel marketing. Traveling for religious reasons is becoming increasingly important these days. It is a long-standing tradition in many regions of the world. People in traditional Indian living place high value on spiritual things and frequently visit sacred locations. Various ceremonies and offerings are conducted at pilgrim centers to atone for sins. Over the last three decades, a fresh focus on pilgrimage has evolved from tourism scholars who have investigated fascinating political, cultural, behavioral, economic, and geographical study themes. Pilgrimage tourism is important for socio-economic growth.

INTRODUCTION

Pilgrimage tourism is one of the very important types of tourism which is practiced worldwide by almost all the existing religions. Pilgrimage tourism is a result of religious beliefs & faiths of a particular religion towards visiting their holy land/place. Pilgrimage tourism is one of the oldest forms of tourism & is classified into short-term pilgrimage tourism & long-term pilgrimage tourism. In short-term pilgrimage tourism, the pilgrims visit nearby pilgrimage centers & return to their native place the same day. However, in long-term pilgrimage tourism, the pilgrims visit far-off pilgrimage centers & return

DOI: 10.4018/978-1-6684-4817-5.ch011

to their native place in some/few days. Pilgrimage tourism contributes heavily towards increasing the Gross Domestic Product (GDP) of the country to which pilgrims visit for their worships & prayers. For instance, the most populated community of the world is Islamic & every Islamic must visit Mecca (the holy city in Saudi Arabia) once in their lifetime for the Haj ceremony. As a result of this, Saudi Arabia is visited by Muslims from all over the world for the Haj ceremony. This mass movement of pilgrims to Saudi Arabia brings enormous foreign exchange every year & today, Saudi Arabia is one of the most developed & financially reliable countries in the world with a strong economy. Many issues & challenges are associated with pilgrimage tourism globally in the current era. The optimum infrastructure development at the pilgrimage site/center is the need of the hour. It is the duty & responsibility of the pilgrimage board/trust/state government/center government (as applicable) to establish & maintain the required infrastructure level/standard of each & every pilgrimage site/center falling under their geographical jurisdiction. The infrastructure should be such that a large number of pilgrims can be accommodated & catered to in a systematic & peaceful way. The political state of the country which is to be visited for pilgrimage tourism needs to be stable & amicable to win the confidence of the pilgrims. The various embassies in the country which is to be visited for pilgrimage tourism must help grant visas to the pilgrims worldwide in the prescribed duration of time in case of international pilgrimage tourism. The government of the country which is to be visited for pilgrimage tourism must provide subsidies to the economically weak pilgrims in terms of travel/transportation charges, accommodation charges, food & beverage charges, etc. as a gesture of nobility. It must be ensured that each & every pilgrimage site/center is well maintained, hygienic & clean. The sanctity of the holy land/place must always be maintained so that the pilgrims can offer their respective worships & prayers happily & satisfactorily. The misguiding of pilgrims & lack of sustainable development at the pilgrimage site/center are also growing concerns. It must be ensured that tourists do not get misguided under any circumstance & adequate measures must be taken to avoid such situations. It must also be ensured that sustainable practices are adopted & implemented at the pilgrimage site/center for all-around sustainable development.

BACKGROUND

The boundaries between tourism & pilgrimage have become blurred. The phenomena of pilgrimage tourism are changing in the 21st Century (Collins-Kreiner, N., 2018). The pilgrimage sites and centers are like tourist destinations. The pilgrimage sites and centers are the products that cater to the need of religious believers & holiday planners (Shuo, Y. S. S., Ryan, C., & Liu, G. M., 2009). Both, traditional religious pilgrimage tourism & modern secular pilgrimage tourism, are experiencing a resurgence globally. The pilgrimage sites and centers are visited by various pilgrims such as orthodox pilgrims, traditional pilgrims, secular pilgrims, etc. (Collins-Kreiner, N., 2010). The nexus of pilgrimage sites and centers & holy lands and places are enormously expanding. The impacts of pilgrimage tourism should not be underestimated & neglected. Pilgrimage tourism has helped in the socio-economic development of countries worldwide. Pilgrimage tourism has helped in increasing the Gross Domestic Product (GDP) of countries worldwide. Pilgrimage tourism has helped in increasing the export of products worldwide. Pilgrimage tourism has helped in increasing foreign exchange worldwide. Pilgrimage tourism has helped in increasing the employment opportunities for skilled & unskilled people worldwide. Pilgrimage tourism has helped in uplifting the standards of living in countries worldwide. Pilgrimage tourism has helped in lowering the levels of poverty in countries worldwide (Vijayanand, S., 2012). The holy sites & holy land are of prime importance to Christian pilgrims. The behavioral aspects of Christian pilgrims are related to their visit to their respective holy sites & holy land. The old-fashioned pilgrimage tourism is drastically changing (Collins-Kreiner, N., & Kliot, N., 2000). At the shore of the Atlantic Ocean, there is a town named Cape Coast in Southern Ghana which is famous for pilgrimage tourism (Schramm, K., 2004). Pilgrimage tourism is the oldest form of tourism. Pilgrimage tourism is the most widespread tourism globally. The shortcomings in marketing communication & infrastructure are the two main factors that have hampered the growth and development of pilgrimage tourism in Slovakia. The pilgrimage travel market has broadened significantly in Slovakia. The pilgrimage travel market of Slovakia consists of Christians, Jews, Buddhists & Muslims. The marketing communication strategy is a very important strategy for the growth & development of each & every pilgrimage site/center worldwide. Globalization has increased the level of competition among pilgrimage sites/centers worldwide. There should be an increased awareness & marketing of each & every pilgrimage site/center worldwide for enhancing the number of visits by pilgrims (Štefko, R., Kiráľová, A., & Mudrík, M., 2015). A lot of changes have taken place in pilgrimage tourism. Pilgrimage tourism has developed new forms such as secular pilgrimage tourism, dark pilgrimage tourism, and eco pilgrimage tourism (Collins-Kreiner, N., 2016). Pilgrimage tourism in India is very popular. The central government & state governments have made huge investments to develop the infrastructure of pilgrimage sites and centers. The need of the hour is to preserve & maintain sacred temples, mosques, churches, gurudwaras, shrines, etc. The increased construction of hotels, dharamshalas, sarais, inns, etc. at the pilgrimage sites and centers has led to environmental threats such as dumping of waste in water bodies, lowering of underground water, increased use of plastic, etc. Thus, sustainable practices need to be developed and followed to deal effectively with the above-mentioned environmental threats (Hole, Y., & Snehal, P., 2019). Pilgrimage and tourism are two different phenomena. Tourism is more secular than a pilgrimage. The Santiago de Compostela is a famous pilgrimage site for Christians. The experience of visiting Santiago de Compostela lasts forever in memories of devoted pilgrims (Lopez, L., 2013). 87.18 percent of the Indonesian population are Muslims. The most famous pilgrimage destination for Indonesian Muslim pilgrims is the Tomb of Wali. The Indonesian Muslim pilgrims feel satisfied whenever they visit the Tomb of Wali (Handriana, T., Yulianti, P., & Kurniawati, M., 2019). Modern information and communication technologies contribute significantly towards the development of pilgrimage tourism in Uzbekistan (Dadamuhamedov, A., 2019). Domestic pilgrimage tourism in Ireland has declined significantly whereas inbound and outbound pilgrimage tourism has experienced enormous growth (Griffin, K. A., 2007). Pilgrimage tourism is in the stage of rejuvenation. Pilgrimage tourism is in the process of losing its unique religious identity. Pilgrimage tourism is developing new identities such as secular tourism, spiritual tourism, dark tourism, and transformational tourism (Collins-Kreiner, N., 2019). A change is noticed in the behavior of host countries & communities towards pilgrimage tourism. The host countries & communities see pilgrimage tourism as a medium for regional development & cultural exchange (Raj, R., & Griffin, K. A., 2015).

MAIN FOCUS OF THE CHAPTER

Tourism and Pilgrimage: A Discussion

Pilgrimage is one of the most fundamental and ancient forms of human population mobility, with far-reaching political, social, cultural, and economic ramifications. The researcher tried to evaluate geographical studies on pilgrimage, paying particular emphasis to pertinent data from related fields. The researcher's goal is to investigate significant topics, arguments, and conceptualizations in pilgrimage research. This is to demonstrate the changes that the study of pilgrimage has undergone. As part of this purpose, the research will seek to highlight the differences between the various types of pilgrimage researchers. The study of pilgrimage has evolved toward the blurring of tourism and pilgrimage, especially, secular pilgrimage and religious pilgrimage. Dedifferentiation has permeated this study in terms of its aspects and the researchers' interdisciplinary approach to it. Today, the study of the significance of pilgrimage extends beyond geographical and socioeconomic boundaries. It entails a previously unexplored interpretive method to obtaining an alternate meaning. Current studies presume that pilgrimages are products of the culture in which they were founded, and hence they give us "stories" from a geopolitical, spiritual, cultural, and social point of view. These pilgrims are the results of social tradition and order norms and ideals, as well as the makers of such tradition and culture (COLLINS-KREINER, 2010). However pilgrimage and spiritual tourists have a significant impact in rural regions, but the social aspects of pilgrimage and religious tourism have had the greatest impact on rural households. Furthermore, the study showed that communities associated with "Religious tourism" had statistically significantly higher impacts than villages associated with "Pilgrimage tourism" (Pourtaheri, 2012). In the component of international economic reform, it is critical to investigate the analytical and practical elements of this subject to identify trends and features of pilgrimage tourism and to improve the organizational and economic mechanisms of its successful growth (Xudoyarov, 2019). The borderzone idea focuses on a confined event, restricted in location and time, such as a meeting between foreign guests and locals. The two groups approach it from quite different angles. Tourists encounter peak moments; residents and producers organize these experiences for tourists but largely live in the daily. Although the borderzone is concerned with the local, what is done there takes into account global and international movements, as tourist growth in the Third World necessitates the participation of nation-states and transnational networks. To comprehend, it is vital to comprehend questions such as how various people migrate and what kinds of information, tales, embodiment, and ideas they generate in the process. The primary flaw in Bruner's model is that it implies tourists and natives are mutually exclusive: the former visit a location just briefly, while the latter stay for an extended period (Reed, 2014).

Religious Landscape Characteristics: Value, Identity, Customs

The researcher also found in the review of literature that Cathedral visits are one type of cultural tourism. The reasons for these travels are global. The primary attractions are cultural heritage and architecture. Visitors to the place, on the other hand, are generally moved by religious impulses. Religion appears to be a reasonable motivation for visiting England's cathedrals. The predominant focus of visitors in the cathedral was religious in nature, indicating the popularity of pilgrimage tourism as a causative component in visits to English cathedrals. It is critical to recognize the value of preserving religious landscape characteristics from the past since it adds to the beauty of a sacred site by implying to the visitor the sacrifice needed of its builders, the glories of ages past, or simply the architectural styles that distinguished the era in which they were erected. The dominance of architecture in answers suggests that there is a strong desire to have a better knowledge of cathedrals as landscape characteristics, which geographers may be able to assist in explaining. Geographers, without a doubt, should be at the forefront of efforts to conserve and restore the religious structures that have survived intact to the present (Jackson & Hudman, 1995). Turning sacred sites into pilgrimage clusters involves four stages: i) making holy sites a location of pilgrimage; ii) publicizing holy sites as enticing pilgrimage tourist attractions, and iii) forming pilgrimage groups with a high level of performance based on appealing pilgrimage locations. It is necessary to examine the quality and level of connectivity in locations to develop regional pilgrimage tourism clusters (Navruz-zoda, 2020). The pilgrimage also gives a chance for local inhabitants of the specific place to perform their identity. The unique appearance of the holy sites and appealing local customs linked with pilgrimage attracts tourists to the holy sites. Some visitors and pilgrims want to re-trace the history of holy sites and try to figure out the origins of history which are popularised in the touristic site or sacred land and related works of literature (Badone, 2008).

Pilgrimage Tourism and Post-Secular Culture: Open, Unfolding, and Constantly Developing

Additionally, there has been an emergence of a post-secular culture that has created new opportunities for cross-disciplinary study and viewpoints. Academic and research discussions on the post-secular have failed to make inroads into tourist studies, which is odd given tourism's inherent religiosity. Tourism, we argue, is at the heart of the post-secular trend and should be examined as such. Similarly, the scholarly discourse on post-secularism has not given tourism its due. There are four primary characteristics of post-secular tourism: (self-)reflexivity, the yearning for existential security, ethical relativism, and topos-disloyalty. Researchers also found in the review that the post-secular tourist transcends traditional dualism between tourism and religious pilgrimage, as well as traveling as an assertion of denominational religion and pure recreation. Post-secular tourism is concerned with ontogenesis, which is why it is performing rather than declarative. In this way, post-secular tourism might be described as something open, unfolding, and constantly developing (Nilsson & Tesfahuney, 2017). Pilgrimages, particularly religiously or culturally, are becoming increasingly famous worldwide. The phenomena of pilgrimage to the graves of saintly Jews in Israel is an intriguing and instructive tourism occurrence for studying and analyzing the features of Jewish pilgrims to holy burial sites in Israel at the current moment, as well as assessing the phenomenon of pilgrimage. Along with these causes, other variables such as pilgrimage motivations, pilgrimage activities, and the impact of tourism on it are also essential components of future inquiry and study (Collins-Kreiner, 2009). Every year, Islamic people go to Saudi Arabia to do pilgrim worship, and the fact that Saudi Arabia is such a holy site has several ramifications. In this context, researchers examined the landscape in terms of the relationship between CO2 emissions, economic expansion, energy use, and pilgrimage tourism. Researchers discovered that power consumption, the frequency of pilgrims, and oil prices all have a positive influence on CO2 emissions, but GDP has a negative effect, although DOLS data indicate that only energy consumption has a favorable effect on carbon emissions. Furthermore, there is linear causation from CO2 to pilgrimage tourism and unidirectional causality from pilgrimage tourism to oil, as well as bi-directional causality between pilgrimage GDP and tourism. As a result, the environmental impact of pilgrimage tourism is unavoidable (Ozturk, Aslan, & Altinoz, 2021). Religious tourism is a custom that is seen in all religions. However, pilgrimages and how they are carried out vary according to faith. A person is called a participant in religious tourism if he visits sacral and religious sites not only for a spiritual experience, but also to learn about them. Pilgrims, regardless of faith, travel to a sacred location, such as a church, mosque, or shrine, to mention a few. In terms of pilgrims' demands, these artifacts represent the foundation of religious tourism infrastructure, as opposed to regular visitors, for whom they serve as a means of localization (Čábyová, 2016).

GLOBAL PILGRIMAGE TOURISM GAP AREAS

The influence of tourism on the local inhabitants is frequently given more emphasis in tourist literature than the effect on the visitors themselves. This refers to the length, strength, and level of the effect. The visitor's varying perspectives, whether pilgrim or vacationer, should thus be displayed on a scale based on their duration and strength: to what extent the visitor was affected after his or she come back home from the visit, regardless of whether he or she was classified as a tourist or pilgrim at the outset (COLLINS-KREINER, 2010). The same causes should have an impact on the preservation of other important religious structures in other nations. Most tourists to religious places are drawn in by some form of tourism appeal. Researchers and government travel authorities must begin to obtain a better knowledge of what landscape aspects, activities, or other phenomena attract tourism pilgrims (Jackson & Hudman, 1995). It is proposed that pilgrims be divided into three categories: "Holy Sights," "Pilgrims Place," and "Appealing Sanctuary". It is suggested that pilgrim monuments be assessed using the screening indicators: sophisticated pilgrim infrastructures, pilgrim reputation, and pilgrim historical evidence (Navruz-zoda, 2020).

DESIGN/METHODOLOGY/APPROACH

This chapter will be analyzing the extensive literature by using the PRISMA approach. Using Preferred Reporting Items for Systematic Reviews and Meta-Analyses (PRISMA). The scale was developed with 11 items on a 5-point Likert-type scale from 1 to 5, 1 being very low & 5 being very high. Survey questions were kept close and largely aimed at the 03 elements:

- Differing experiences of the visitor, including the visit of the pilgrim place, time spent for a better experience, and association of attraction.
- Preservation of other significant religious structures, including the decision to travel based on its preservation.
- Tourist-related attractions to attract tourist pilgrims, including the importance of landscape elements, advanced infrastructure historicity, in choosing a pilgrimage place.

The research items in the questionnaire were selected based on the literature review-based research gap areas. A robust value of 0.809 is also derived for the Reliability test based on Cronbach's alpha coefficient for the scale used using SPSS software.

The research study has assessed & analyzed the prospects and challenges of global pilgrimage tourism and hospitality on changing facts and facets of pilgrimage tourism, community integration, tolerance, peace, etc. for a secular and sustainable future with a dedicated focus to pilgrimage tourism. The study has explored the perspectives and the possible effects of new issues, scopes, challenges, entrepreneurial opportunities on pilgrimage tourism and hospitality, and what measures we can take to preserve pilgrimage tourism and hospitality for our future. As is now a challenging task to protect our environment because of the recent pandemic caused by COVID-19.

ARTICLES SELECTION WITH KEYWORDS FROM THE ONLINE DATABASE

The research for apt literature and subsequent review was conducted using Google Scholar, one of the most well-known free research tools. It is recognized as one of the most reliable sources of reputable articles on subjects and themes of study. Following database selection, advanced research was undertaken on Google for the selected key phrases using the required keyword combination while keeping the term 'Global Pilgrimage Tourism' constant in all search results. Other word combinations besides Global Pilgrimage Tourism include pilgrimage tourist, hospitality, and tourism industry, changing facts and facets of pilgrimage tourism, sustainable pilgrimage tourism, & pilgrimage tourism development. The discovered papers were then used to determine citations using, and the 61 articles that were chosen were double screened using the PRISMA approach, as shown in the figure below.

The prior stages followed in the selection of 26 papers for study purposes. The screening of each paper resulted in the elimination of 20 duplicate papers. The remaining 41 publications were read and assessed in the next chapter based on their abstract, title, and literature review. Articles were removed from the study during the preceding phase. The researcher's knowledge and the information presented in their abstract led to the exclusion of such. Exclusions based on article titles that match the criteria and their abstract. The Exclusion based on article title (n=5), exclusion based on abstract (n=4), excluded duplicate articles (n=3). Further, 03 more articles were excluded articles according to the researcher's understanding of the domain (n=3). Finally, the systematic literature review was conducted for 26 research articles.

RESEARCH LIMITATION/ IMPLICATIONS

The study has limitations as data available on various indexing portals may not be as suitable as required. So, the researcher will try to gather the relevant data from online portals, journals, and reputed online research data portals such as Google Scholar, etc. As we are aware that we have not recovered fully from the pandemic caused by Covid-19 and people are still hesitant to come in contact due to this constant fact it may be a limiting factor for gathering responses for the survey. Hence researcher will focus on using the online survey in the form of a google questionnaire because it will be difficult to take the survey from most of the respondents in a positive way.

SOLUTIONS AND RECOMMENDATIONS

Geologists have something to say about contemporary discussions about pilgrimage, tourism, place, and experience that are surfacing across a variety of disciplines. In reviewing these central themes, which highlight the importance of a different direction in pilgrims' geography, the researchers not only framed recent efforts in geographic location but also recognized how well this field of research is and should be pursued, drawing on theoretical advancements that have been impactful. Researchers have found that the visiting demographic ranged from intensely religious orthodox pilgrims to "typical" pilgrim-visitors to secular tourists. The characteristics of modern pilgrims may be classified into two categories: secular vs spiritual, and tourism versus pilgrimage. This typology also provides a framework for the building of pilgrimage sites. The study findings emphasize the growing confluence of traditional pilgrimage and modern tourist, which share many similarities. Furthermore, this study underlines the growing link between holy places, society, politics, ideology, and culture. Faith tourism is studied in various disciplines, including sociological, anthropology, psychological, and economic. The literature, on the other hand, reveals that research in the disciplines of economics typically focuses on tourism and has a relatively low interest in belief tourism. However, pilgrimage tourism is comparable to other types of tourism in many respects, and a pilgrim nearly entirely resembles a tourist. Thus, the crowding of holy sites and locations in the main cities during the pilgrimage time has both societal and economic consequences. Because of the accelerated pace of climate change research, its influence on environmental contamination, which cannot be compensated for, should be considered significant.

META-ANALYSIS & RESULTS

A robust value of 0.809 is also derived for the Reliability test based on Cronbach's alpha coefficient for the scale used using SPSS software. The below-mentioned meta-analyses are carried out with the help of SPSS software.

The survey questionnaire was filled by a total of 72 respondents including 73.6% males and 26.4% females.

Table above lists the age categories of the respondents. There the greatest number of respondents was 81.9% of the 21-30 years age group followed by 11.1% from 31-40 years age group.

Table 1. Gender

		Frequency	Percent	Valid Percent	Cumulative Percent
Valid	Male	53	73.6	73.6	73.6
	Female	19	26.4	26.4	100.0
	Total	72	100.0	100.0	

Table 2. Age

		Frequency	Percent	Valid Percent	Cumulative Percent
Valid	21-30 years	59	81.9	81.9	81.9
	31-40 years	8	11.1	11.1	93.1
	41-50 years	3	4.2	4.2	97.2
	51-60 years	2	2.8	2.8	100.0
	Total	72	100.0	100.0	

Table 3. Q1. Do you agree that visiting or traveling to a sacred sights, pilgrimage place, and attractive sanctuary has the same objective?

		Frequency	Percent	Valid Percent	Cumulative Percent
Valid	Yes	38	52.8	52.8	52.8
	No	34	47.2	47.2	100.0
	Total	72	100.0	100.0	

For questions number 1, the majority of the respondents agreed that visit to a Sacred Sights, Pilgrimage Place, and Attractive Sanctuary has the same objective, however, 47.2% of respondents also replied with 'No'. This also draws attention to the need for appropriate education and literature through clarity on the subject matter.

For question 2 responses are aligned in the range of being neutral, for question3 to 6 respondents agreed with the statement, for question number 07 respondents were neutral with the statement, and for question number 8,9 and 10 respondents agreed with the statement. However, for question number 11 respondents were neutral and for question number 12 they strongly agreed with the statement.

FUTURE RESEARCH DIRECTIONS

Discuss future and emerging trends. Provide insight about the future of the book's theme from the perspective of the chapter focus. Viability of a paradigm, model, implementation issues of proposed programs, etc., may be included in this section. If appropriate, suggest future research opportunities within the domain of the topic.

CONCLUSION

Pilgrimage tourism is a global phenomenon. There is a global resurgence for pilgrimage tourism. Pilgrimage tourism is the oldest and most widespread form of tourism. The pilgrims see their pilgrimage sites and centers as attractive tourist destinations in the 21st Century. Pilgrimage tourism contributes significantly towards the growth and development of countries worldwide by increasing their respective Gross Domestic Product (GDP), foreign exchange, employment opportunities, exchange of products, standards of living & lowering levels of poverty. The major issues & challenges associated with global

Table 4. Descriptive statistics

	N	Mean	Std. Deviation
Q2. Visiting a pilgrim place is mandatory.	72	3.3750	1.19196
Q3. The more time you spent on pilgrimage is better for experiencing the travel.	72	3.7361	1.08761
Q4. You want to travel to a pilgrimage place due to an associated attraction with it.	72	3.7500	1.05817
Q5. After you have traveled pilgrimage place, you believe that visiting a pilgrim place is mandatory.	72	3.5000	1.23334
Q6. After you have traveled pilgrimage place, you want more time to spend in the place you visited.	72	3.8056	1.14620
Q7. After you have traveled pilgrimage place, you want to travel only to such a place if it has some of the attractions associated with it.	72	3.4722	1.26683
Q8. The decision to travel to a pilgrimage place is based on how it is preserved.	72	3.5833	1.09737
Q9. Even if the pilgrimage place is not well preserved, you travel to it.	72	3.6111	1.05558
Q10. landscape elements are an important reason for you to choose a pilgrimage place.	72	3.5417	1.23263
Q11. The advanced infrastructure of pilgrimage place will make you choose it.	72	3.2639	1.35307
Q12. The historicity of pilgrimage place plays an important role in choosing it to travel.	72	4.0972	1.07677
Valid N (listwise)	72		

pilgrimage tourism are infrastructure development, cleanliness, sanitation, hygiene, political stability, pilgrimage subsidy, incidents of pilgrim misguiding, incidents of the stampede, sustainable practices & development, convenient grant of visas to pilgrims, marketing communication, information, and communication technology, etc. A lot of changes have been noticed & identified in present-day pilgrimage tourism & pilgrims. The present-day pilgrimage tourism has developed new forms such as secular pilgrimage tourism, spiritual pilgrimage tourism, dark pilgrimage tourism, transformational pilgrimage tourism, eco pilgrimage tourism, etc. The present-day pilgrims are classified as orthodox pilgrims, traditional pilgrims, modern pilgrims, secular pilgrims, etc. Some environmental threats have also crept in due to increased movement of pilgrims worldwide namely cutting of forests for constructing accommodation facilities for pilgrims near the pilgrimage site, increased use of plastic at the pilgrimage site, dumping of waste into water bodies at the pilgrimage site, lowering of underground water level at the pilgrimage site, etc.

Today's researchers dispute established hypotheses and reject the clear-cut divisions of the current study. The tendencies of deconstruction and breaking down old ideas, as well as a propensity to emphasize the subjective rather than the objective, and a rising emphasis on one's experience, all fit in with the new perspective. All of these tendencies, together with the increasing decoupling of pilgrimage, tourism, and secular tourism, and even the diminishing gap between people's desire to find a different meaning in their everyday lives, indicate that the study of pilgrims has altered. This shows that a theoretical shift, rather than merely a procedural shift, has happened. According to the researchers, pilgrims and religious tourists have a substantial influence in rural regions, but the social aspects of pilgrimage and religious tourism have the greatest impact on rural households. Also, social effects such as increased social engagement with pilgrims and visitors, reduced migration, particularly among young people, increased sense of security, friendly connection with neighboring communities, and reduced social tensions in the community. This type of rural tourism provides not only social profits, but also physical and economic benefits, such as improving housing quality, improving rural transportation systems, increasing access to services (educational, administrative, etc.), increasing service quality, promoting health fields such as waste and sewage disposal, increasing income, job creation, creating commercial and retail centers, handicraft development, and increasing land prices. According to Ľudmila Čábyová (2015), for 68 percent of pilgrims, the purpose for making a pilgrimage is faith, while 13 percent are seeking some type of spirituality. Surprisingly, up to 90% of pilgrims experienced a significant emotional experience, primarily related to the holy site - the destination of their travel (55%), prayer (21%), or the journey itself (17 percent).

ACKNOWLEDGMENT

This research received no specific grant from any funding agency in the public, commercial, or not-for-profit sectors.

REFERENCES

Badone, E. (2008). Pilgrimage, tourism and The Da Vinci Code at Les-Saintes-Maries-de-la-Mer, France. *Culture and Religion*, *9*(1), 23–44. doi:10.1080/14755610801954847

Čábyová, Ľ. (2016). Pilgrimage tourism in Slovakia, the potential for development. *European Journal of Science and Theology*, *12*(1), 231–244.

Collins-Kreiner, N. (2007). Graves as attractions: Pilgrimage-tourism to Jewish holy graves in Israel. *Journal of Cultural Geography*, *24*(1), 67–89. doi:10.1080/08873630709478217

Collins-Kreiner, N. (2010). Current Jewish pilgrimage tourism: Modes and models of development. *Tourism: An International Interdisciplinary Journal*, *58*(3), 259–270.

Collins-Kreiner, N. (2010). Geographers and pilgrimages: Changing concepts in pilgrimage tourism research. *Tijdschrift voor Economische en Sociale Geografie*, *101*(4), 437–448. doi:10.1111/j.1467-9663.2009.00561.x

Collins-Kreiner, N. (2016). The lifecycle of concepts: The case of 'Pilgrimage Tourism'. *Tourism Geographies*, *18*(3), 322–334. doi:10.1080/14616688.2016.1155077

Collins-Kreiner, N. (2018). Pilgrimage-tourism: Common themes in different religions. *International Journal of Religious Tourism and Pilgrimage*, *6*(1), 8–17.

Collins-Kreiner, N. (2019). Pilgrimage tourism-past, present and future rejuvenation: A perspective article. *Tourism Review*.

Collins-Kreiner, N., & Kliot, N. (2000). Pilgrimage tourism in the Holy Land: The behavioural characteristics of Christian pilgrims. *GeoJournal*, *50*(1), 55–67. doi:10.1023/A:1007154929681

Dadamuhamedov, A. (2019). The role of information and communications technologies in pilgrimage tourism in Uzbekistan. *The Light of Islam*, *2019*(1), 17.

Griffin, K. A. (2007). The globalization of pilgrimage tourism? Some thoughts from Ireland. *Religious tourism and pilgrimage management: An international perspective*, 15-34.

Handriana, T., Yulianti, P., & Kurniawati, M. (2019). Exploration of pilgrimage tourism in Indonesia. *Journal of Islamic Marketing*.

Hole, Y., & Snehal, P. (2019). The significance of pilgrimage tourism to sustainable development with special reference to the Indian context. *African Journal of Hospitality, Tourism and Leisure*, *8*(3), 1–11.

Jackson, R. H., & Hudman, L. (1995). Pilgrimage tourism and English cathedrals: The role of religion in travel. *Tourism Review*.

Lopez, L. (2013). How long does the pilgrimage tourism experience to Santiago de Compostela last? *The international journal of religious tourism and pilgrimage, 1*(1), 1-14.

Navruz-zoda, Z. (2020). Evaluation of Holy Places of the Regions for the Development of Pilgrimage Tourism. *Indonesian Journal of Law and Economics Review*, *6*, 10–21070. doi:10.21070/ijler.2020.V6.493

Nilsson, M., & Tesfahuney, M. (2018). The post-secular tourist: Re-thinking pilgrimage tourism. *Tourist Studies*, *18*(2), 159–176. doi:10.1177/1468797617723467

Ozturk, I., Aslan, A., & Altinoz, B. (2021). Investigating the nexus between CO2 emissions, economic growth, energy consumption and pilgrimage tourism in Saudi Arabia. *Economic Research-Ekonomska Istraživanja*, 1-16.

Pourtaheri, M., Rahmani, K., & Ahmadi, H. (2012). Impacts of religious and pilgrimage tourism in rural areas: The case of Iran. *Journal of Geography and Geology*, *4*(3), 122. doi:10.5539/jgg.v4n3p122

Raj, R., & Griffin, K. A. (Eds.). (2015). *Religious tourism and pilgrimage management: An international perspective*. Cabi. doi:10.1079/9781780645230.0000

Reed, A. (2014). *Pilgrimage tourism of diaspora Africans to Ghana*. Routledge. doi:10.4324/9781315771304

Schramm, K. (2004). Coming home to the motherland: Pilgrimage tourism in Ghana. In *Reframing pilgrimage* (pp. 143–159). Routledge.

Shuo, Y. S. S., Ryan, C., & Liu, G. M. (2009). Taoism, temples and tourists: The case of Mazu pilgrimage tourism. *Tourism Management*, *30*(4), 581–588. doi:10.1016/j.tourman.2008.08.008

Štefko, R., Kiráľová, A., & Mudrík, M. (2015). Strategic marketing communication in pilgrimage tourism. *Procedia: Social and Behavioral Sciences*, *175*, 423–430. doi:10.1016/j.sbspro.2015.01.1219

Vijayanand, S. (2012). Socio-economic impacts in pilgrimage tourism. *International Journal of Multidisciplinary Research*, *2*(1), 329–343.

Xudoyarov, A. (2019). A. (2019). Specific organizational and economic aspects of the development of pilgrimage tourism in uzbekistan. *The Light of Islam*, (4), 39.

Chapter 12
Perspectives, Impact, and Challenges of Pilgrimage Tourism in Chidambaram

M. Kalaivani
SRM Institute of Science and Technology, Vadapalani, India

V. Suganya
https://orcid.org/0000-0001-5301-8317
SRM Institute of Science and Technology, Vadapalani, India

ABSTRACT

The travel and tourism industry is one of the fastest-growing industries in today's Indian global economy. The fast growth of the industry is not only generating jobs for millions, but also conducive on a large scale to the entire economy of the world. Pilgrimage tourism in Chidambaram is a popular form of spiritual tourism that attracts thousands of devotees every year. Chidambaram is a holy town located in the southern state of Tamil Nadu, India, and is considered one of the most important pilgrimage destinations in the country. The town is renowned for its ancient temple dedicated to Lord Nataraja, a form of Lord Shiva, and is also known for its rich cultural heritage, traditional arts, and dance forms. In Tamil Nādu, there are a number of tourist places but pilgrimage tourism in Chidambaram has its own importance. Therefore, it attracts a large number of pilgrims from various parts of the world. This chapter is an attempt to study the pilgrimage tourism on the overall development of the place and moreover aims at the challenges which the tourism industry is facing in Chidambaram.

INTRODUCTION

Chidambaram is a pilgrimage destination situated in the state of Tamil Nadu, India. The town is known for its famous Chidambaram Nataraja temple where Lord Shiva is worshipped. The temple is one of the five Pancha Bootha Sthalams, which represents the five elements of nature. Chidambaram is considered a significant place for Shaivism and attracts a large number of devotees every year.

DOI: 10.4018/978-1-6684-4817-5.ch012

Apart from the temple, the town also has other places of religious importance such as Thillai Kali Temple, Thirukovilur, and Pitchavaram Mangrove Forest. The town is also known for its rich cultural heritage and traditional art forms such as Bharatanatyam and Nadi astrology.

The pilgrimage tourism in Chidambaram attracts both domestic and international tourists who visit the town to seek blessings and immerse themselves in the spirituality and culture of the place. The town is also a popular destination for spiritual seekers who are interested in learning and practicing yoga and meditation. Overall, Chidambaram is a unique destination that offers a perfect blend of religion, spirituality, and culture to tourists.

Pilgrimage Tourism in Chidambaram

Chidambaram, also known as Thillai, is a pilgrimage destination located in Tamil Nadu, India. It is famous for its ancient temple dedicated to Lord Nataraja, the cosmic dancer of Hindu mythology. Here are some of the pilgrimage sites to visit in Chidambaram:

- Nataraja Temple: The Nataraja Temple is a famous pilgrimage site that attracts thousands of devotees every year. It is one of the few temples dedicated to Lord Nataraja, the cosmic dancer of Hindu mythology. The temple is known for its magnificent architecture, with intricate carvings and sculptures depicting various scenes from Hindu mythology.
- Thillai Kali Temple: The Thillai Kali Temple is another important pilgrimage site in Chidambaram. It is dedicated to the goddess Kali, who is believed to have appeared here to protect the town from evil forces. The temple is known for its unique architecture and the annual Kali festival, which draws large crowds of devotees.
- Pichavaram Mangrove Forest: The Pichavaram Mangrove Forest is located near Chidambaram and is a popular destination for pilgrims. The forest is home to several species of birds and other wildlife and is believed to have healing properties.
- Annamalai University: Annamalai University is located in Chidambaram and is a renowned educational institution in India. The University is also a pilgrimage site, as it houses the Sri Ramana Maharshi Study Centre and the Sri Aurobindo Ashram.
- Pitchavaram Temple: Pitchavaram Temple is famous for its annual festival that happens during the Tamil month of Thaipoosam. Thousands of pilgrims visit the temple to witness the grand procession of the deity that starts at Pitchavaram Lake and ends at the temple.

Pilgrimage tourism in Chidambaram offers a unique experience for those seeking spiritual enlightenment and a deeper understanding of Hindu mythology. The town's rich cultural heritage and ancient temples are sure to leave a lasting impression on every visitor.

Review of Literature

Like many other papers in the special issue collection, this one emphasizes the importance of applying sustainable tourism concepts and practices to the management and growth of Indigenous tourism. However, there is a contrarian viewpoint to that debate as well. A closer look at and interaction with Indigenous tourism could teach sustainable tourism as a larger concept and practice many valuable lessons. It involves so many important problems. Generally speaking, the majority of the work on sustainable tourism up to this point has been focused on comprehending and controlling the effects of tourism on the physical environment. The management of the intangible cultural heritage is particularly challenging (McKercher & du Cross, 2002), making it a far more challenging subject to analyze. While not all of Indigenous people's history.

In order to maintain variety and choices in a globalizing world, it is crucial to widen the concept of sustainable tourism away from its roots in the western industrialized world. The focus of previous discussions on this topic has been the requirement for developing a kind of sustainable tourism in Asia, particularly China (Honggang, Dan, & Jigang, 2016; Sofield & Li, 2011). Alternative methods to company ownership, governance, and capacity building are also necessary for indigenous tourism. These difficulties are highlighted by the concepts and problems raised by Higgins-Desbiolles (2016), Hillmer-Pegram (2016), and Pereiro (2016). These substitutes might be useful for various non-indigenous types of tourism. There may be much to be learned from CBT, whose value is being hotly debated (Salazar, 2012).

Research on Indigenous tourism, which moves CBT studies to a new geographical foundation and examines numerous communities brought together by their Indigenous people, may also have a lot to teach us. Particularly meriting more consideration and testing abroad is the Guna sustainable tourism model (Pereiro, 2016).

Finally, we hope that publications will include more research conducted by or in partnership with Indigenous scholars in the future. Indigenous authors and editors of scholarly journals should become standard practice in academia based on their own intellectual merits. In the end, Indigenous academics and tourist providers will make significant contributions to and commentary on mainstream and specialized methods to tourism management and tourism studies, while communities acquire recognition as world leaders in tourism and allied fields rather than just as the "Other" that is visited.

Impacts of Pilgrimage Tourism in Chidambaram

Pilgrimage tourism in Chidambaram has a significant impact on the local economy, culture, and environment. Here are some of the impacts of the pilgrimage tourism in Chidambaram:

Economic Impact: Pilgrimage tourism in Chidambaram has created job opportunities for local people. The influx of visitors has led to the growth of the hospitality industry, with many hotels, restaurants, and shops catering to the needs of pilgrims. The money spent by pilgrims also benefits the local economy and helps in the growth of the tourism sector.

Cultural Impact: The influx of pilgrims has helped preserve the cultural heritage of Chidambaram. The ancient temples, rituals, and traditions associated with pilgrimage sites have been passed down from generation to generation, and the town has retained its unique cultural identity.

Environmental Impact: The increase in the number of visitors to Chidambaram has led to environmental degradation. The pilgrims contribute to air, water, and noise pollution, which can have a negative impact on the ecosystem. However, efforts are being made to mitigate the environmental impact, and many government and non-governmental organizations are working towards sustainable tourism in the region.

Social Impact: Pilgrimage tourism has led to social development in Chidambaram. The influx of visitors has led to the development of infrastructure such as roads, transportation, and communication, which has improved the standard of living for local people. Additionally, the interaction between visitors and local people has led to cultural exchange and mutual understanding.

Figure 1. Chidambaram (must watch)

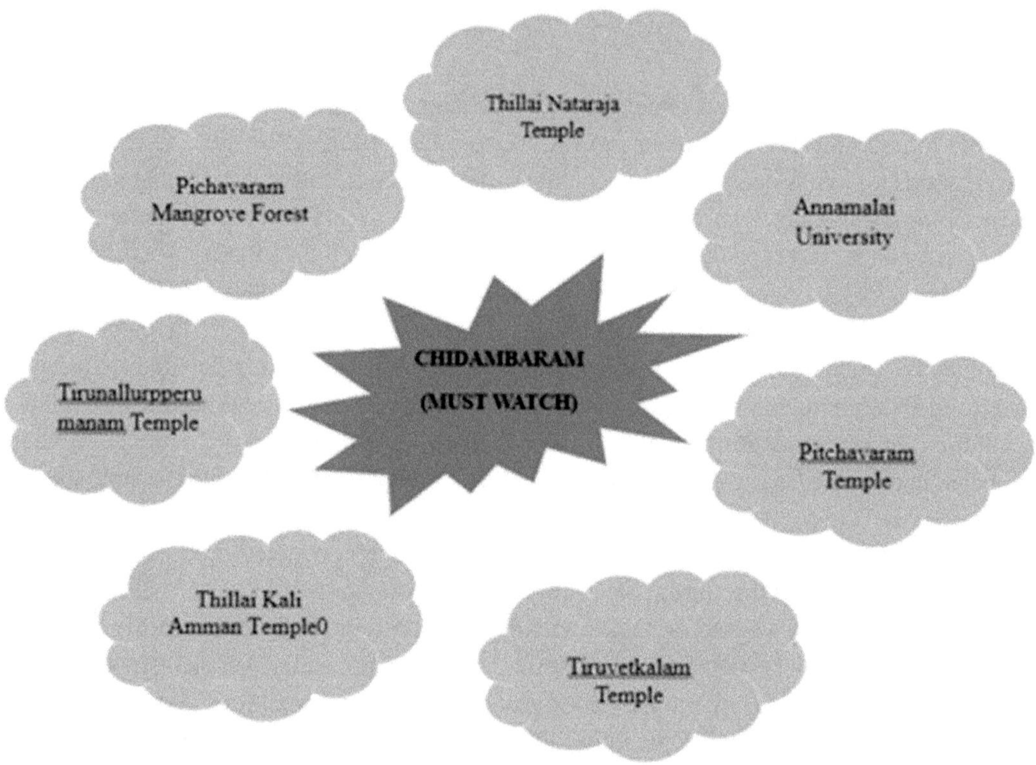

Overall, pilgrimage tourism in Chidambaram has had a positive impact on the local economy, culture, and people. However, it is essential to ensure that the environmental impact is minimized, and that tourism is sustainable for the long term.

Challenges of the Pilgrimage Tourism in Chidambaram

Pilgrimage tourism in Chidambaram, like any other tourism activity, faces several challenges. Here are some of the challenges of the pilgrimage tourism in Chidambaram:

Despite being a popular pilgrimage site, Chidambaram lacks adequate infrastructure. The town's roads, railways, and airport connectivity need improvement to accommodate the growing number of visitors. The increase in the number of pilgrims has led to a rise in waste, causing environmental pollution. The town lacks adequate waste management facilities, and waste disposal is a significant issue. Chidambaram is home to several ancient temples and heritage sites. However, the large influx of visitors has led to damage to these ancient temples and historical sites. Preserving these sites without compromising the cultural experience of pilgrims is a challenge. The safety and security of tourists, particularly women and foreign travelers, is a significant concern. Ensuring the safety of people visiting Chidambaram should be a priority for the local authorities. The number of tourists visiting Chidambaram is seasonal, with the peak season being during festivals and auspicious days. This can lead to overcrowding and long queues at temples. Tourists visiting Chidambaram must be wary of illegal activities like theft, pickpocketing, and cheating. Pilgrims are often targeted by local vendors who sell counterfeit items. With the rise in the number of visitors, there is a need for effective regulation of tourism activities. This will help in ensuring the safety of tourists and avoid overcrowding during special events and festival seasons.

Overall, addressing these challenges is crucial to achieving sustainable tourism in Chidambaram. The local authorities and other stakeholders must work together to address these issues and promote responsible tourism.

Recommendations for Promoting Pilgrimage Tourism in Chidambaram

The local authorities should work towards improving the infrastructure in Chidambaram, including better roads, railway connectivity, and airport development. Effective waste management practices should be put in place to ensure cleanliness and hygiene in the town. This could involve setting up waste collection centers and conducting awareness campaigns. There should be a focus on preserving the ancient temples and historic sites in Chidambaram. Proper measures, including restoration, preservation, and documentation, should be taken to maintain and protect these valuable heritage sites. Measures should be taken to ensure the safety and security of pilgrims. Training local police and security staff in emergency response and crowd control will be helpful. The town planning of Chidambaram should be done in such a way that it accommodates a large number of pilgrims, especially during festivals and auspicious days. The local administration should conduct awareness campaigns and information centers for tourists. Educational programs for vendors about responsible tourism should also be organized. A proper regulatory framework should be in place to ensure quality control of tourism services, including the licensing of hotels, restaurants, and tour operators. Eco-friendly tourism and measures should be promoted, such as promoting local handicrafts and products from the area. It is vital to promote responsible tourism by making use of bio-friendly materials.

By implementing the above recommendations, Chidambaram can develop not only as a popular pilgrimage destination but also as a sustainable tourism center.

CONCLUSION

In conclusion, pilgrimage tourism plays a significant role in the economy and culture of Chidambaram. The town's renowned Nataraja Temple is a crucial religious site for Hindus, and the festivals and rituals held at the temple attract countless devotees yearly, generating an enormous income for the town's economy. Chidambaram also has several other temples and holy sites, such as Thillai Kali Temple, which is also a major pilgrimage site. The continued growth of the tourism sector in Chidambaram will not only generate revenue for the town but will also provide opportunities for the development of local businesses and industries. Moreover, pilgrimage tourism promotes a deeper understanding and appreciation of the religious and cultural traditions of Chidambaram. As such, the local government and stakeholders should invest in the development of tourism infrastructure and facilities in Chidambaram to attract more tourists and sustain the town's tourism industry.

REFERENCES

Higgins-Desbiolles, F. (2016). Sustaining spirit: A review and analysis of an urban Indigenous Australian cultural festival. *Journal of Sustainable Tourism*, *24*(8–9), 1280–1297. doi:10.1080/09669582.2016.1149184

Hillmer-Pegram, K. (2016). Integrating indigenous values with capitalism through tourism: Alaskan experiences and outstanding issues. *Journal of Sustainable Tourism*, *24*(8–9), 1194–1210. doi:10.1080/09669582.2016.1182536

Honggang, X., Dan, Z., & Jigang, B. (2016). Sustainability and nature-based mass tourism: Lessons from China's approach to the Huangshan Scenic Park. *Journal of Sustainable Tourism*, *24*(2), 182–202. doi:10.1080/09669582.2015.1071381

McKercher, B., & du Cros, H. (2002). *Cultural tourism: The partnership between tourism and cultural heritage management*. The Howarth Hospitality Press. [Google Scholar]

Pereiro, X. (2016). A review of indigenous tourism in Latin America: Reflections on an anthropological study of Guna Tourism (Panama). *Journal of Sustainable Tourism*, *24*(8–9), 1121–1138. doi:10.1080/09669582.2016.1189924

Salazar, N. B. (2012). Community-based cultural tourism: Issues, threats and opportunities. *Journal of Sustainable Tourism*, *20*(1), 9–22. doi:10.1080/09669582.2011.596279

Sofield, T., & Li, S. (2011). Tourism governance and sustainable national development in China: A macro-level synthesis. *Journal of Sustainable Tourism*, *19*(4–5), 501–534. doi:10.1080/09669582.2011.571693

ADDITIONAL READING

Abascal, T., Fluker, M., & Jiang, M. (2016). Domestic demand for indigenous tourism in Australia: Understanding intention to participate. *Journal of Sustainable Tourism*, *24*(8—9), 1350–1368. doi:10.1080/09669582.2016.1193187

Almagor, U. (1985). A tourist's 'vision quest' in an African game reserve. [Crossref], [Web of Science ®], [Google Scholar]. *Annals of Tourism Research*, *12*(1), 31–47. doi:10.1016/0160-7383(85)90038-6

Altman, J. C., & Finlayson, J. (1993). Aborigines, tourism and sustainable development. [Google Scholar]. *Journal of Tourism Studies*, *4*(1), 38–48.

Amoamo, M. (2008). *Decolonising Māori tourism: Representation and identity*. [Unpublished thesis, University of Otago, Dunedin]. Google Scholar.

Amoamo, M., & Thompson, A. (2011). (Re)imaging Māori tourism: Representation and cultural hybridity in postcolonial New Zealand. *Tourist Studies*, *10*(1), 35–55. doi:10.1177/1468797610390989

Barnett, S. J. (1997). Māori tourism. *Tourism Management*, *18*(7), 471–473. doi:10.1016/S0261-5177(97)00050-2

Bunten, A., & Graburn, N. (2009). Guest editorial: Current issues in indigenous tourism. [Google Scholar]. *London Journal of Tourism, Sport and Creative Industries*, *2*, 102–111.

Butler, R., & Hinch, T. (Eds.). (1996). *Tourism and indigenous peoples*. International Thomson Business Press. [Google Scholar]

Carr, A. M. (2004). Mountain places, cultural spaces – interpretation and sustainable visitor management of culturally significant landscapes: A case study of Aoraki/Mount Cook National Park. [Taylor & Francis Online], [Google Scholar]. *Journal of Sustainable Tourism*, *12*(5), 432–459. doi:10.1080/09669580408667248

Chan, J. H., Iankova, K., Zhang, Y., McDonald, T., & Qi, X. (2016). The role of self-gentrification in sustainable tourism: Indigenous entrepreneurship at Honghe Hani Rice Terraces World Heritage Site, China. *Journal of Sustainable Tourism*, *24*(8–9), 1262–1279. doi:10.1080/09669582.2016.1189923

Cohen, E. (1979). A phenomenology of tourist experiences. *Sociology*, *13*(2), 179–202. doi:10.1177/003803857901300203

Espeso-Molinero, P., Carlisle, S., & Pastor-Alfonso, M. J. (2016). Tourism product design from within: A collaborative research process with the Lacandon of Chiapas, Mexico. *Journal of Sustainable Tourism*, *24*(8–9). doi:10 doi:.1080/09669582.2016.193188

Fletcher, C., Pforr, C., & Brueckner, M. (2016). Factors influencing indigenous engagement in tourism development: An international perspective. *Journal of Sustainable Tourism*, *24*(8–9), 1100–1120. doi: 10.1080/09669582.2016.1173045

Gochring, B. (1993). *Indigenous peoples of the world: An introduction to their past and future*. Purich Publishing.

Graburn, N. (Ed.). (1976). *Ethnic and tourist arts: Cultural expressions from the fourth world*. University of California Press. doi:10.1525/9780520316775

Hinch, T., & Butler, R. (1996). Indigenous tourism: A common ground for discussion. In R. Butler & T. Hinch (Eds.), *Tourism and indigenous peoples* (pp. 3–21). International Thomson Business Press.

Holmes, A. P., Grimwood, B. S. R., & King, L. J.The Lutsel K'e Dene First Nation. (2016). Creating an indigenized visitor code of conduct: The development of Denesoline self-determination for sustainable tourism. *Journal of Sustainable Tourism*, *24*(8–9), 1177–1193. doi:10.1080/09669582.2016.1158828

Ingram, T. (1990). *Indigenous entrepreneurship in the tourism industry in the pacific Islands*. [Unpublished PhD thesis, Massey University, Palmerston North]. Google Scholar.

Koot, S. P. (2016). Contradictions of capitalism in the South African Kalahari: Indigenous Bushmen, their brand and baasskap in tourism. *Journal of Sustainable Tourism*, *24*(8–9), 1211–1226. doi:10.108 0/09669582.2016.1158825

Lemelin, H., Thompson-Carr, A., Johnston, M., Stewart, E., & Dawson, J. (2013). Indigenous people: Discussing the forgotten dimension of dark tourism and battlefield tourism. In D. K. Müller, L. Lundmark, & R. H. Lemelin (Eds.), *New issues in polar tourism: Communities, environments, politics* (pp. 205–212). Springer Publications. [Crossref], [Google Scholar] doi:10.1007/978-94-007-5884-1_14

MacCannell, D. (1976). *The tourist: A new theory of the leisure class*. Schocken. [Google Scholar]

Mkono, M. (2016). Sustainability and indigenous tourism insights from social media: Worldview differences, cultural friction and negotiation. *Journal of Sustainable Tourism*, *24*(8–9), 1315–1330. doi:10.1080/09669582.2016.1177066

Notzke, C. (1999). Indigenous tourism development in the Arctic. [Crossref], [Web of Science ®], [Google Scholar]. *Annals of Tourism Research*, *26*(1), 55–77. doi:10.1016/S0160-7383(98)00047-4

Notzke, C. (2004). Indigenous tourism development in Southern Alberta, Canada: Tentative engagement. *Journal of Sustainable Tourism*, *12*(1), 29–54. doi:10.1080/09669580408667223

Pereiro, X., Martínez, M., Ventocilla, J., De León, C., & Del Valle, Y. (2012). *Los turistores kunas. Antropología del turismo étnico en Panamá* [Kuna tourism: The anthropology of ethnic tourism in Panama]. University of the Balearic Islands. [Google Scholar]

Pihema, W. T. (2002). Boundaries of desire: How hapu and whanau mitigate the vectors of power during the establishment of a Māori tourism attraction (Unpublished thesis). Victoria University, Wellington. [Google Scholar]

Reggers, A., Grabowski, S., Wearing, S. L., Chatterton, P., & Schweinsberg, S. (2016). Exploring outcomes of community-based tourism on the Kokoda Track, Papua New Guinea: A longitudinal study of participatory rural appraisal techniques. *Journal of Sustainable Tourism*, *24*(8–9), 1139–1155. doi:10.1080/09669582.2016.1145229

Ryan, C., & Aicken, M. (2005). *Indigenous tourism; the commodification and management of culture (advances in tourism research)*. Elsevier. [Google Scholar]

Seiver, B., & Matthews, A. (2016). Beyond whiteness: A comparative analysis of representations of aboriginality in tourism destination images in New South Wales, Australia. *Journal of Sustainable Tourism*, *24*(8–9), 1298–1314. doi:10.1080/09669582.2016.1182537

Shackley, M. (2001). *Managing sacred sites*. Continuum. [Google Scholar]

Shultis, J., & Heffner, S. (2016). Hegemonic and emerging concepts of conservation: A critical examination of barriers to incorporating indigenous perspectives in protected area conservation policies and practice. *Journal of Sustainable Tourism*, *24*(8–9), 1227–1242. doi:10.1080/09669582.2016.1158827

Smith, M., & Richards, G. (Eds.). (2013). *Handbook of cultural tourism*. Routledge. [Crossref], [Google Scholar] doi:10.4324/9780203120958

Smith, V. (1989). *Hosts and guests: The anthropology of tourism*. University of Pennsylvania Press. doi:10.9783/9780812208016

Smith, V. (1996). Indigenous tourism: The 4 Hs. In R. Butler & T. Hinch (Eds.), *Tourism and indigenous peoples* (pp. 283–307). International Thomson Business Press. [Google Scholar]

Smith, V., & Brent, M. (Eds.). (2001). *Hosts and guests revisited: Tourism issues of the 21st century*. Cognizant Communications. [Google Scholar]

Sofield, T. H. B. (1991). Sustainable ethnic tourism in the South Pacific: Some principles. *Journal of Tourism Studies*, *2*(1), 56–72.

Sofield, T. H. B. (1993). Indigenous tourism development. *Annals of Tourism Research*, *20*(4), 729–750. doi:10.1016/0160-7383(93)90094-J

Sofield, T. H. B. (2003). *Empowerment for sustainable tourism development*. Pergamon.

Sofield, T. H. B., & Birtles, R. A. (1996). Indigenous peoples' cultural opportunity spectrum for tourism (IPCOST). In R. Butler & T. Hinch (Eds.), *Tourism and indigenous peoples: Issues and implications* (pp. 396–434). International Thomson Business Press.

Spark, C. (2002). Brambuk living cultural centre: Indigenous culture and the production of place. [Crossref], [Google Scholar]. *Tourist Studies*, *2*(1), 23–42. doi:10.1177/146879760200200l095

Stone, J. (2003). Max weber on race, ethnicity and nationalism. In J. Stone & R. Dennis (Eds.), *Race and ethnicity: Comparative and theoretical approaches* (pp. 28–42). Blackwell Publishing.

Tahana, N., & Opperman, M. (1998). Māori cultural performances and tourism. *Tourism Recreation Research*, *23*(1), 23–30. doi:10.1080/02508281.1998.11014816

Thompson, A. (2013). Māori Tourism: A case study of managing indigenous cultural values. In M. Smith & G. Richards (Eds.), *Handbook of cultural tourism* (pp. 227–235). Routledge.

Thompson, A. M. (2007). Strengthening of identity through participation in tourism guiding. In R. Butler & T. Hinch (Eds.), *Tourism and indigenous peoples* (2nd ed., pp. 113–127). Butterworth-Heinemann. [Google Scholar]

Tuhiwai Smith, L. (1999). *Decolonising methodologies: Research and indigenous peoples*. Otago University Press.

Tuhiwai Smith, L. (2012). *Decolonizing methodologies: Research and indigenous peoples* (2nd ed.). Zed Books.

United Nations. (2016). 2007 UN declaration on the rights of indigenous peoples. Retrieved from https://www.un.org/development/desa/Indigenouspeoples/declaration-on-the-rights-of-Indigenous-peoples.html

Walker, K., & Moscardo, G. (2016). Moving beyond sense of place to care of place: The role of indigenous values and interpretation in promoting transformative change in tourists' place images and personal values. *Journal of Sustainable Tourism*, *24*(8–9), 1243–1261. doi:10.1080/09669582.2016.1177064

Whitford, M., & Ruhanen, L. (2016). Indigenous tourism research, past and present: Where to from here? *Journal of Sustainable Tourism*, *24*(8–9), 1080–1099. doi:10.1080/09669582.2016.1189925

Whitney-Squire, K. (2016). Sustaining local language relationships through indigenous community-based tourism initiatives. *Journal of Sustainable Tourism*, *24*(8–9), 1156–1176. doi:10.1080/09669582.2015.1091466

Wikitera, K.-A. (2006). *Whakarewarewa Tourism development: A critical analysis of place and space* [Unpublished Master's thesis, Auckland University of Technology, Auckland].

Zeppel, H. (2006). *Indigenous ecotourism: Sustainable development and management*. CABI. doi:10.1079/9781845931247.0000

Chapter 13
Assessing Water and Sanitation Service Quality at Pilgrimage Sites After COVID-19:
An Empirical Investigation

Jeganathan Gomathi Sankar
https://orcid.org/0000-0002-5077-5109
BSSS Institute of Advanced Studies, India

Arokiaraj David
https://orcid.org/0000-0002-9591-2410
Al Tareeqah Management Studies, UAE

Rajesh Ramasamy
Mizoram University, India

ABSTRACT

This empirical investigation focuses on assessing the quality of water and sanitation services provided by hotels located on pilgrimage sites, especially after the COVID-19 pandemic. The study examines the factors that contribute to customers' perception of service quality and the associated hygiene practices implemented by the hotels. Data were collected through personal interaction using the SERVQUAL questionnaire which measures service quality in accordance with five dimensions. It highlights the importance of improving the tangibility of service and increasing empathy between hotel staff and guests, especially in the post-COVID-19 era. The study suggests recommendations for improving the quality of hospitality services and increasing tourist satisfaction. Overall, the study highlights the importance of providing high-quality water and sanitation services at pilgrimage sites, especially in the current context of the pandemic, to ensure that tourists have a positive experience and feel confident in using these facilities.

INTRODUCTION

India is rich and diversified in natural, historical, cultural, and spiritual values. Millions of domestic and international tourists travel to India each year to witness its spiritual grandeur and Indian heritage.

DOI: 10.4018/978-1-6684-4817-5.ch013

Hinduism, Buddhism, Jainism, Sikhism, and Sufism are just a few of the countless religions that call India home (Jeganathan & Srinivasulu, 2014). The country has long been a popular destination for pilgrims of all faiths. The development of Indian heritage values has been significantly affected by these centers. In fact, associations with holy or historical sites, figures, or events have contributed significantly to the growth of many well-known tourist destinations, making religion, spirituality, and heritage key travel motivations (Kumar et al., 2020). The growth of pilgrimage and historical tourism is crucial to developing domestic tourism in India. To fully realize this potential, it is required to collaborate with other stakeholders to integrate the development of the selected heritage cities and pilgrimage sites (Yousaf, 2021). Integrated development encompasses all aspects of a city's growth, including planning, such as basic services, economy, livelihoods, community quality of life, and also cleanliness, security, soul-stirring renewal, and outward displays of a city's personality. It goes beyond just the creation and upkeep of a few monuments.

Unlike holiday or leisure tourism destinations, which are predominantly marketed to the segment with a propensity to travel and spend money, religious tourism destinations are not intended for this group. Community involvement in the chosen location/destination is crucial when creating a spiritual tourism package. Furthermore, it is crucial to consider how visitors engage and seek novel experiences while designing modern amenities for religious institutions (David et al., 2022). To pave the way for the growth and promotion of religious tourism in India, it is necessary first to identify the fundamental shortcomings, then the facilitation of services (Yousaf, 2021).

Pilgrimage tourism would mostly draw a different demographic. But the problems are the same: the heritage regions are overcrowded, neglected, and lack essential infrastructure like clean water, sanitary facilities, and roads. Streetlights, signage, and other necessities are absent. In order to ensure a quick pace and successful implementation on the ground, a more systematic strategy for managing the construction of such a city based on tourism infrastructure is also required. Pilgrimage and heritage tourism boost foot traffic in towns and destinations (Jeganathan et al., 2022). Religious/pilgrimage tourism is more of a destination-oriented theme. Rejuvenation through pilgrimage and spiritual augmentation satisfies developments within constrained touristic footprints surrounding the sites (Collins-Kreiner, 2020).

Moreover, city interventions affect several areas, including water supply, sewage, urban development, water body preservation, roads & transportation, urban transit, urban regeneration, and cultural conservation, among others, necessitating an integrated and coherent strategy for development (David, 2020). The success of pilgrimage locations depends heavily on travelers' satisfaction, so there has been an increase in interest in the caliber of services offered to pilgrims in recent years. The degree to which services given in pilgrimage tourism match or surpass the expectations of pilgrims can be referred to as service quality (Ganeshkumar et al., 2023). The experience of pilgrims is significantly impacted by the receivers' service quality, with accommodation, transit, and places to eat and drink having an important impact (Alnsour & Al-Abdallah, 2020). Service quality greatly impacts pilgrims' loyalty; hotels, transportation, and eateries must deliver high-quality services (Mishra & Gupta, 2021).

THEORETICAL FRAMEWORK

Tourism, particularly in the context of pilgrim tourism, heavily relies on the availability of suitable water and sanitation facilities. As many individuals visit religious sites for extended periods, water quality and sanitation services play a vital role in shaping the overall experience of pilgrim tourists. Thus, this study aims to conduct a comprehensive literature review to analyze the service quality experience of pilgrim tourists regarding water and sanitation.

Service Quality

SERVQUAL is a widely used framework for measuring and evaluating service quality. Parasuraman, Zeithaml, and Berry (1988) developed SERVQUAL as a comprehensive approach to understanding customer perceptions of service quality. The framework is based on the premise that service quality is a function of the customer's expectations and the actual service delivered.

SERVQUAL has five dimensions used to measure service quality: reliability, responsiveness, assurance, empathy, and tangibles. Reliability refers to the ability of a service provider to deliver the service as promised. Responsiveness refers to the willingness of a service provider to help customers promptly (Spreng & Mckoy, 1996). Assurance refers to service providers' knowledge and competence and ability to inspire trust and confidence (Clow & Beisel, 1995). Empathy refers to the degree of caring and individualized attention service providers show their customers (Grönroos, 1984). Finally, tangibles refer to the physical and visual aspects of the service environment, such as facilities, equipment, and appearance.

Parasuraman et al. (1988) introduced the SERVQUAL instrument, a five-dimensional measurement scale designed to assess perceived service quality. The scale includes reliability, responsiveness, assurance, empathy, and tangibles. While SERVQUAL has faced criticism, it remains widely used by researchers, and service quality proponents have developed modified versions tailored to specific industries and situations (Sankar et al., 2020). Various service industries have used the original or modified SERVQUAL instrument (Ladhari, 2008).

Pilgrimage tourism is a form of tourism that involves traveling to religious sites or destinations for spiritual or religious purposes. Service quality is a critical factor in the success of pilgrimage tourism, as pilgrims expect a high level of service and hospitality during their journeys (Parasuraman et al., 1988). Service quality in pilgrimage tourism is important for the pilgrims' satisfaction and the long-term sustainability of the destination and the tourism industry (Jeganathan et al., 2022_a).

Water and Sanitation Service Quality

Al-Shaer et al. (2020) have found that the quality of water and sanitation services is a critical component of the pilgrim tourist experience, as inadequate facilities can adversely affect health, hygiene, and overall satisfaction (Madaan et al., 2021). This research aimed to evaluate the quality of water and sanitation services in Bali, Indonesia, a well-known tourist spot. The SERVQUAL model was employed to measure service quality, and data was gathered from local inhabitants, tourists, and service providers. The results showed that although Bali has significantly improved water and sanitation services, there is still room for improvement in infrastructure, facilities, and management. The authors recommend enhancing service quality by improving water treatment quality, increasing wastewater treatment plant capacity, increasing public awareness about water and sanitation issues, and strengthening regulatory frameworks to ensure compliance with standards. The significance of assessing the quality of water and sanitation services in tourist spots was highlighted by this research. It offers valuable perspectives to policymakers, service providers, and other concerned parties on enhancing service quality and fostering sustainable tourism development (Pratama et al., 2020).

In their study, Pradhan and Mishra extensively analyze the water and sanitation services in tourist destinations. They shed light on challenges such as inadequate infrastructure, poor management, and lack of funds that impact the sector. The authors also emphasize the impact of tourism on the environment and the crucial role of sustainable practices in the water and sanitation sector (Parne et al., 2021). They suggest various best practices, including innovative technologies, public-private partnerships, and community participation. The research provides valuable insights for policymakers, managers, and practitioners in the water and sanitation sector to improve services in tourist destinations (Pradhan & Mishra, 2019).

The authors, Ceesay, Jallow, and Njie, comprehensively examine the obstacles encountered by water and sanitation services in the Gambia and their effects on the tourism sector. They highlight the importance of sustainable practices in the water and sanitation sector to support tourism development and promote environmental conservation. The authors suggest several best practices, including community engagement, public-private partnerships, and innovative technologies. They provide recommendations for policymakers, managers, and practitioners in the water and sanitation sector to improve services in the Gambia and promote sustainable tourism development (Ceesay et al., 2019). In another study, Boonchuay and Pinit analyze the challenges faced by water and sanitation services in Phuket, including insufficient infrastructure, limited resources, and ineffective management. The authors stress the importance of improving services to support sustainable tourism development and protect the environment. They provide several recommendations, including increased investment in infrastructure, improved management practices, and community engagement (Garwal et al., 2020). The paper offers valuable insights for policymakers, managers, and practitioners in the water and sanitation sector to improve services in tourist destinations (Boonchuay & Pinit, 2018).

Furthermore, Alam and Choudhury present a detailed analysis of policies and strategies at the national and international levels, emphasizing the importance of sustainable practices in the water and sanitation sector to support tourism development and promote environmental conservation. They discuss the need for effective implementation and monitoring of policies and strategies. The research reviews several best practices, including public-private partnerships, community participation, and innovative technologies. The study offers significant perspectives for policymakers, managers, and practitioners in the water and sanitation sector on enhancing services in tourist hotspots (Alam & Choudhury, 2022). According to Salem et al. (2019), cultural factors play a vital role in shaping pilgrim tourists' quality of service experience. The authors highlight how pilgrim tourists from different cultural backgrounds have different perceptions and expectations regarding water quality and sanitation services (Haverila & Haverila, 2018). These cultural differences can significantly impact the expectations of pilgrim tourists and ultimately influence their satisfaction levels (Ramasamy et al., 2021). Therefore, it is crucial to consider cultural differences when designing and providing water and sanitation services for pilgrim tourists.

Water and sanitation service quality are crucial determinants of the overall experience of pilgrim tourists. The quality of service factors, including responsiveness, assurance, empathy, reliability, and tangibility, influences the satisfaction of pilgrim tourists. The quality of water and sanitation services is particularly significant, as it can significantly affect pilgrim tourists' satisfaction levels and behavioral intentions. Therefore, it is important to prioritize service quality in these areas and ensure that cultural differences are considered when providing services to pilgrim tourists. By doing so, service providers can create a positive experience for pilgrim tourists, leading to increased satisfaction levels and favorable behavioral intentions (Datrika et al., 2022).

Service Quality Experience and Behavioral Intentions

The service quality experienced by pilgrim tourists plays a significant role in determining their behavioral intentions, such as loyalty, communication through word-of-mouth, and willingness to return. Kim et al. (2020) found that a positive service quality experience can significantly influence the future success of pilgrimages by positively affecting loyalty and communicated word-of-mouth. Meanwhile, Kandel and Adhikari (2018) conducted an in-depth analysis of the sanitation service quality in Kathmandu's tourist areas. They identified various challenges the sector faces, such as inadequate infrastructure, poor management, and lack of awareness among tourists and service providers. They emphasized the significance of sanitation service quality in promoting tourism and suggested that sustainable practices must be implemented in the sector (Ganeshkumar et al., 2020). The paper provides useful recommendations for policymakers, managers, and practitioners to improve the service quality of tourist destinations.

Kumar and Singh (2019) conducted a detailed analysis of sanitation service quality in tourist areas of Goa, India, and highlighted the same challenges as in the Kandel and Adhikari study. They proposed several recommendations, including improved infrastructure, increased awareness among tourists and service providers, and better management of sanitation services (David et al., 2019). The study provides valuable insights for policymakers, managers, and practitioners in the sanitation sector and is a useful example for other tourist destinations facing similar challenges.

Valencia et al. (2018) also identified various challenges that water and sanitation services face in Cancun, such as inadequate infrastructure, population growth, and the impact of tourism on natural resources. They stressed the importance of sustainable practices in the water and sanitation sector. They proposed several recommendations, including increased investment in infrastructure, water conservation measures implementation, and innovative technologies (David, 2022). The study provides valuable visions for policymakers, practitioners, and managers in the water and sanitation sector to improve services in tourist places. Additionally, Mkenda et al. (2020) conducted an in-depth analysis of the challenges and difficulties faced by Zanzibar, Tanzania, regarding the water and sanitation services similar to those in other tourist destinations. They also stressed the importance of sustainable practices. They proposed several recommendations, including increased investment in infrastructure, effective management of water and sanitation services, and increased awareness among tourists and service providers.

Finally, the authors provide a comprehensive review of innovative solutions and best practices adopted in various tourist destinations worldwide to address the challenges faced by water and sanitation services. The paper emphasizes the importance of sustainable practices, including water conservation, recycling, and reuse, in promoting tourism development and protecting the environment. It identifies several key factors that contribute to the success of water and sanitation services in tourist places, including stakeholder engagement, effective management, and investment in infrastructure (Jeganathan et al., 2022_b). The review provides valuable insights for researchers, practitioners, and policymakers in water and sanitation.

Based on these previous findings by various authors who have employed and validated SERVQUAL, the present study posits the following hypotheses for SERVQUAL dimensions' impact on satisfaction (as shown in figure 01) and revisits intention to the place they have stayed. The SERVQUAL dimensions focus on the water and sanitation perspective.

H1: The reliability factor has a significant impact on customer satisfaction.
H2: The responsiveness factor has a significant impact on customer satisfaction.

Figure 1. Postulated model

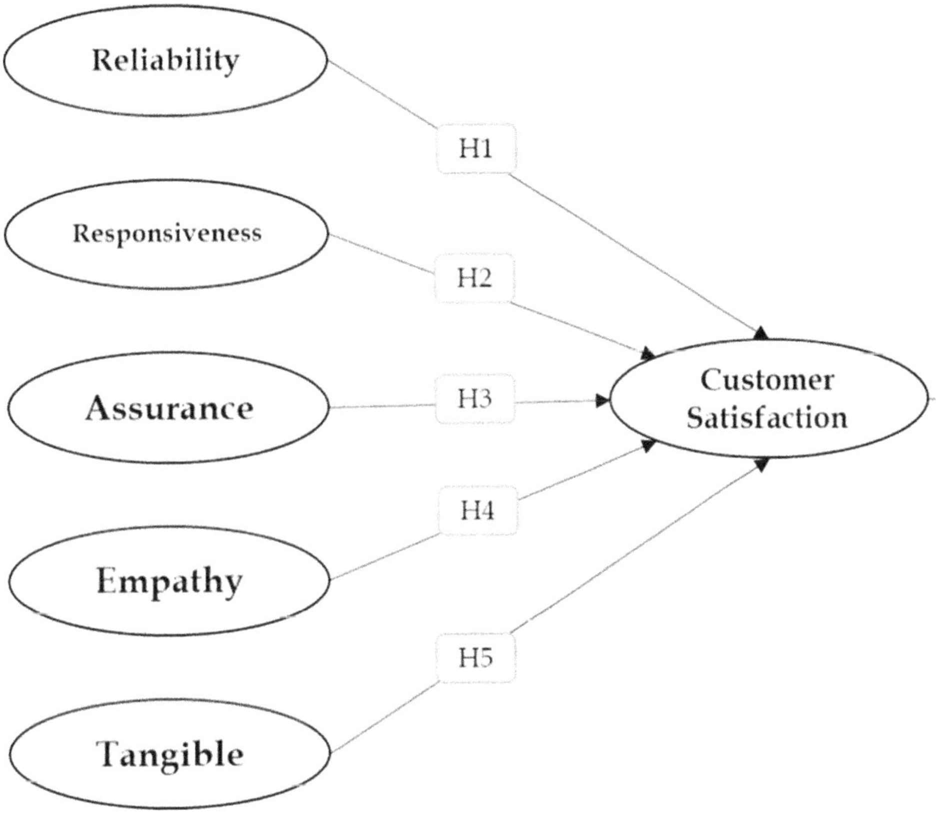

H3: The assurance factor has a significant impact on customer satisfaction.
H4: The empathy factor has a significant impact on customer satisfaction.
H5: The tangible factor has a significant impact on customer satisfaction

Research Methodology

The SERVQUAL scale was used to create a self-administrated questionnaire for this study. The questionnaire had two primary elements, the first of which was used to gauge hotel guests' impressions of service quality. A further item was included for estimating the hotels' overall degree of service quality from their guests' perspectives. The questionnaire's second section asks questions regarding respondents' demographic information (nationality, gender, etc.). A five-point scale (5=very high; and 1=very low) was used to assess the variables taken for the questionnaire.

The present study collected data from Kumbakonam, a beautiful temple town in Tamil Nadu, India, known for its rich cultural heritage and numerous Hindu temples. Most temples in Kumbakonam are dedicated to Lord Vishnu and Lord Shiva and play a significant role in the town's religious and cultural life (Thummula et al., 2019). The Shiva temples in Kumbakonam are associated with the Mahamaham festival, which occurs every twelve years (Ganeshkumar et al., 2022). There are twelve Shiva temples in total, with ten of them located in Kumbakonam town itself. The remaining five Vishnu temples are also situated in Kumbakonam and connected to the Mahamaham festival. In addition to the temples, there are also nine Navagraha temples within a 60 km radius around Kumbakonam in the Thanjavur district, each representing one of the nine planets (Ravi et al., 2018). In December 2022, questionnaires were administered to patrons departing from multiple hotels in Kumbakonam to gather data for the study. Convenient sampling was utilized; respondents typically completed the survey within five minutes while standing outside the hotels. Out of the 320 questionnaires collected, 16 were deemed incomplete or unusable, leaving 304 questionnaires for data analysis (Yuan & Wu, 2008). The usable rate was 95%.

RESULTS AND DISCUSSION

Reliability and Validity Analyses

To ensure the accuracy and dependability of the analysis, it is crucial to perform preliminary verifications using the measurement instrument and collected data before testing the hypotheses, as noted by Ganeshkumar et al. (2023). Various assessments were implemented to identify common method variance on the scale, appraise the psychometric properties of the measurement instrument, and evaluate the data distribution. After validating the data, Confirmatory Factor Analysis (CFA) was utilized to verify the dimensions of service quality described in the literature (measurement theory). Finally, the study examined the influence of service quality on customer satisfaction. Conducting these preliminary verifications is necessary to ensure the study's findings are valid and reliable.

Exploratory Factor Analysis

Exploratory factor analysis (EFA) is a statistical method used to identify latent or underlying constructs that explain the correlations among observed variables. EFA involves examining the factor structure of a set of variables to determine the number of factors that account for the observed correlations. Factor extraction involves using mathematical algorithms, such as principal component analysis, to identify the initial set of factors that account for the majority of the variance in the data. Factor rotation, on the other hand, involves reorienting the factor axes to maximize the interpretability of the factor structure.

Based on the above results of Bartlett's sphericity test and the KMO measure are represented in Table 1. These tests are frequently used in factor analysis, determining whether the data taken are appropriate to the statistical method. The correlation matrix's suitability for factor analysis is determined by the KMO measure, which also assesses how well the data were sampled. The outcome of 0.883 shows that the data are sufficient for factor analysis. Bartlett's test of sphericity is utilized to test the null hypothesis that the variables in the correlation matrix are not interrelated in the population. The variables are

Table 1. KMO and Bartlett's test

Kaiser-Meyer-Olkin Measure of Sampling Adequacy		0.883
Bartlett's Test of Sphericity	Approx. Chi-Square	12385.816
	df	231
	Sig.	0.000

likely associated, and component analysis may be applicable according to a significant result (p .05). The test's outcome, which shows that the variables are substantially associated, and that factor analysis is appropriate for this data set, is an approximate chi-square of 12385.81 (P value =0.000) with 231 (Df).

Table 2 shows the reliability analysis results for the five factors in the research instrument used in the study. The factors are responsiveness, reliability, empathy, tangible, and assurance, and each factor has a different number of items. The reliability coefficients of each factor were used to assess Cronbach's alpha coefficient for ensuring internal consistency. The results show that the instrument is highly reliable for all factors, with alpha coefficients ranging between 0.805 and 0.952. A coefficient of 0.7 or higher is generally accepted for research instruments, and all factors in this study exceeded this threshold. The factor with the highest reliability is empathy, with a coefficient of 0.952, while the factor with the lowest reliability is tangible, with a coefficient of 0.805. These reliability coefficients suggest that the research instrument is a reliable and valid tool for measuring the five customer satisfaction factors. Researchers can use this instrument to collect customer perceptions of service quality and analyze the connection between service quality and customer satisfaction.

Table 3 shows the exploratory factor analysis with a varimax rotation and Kaiser normalization on a set of variables related to service quality. The table presents the rotated component matrix with five components, labeled 1 through 5, and the loadings of each variable on each component. The components can be interpreted as follows:

- Component 1: Responsiveness, with high loadings on all four responsiveness variables, indicating that they share a common factor. This component represents the extent to which the service provider is perceived as being prompt and helpful in responding to customer needs.
- Component 2: Reliability, with high loadings on all five reliability variables, indicating that they share a common factor. This component represents the extent to which the service provider is perceived as being dependable and accurate in delivering the promised service.
- Component 3: Empathy, with high loadings on all five empathy variables, indicating that they share a common factor. This component represents the extent to which the service provider is perceived as being caring and understanding of customer needs.
- Component 4: Tangible, with high loadings on all four tangible variables, indicating that they share a common factor. This component represents the extent to which the physical aspects of the service environment are perceived as being modern and well-maintained.
- Component 5: Assurance, with high loadings on all four assurance variables, indicating that they share a common factor. This component represents the extent to which the service provider is perceived as being knowledgeable and trustworthy in delivering the promised service.

Table 2. Reliability

Factor	Number of Items	Cronbach Alpha
Responsiveness	04	0.886
Reliability	05	0.937
Empathy	05	0.952
Tangible	04	0.805
Assurance	04	0.893

Table 3. Rotated component matrix

	Component				
	01	02	03	04	05
Responsiveness Item 1			.808		
Responsiveness Item 2			.599		
Responsiveness Item 3			.873		
Responsiveness Item 4			.863		
Reliability Item 1					
Reliability Item 2		.901			
Reliability Item 3		.911			
Reliability Item 4		.898			
Reliability Item 5		.874			
Empathy Item 1	.899				
Empathy Item 2	.901				
Empathy Item 3	.901				
Empathy Item 4	.910				
Empathy Item 5	.791				
Tangible Item 1				.718	
Tangible Item 2				.866	
Tangible Item 3				.776	
Tangible Item 4				.803	
Assurance Item 1					.651
Assurance Item 2					.801
Assurance Item 3					.807
Assurance Item 4					.815
Extraction Method: Principal Component Analysis. Rotation Method: Varimax with Kaiser Normalization. a. Rotation converged in 6 iterations.					

The loadings on each component can be used to create composite scores for each component, which can be used to compare the relative importance of each component in driving overall customer satisfaction or other outcomes, as shown in Figure 02.

Measurement Model

Table 4 provides the factor loading values for a confirmatory factor analysis of variables related to service quality in a pilgrimage context. Factor loading is a statistical measure that represents the direction and strength of the bond between each variable and its underlying factor.

Table 4 shows five factors, each with a set of variables and their corresponding factor loading values:

Figure 2. Measurement model

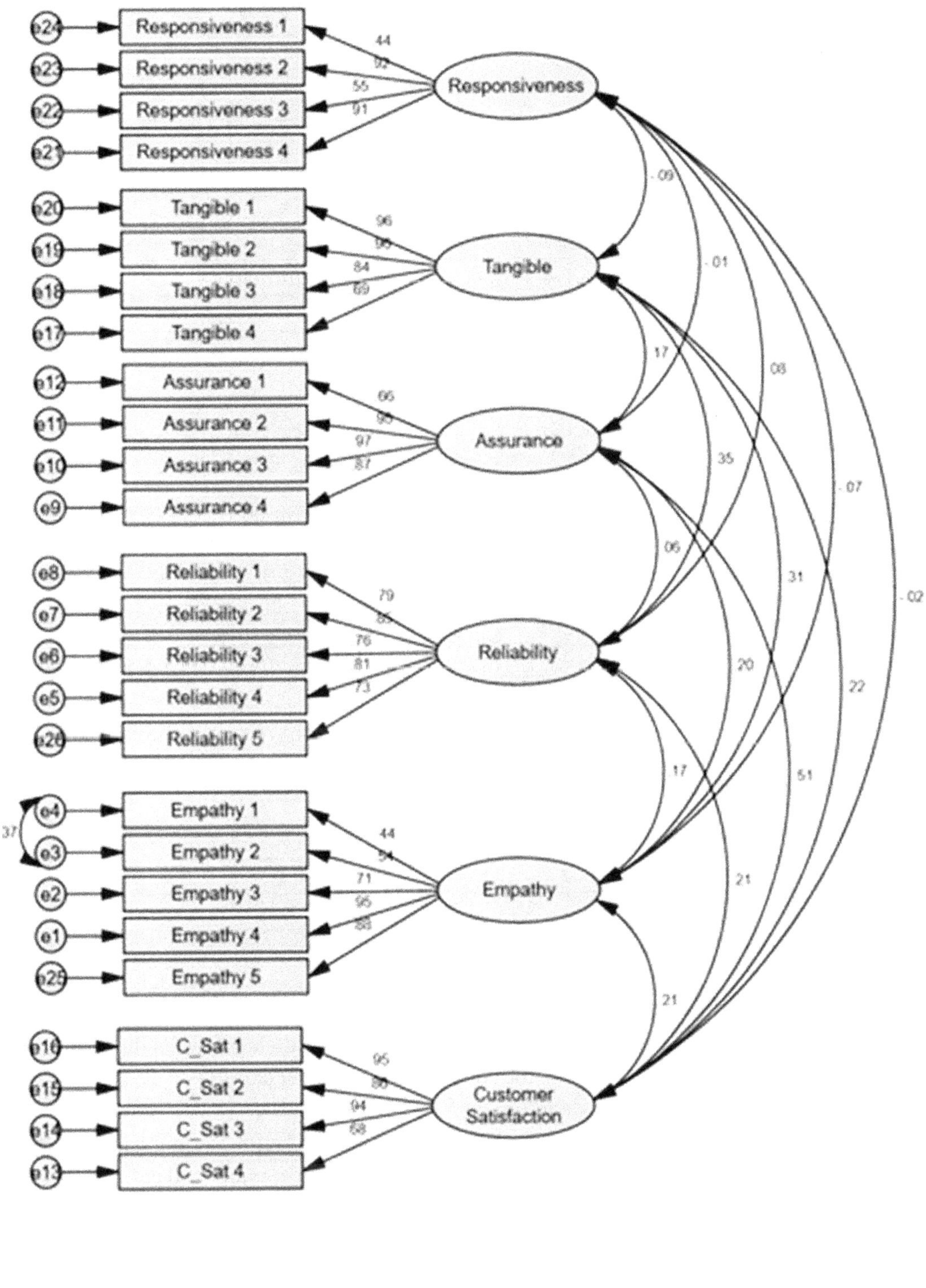

- Responsiveness: This factor has four variables (Responsiveness 1-4), and their factor loading values range from .441 to .925. This factor captures the service provider's competence to respond to

Table 4. Factor loading

Factor	Variables	Factor Loading
Responsiveness	Responsiveness Item 1	.441
	Responsiveness Item 2	.925
	Responsiveness Item 3	.548
	Responsiveness Item 4	.908
Reliability	Reliability Item 1	.794
	Reliability Item 2	.849
	Reliability Item 3	.757
	Reliability Item 4	.808
	Reliability Item 5	.731
Empathy	Empathy Item 1	.441
	Empathy Item 2	.539
	Empathy Item 3	.709
	Empathy Item 4	.952
	Empathy Item 5	.876
Tangible	Tangible Item 1	.964
	Tangible Item 2	.964
	Tangible Item 3	.841
	Tangible Item 4	.693
Assurance	Assurance Item 1	.652
	Assurance Item 2	.946
	Assurance Item 3	.966
	Assurance Item 4	.874
Customer Satisfaction	C_Sat Item 1	.950
	C_Sat Item 2	.856
	C_Sat Item 3	.939
	C_Sat Item 4	.675

the customer's needs and requirements promptly and effectively.

- Reliability: This factor has five variables (Reliability 1-5), and their factor loading values range from .731 to .849. This factor represents the consistency and dependability of the service provider to deliver assured service accurately and timely.
- Empathy: This factor has five variables (Empathy 1-5), and their factor loading values range from .441 to .952. This factor refers to the service provider's ability to comprehend and care about customers' desires and anxieties.
- Tangible: This factor has four variables (Tangible 1-4), and their factor loading values range from .693 to .964. This factor measures the physical aspects of the service, such as amenities, equipment, and appearance, that influence the customer's insight into service quality.

- Assurance: This factor has four variables (Assurance 1-4), and their factor loading values range from .652 to .966. This factor captures the service provider's ability to convey trust and confidence to the customer about their service.

Overall, the factor loading values indicate the direction and strength of the bond between the variables and their underlying factors. Higher factor loading values indicate a stronger relationship between the variable and its factor. Therefore, these results suggest that the five factors identified in the study - responsiveness, reliability, empathy, tangible, and assurance - are relevant and important in measuring service quality in the context of the pilgrimage.

Before testing the hypotheses, it is important to verify the discriminant validity of the constructs using several tests, including the Average Variance Extracted (AVE) test. This test assesses whether the constructs measured in a study are distinct from each other. AVE is considered the most rigorous test for discriminant validity (Fornell & Larcker, 1981). As per Fornell and Larcker (1981), a construct is believed to have sufficient discriminant validity when its correlation with the indicator of the construct (i.e., the square root of AVE) is higher than its correlation with any other constructs. Table 5 presents the results of the discriminant validity analysis, which exhibit the AVE values for each construct and the correlations between them. The diagonal values display the square root of the AVE for each construct, while the off-diagonal values represent the correlations between constructs. The findings indicate that the constructs have discriminant validity because their AVE values are greater than their correlations with other constructs. Specifically, the bold diagonal values are all greater than their corresponding off-diagonal values. It indicates that the measures are adequately distinct from each other and that the constructs can be considered distinct and independent. The correlation between Responsiveness and Reliability is 0.050, which is lower than the square root of the AVE for Responsiveness (0.948) and reliability (0.882), indicating that these constructs are distinct from each other. Similar conclusions can be drawn for the other correlations in the table.

The fit measures used to assess the agreement between the observed data and the theoretical model are presented in Table 6. The table includes four fit measures, such as the achieved value and the suggested value, based on Hair et al.'s (2006) criteria for each measure. The initial measure, CMIN/DF ($\chi 2/df$) (Absolute Fit Indices), compares the chi-square value to the degrees of freedom. The achieved value 2.9 is lower than the recommended value of 3, implying an acceptable fit between the observed data and the theoretical model. The second measure, RMSEA (Absolute Fit Indices), refers to the Root Mean Square Error of Approximation, which evaluates the model's ability to match the covariance ma-

Table 5. Discriminate validity

	Responsiveness	**Reliability**	**Empathy**	**Tangible**	**Assurance**
Responsiveness	**0.897**				
Reliability	0.05	**0.784**			
Empathy	0.035	0.767	**0.785**		
Tangible	-0.101	-0.029	0.014	**0.754**	
Assurance	0.15	-0.02	-0.064	-0.11	**0.852**

Note: Diagonal values are square roots of average variance extracted (AVE), and below the diagonal values are correlations between factors

Table 6. Measurement model fit measures (measurement model)

Fit Measures	Obtained value	Recommended value (Hair et al., 2006)
CMIN/DF (χ^2/df) (Absolute Fit Indices)	2.900	< 3
RMSEA (Absolute Fit Indices)	0.080	< 0.09
CFI (Incremental Fit Indices)	0.938	> 0.9
AGFI (Parsimony Fit Indices)	0.841	> 0.8

trix. The achieved value of 0.080 is below the recommended value of 0.09, indicating a good fit. The third measure, CFI (Incremental Fit Indices), refers to the Comparative Fit Index, which compares the theoretical and null models. The achieved value of 0.938 is higher than the recommended value of 0.9, signifying an acceptable fit. The fourth measure, AGFI (Parsimony Fit Indices), refers to the Adjusted Goodness of Fit Index, which adjusts the fit index for the model's complexity. The achieved value of 0.841 is greater than the recommended value of 0.8, indicating a good fit (Naude et al., 2016). In conclusion, the fit measures in Table 6 demonstrate that the observed data fits the theoretical model well.

In general, based on the obtained values for all four fit measures, it can be concluded that the measurement model fits the observed data satisfactorily, as represented in Figure 03. Nevertheless, it is critical to remember that the fit measures should not be analyzed in isolation and must be interpreted within the context of the research problem and the specific dataset under examination.

Table 7 presents the outcomes of a path analysis that assesses the connection between customer satisfaction and five factors: tangibles, assurance, reliability, responsiveness, and empathy. The path coefficient values indicate the intensity and direction of the association between each factor and customer satisfaction, with higher values indicating a more robust positive relationship between the two variables (Hair et al., 2006; Ganeshkumar et al., 2023). For example, the path coefficient for reliability-customer satisfaction is 0.751, suggesting a significant positive correlation between reliability and customer satisfaction.

The significance level indicates whether the relationship between each factor and customer satisfaction is statistically significant. In this table, all the path coefficients are statistically significant at $P < 0.010$ / $P < 0.050$ level, meaning that the results are unlikely to have occurred by chance (Arokiaraj et al., 2020). After analyzing the data, it can be inferred that the five factors - responsiveness, tangibles, assurance, reliability, and empathy - are positively related to customer satisfaction, and their impact is statistically significant. The analysis further reveals that reliability is the most important factor in customer satisfaction. This information is valuable for organizations to identify and improve customer satisfaction's key drivers. By enhancing these factors, companies can improve customer loyalty and retention, resulting in long-term profitability and sustainability.

Hair et al. (2006), the CMIN/DF value (3.119) is greater than the recommended value of less than 3. It implies that the theoretical model may not fit the observed data well, and there could be some differences between the two. However, the RMSEA value of the model is 0.059, which is below the suggested value of less than 0.09, indicating that the model has a good fit for the data and accurately represents the underlying theoretical construct. Moreover, the CFI value of the model is 0.946, which is above the suggested value of 0.9, indicating a good incremental fit. It means that the model explains more variance than the null model. Additionally, the AGFI value of the model is 0.869, which is above

Figure 3. Structural model

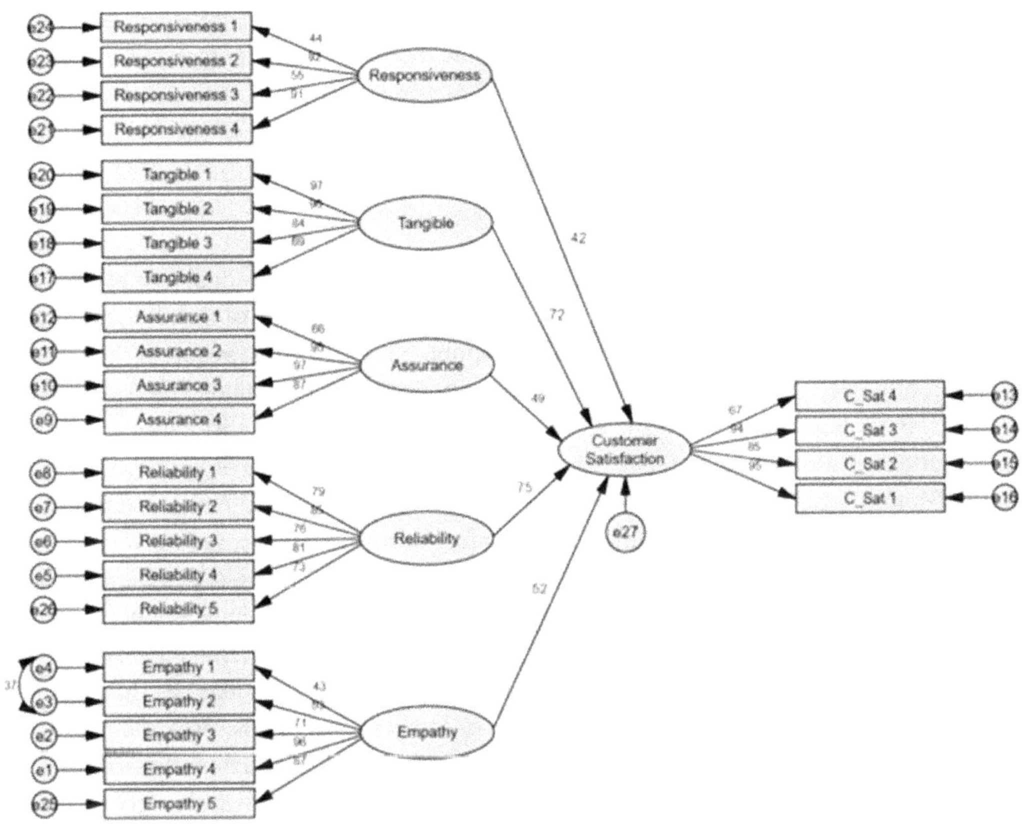

Table 7. Path coefficient values

Path	Path coefficient	Significance level
Responsiveness → Customer Satisfaction	.412	P < 0.010
Tangibles → Customer satisfaction	.719	P < 0.010
Assurance → Customer satisfaction	.496	P < 0.050
Reliability → Customer Satisfaction	.751	P < 0.010
Empathy → Customer Satisfaction	.523	P < 0.010

Table 8. Measurement model fit measures (structural model)

Fit Measures	Obtained value	Recommended value (Hair et al., 2006)
CMIN/DF (χ2/df) (Absolute Fit Indices)	3.119	< 3
RMSEA (Absolute Fit Indices)	.059	< 0.090
CFI (Incremental Fit Indices)	.946	> 0.90
AGFI (Parsimony Fit Indices)	.869	> 0.80

the suggested value of 0.8, suggesting a good parsimony fit. It indicates that the model provides a good balance between model complexity and model fit.

Overall, the results indicate that while the model has some discrepancies between the theoretical model and the actual data, it still has a reasonable fit, and the model is likely to represent the underlying construct accurately.

DISCUSSION

In the aftermath of the COVID-19 pandemic, hospitality and tourism accommodation establishments are responsible for providing a secure and pleasant environment for their guests and all their customers. The guests' stay at an accommodation should be comfortable and safe while receiving their required services. To meet or exceed the service standards, hospitality providers must understand guests' expectations and needs. By doing so, accommodation providers can deliver high-quality services that surpass customer expectations, leading to guest satisfaction. Therefore, accommodation providers must prioritize their guests' and employees' safety and well-being while ensuring their guests' enjoyable stay. It can be accomplished by properly planning and implementing safety protocols and providing various services and amenities catering to guests' diverse needs and preferences.

This study aimed to identify a set of measurement items for SERVQUAL and conduct comparative analyses of quality measurement items based on the type of service and customer in the post-COVID-19 hospitality industry (Lee, 2017). The findings revealed that reliability was ranked as the most crucial quality aspect, followed by tangibles, empathy, assurance, and responsiveness, providing valuable insights into the significance of various quality aspects in the industry. The study emphasizes the crucial need for hospitality and tourism providers to prioritize their guests' and employees' safety and health in the post-COVID-19 era. Providers must ensure that their facilities and services adhere to or exceed the recommended safety protocols to provide their customers with a secure and comfortable environment.

Additionally, the study suggests that empathy and reliability quality aspects, such as cleanliness and hygiene, are paramount in the post-COVID-19 hospitality industry. Providers must empathize with their customers' concerns and needs while offering high-quality facilities and services that meet their expectations and preferences. Furthermore, the study's results emphasize the importance of tailoring services to meet customers' diverse needs and expectations in the post-COVID-19 hospitality industry. Providers must regularly assess their quality standards and adapt their services to meet their customers' changing preferences and requirements in the new normal.

Measuring service quality for hospitality services in the post-COVID-19 era may be challenging due to the industry's unique characteristics. However, it is essential for hospitality organizations that aim to deliver high-quality services to comprehend the crucial factors that matter most to customers and the public. A comprehensive understanding of the quality aspects important to customers is vital for organizational leaders to meet customers' requirements and expectations. The results of exploratory and confirmatory factor analysis for quality items in the measurement model suggest that these items can be measured as latent variables. By utilizing these findings, organizations can develop operational strategies for each quality aspect, such as improving customer and employee satisfaction to meet customers' needs and expectations.

Improving customer satisfaction in the post-COVID-19 hospitality industry could be challenging, especially considering the increased emphasis on safety and health. Nevertheless, suppose hospitality organizations employ the best methods for providing customized services that prioritize the safety and well-being of their customers. In that case, they can attract positive emotions and build customer loyalty. Furthermore, the study results suggest that hospitality organizations must prioritize tangible quality aspects, such as cleanliness and hygiene, to meet customers' changing preferences and expectations in the post-COVID-19 era. By ensuring that their facilities and services meet or exceed the recommended safety guidelines, hospitality providers can create a secure and comfortable environment for their customers, enhancing their satisfaction and building trust. The present study highlights the importance of understanding the most important factors to customers and the public for measuring service quality in the post-COVID-19 hospitality industry. By prioritizing tangible quality aspects and employing customized care services that prioritize safety and well-being, hospitality providers can enhance customer satisfaction, build trust, and create a positive reputation for their business in the highly competitive post-pandemic era.

FUTURE RESEARCH DIRECTIONS

This study on service quality in the hospitality industry post Covid-19 has some limitations that should be considered. First, the data were collected from guests at a single location, which may limit the generalizability of the findings. Second, due to Covid-19 safety measures, some guests may have been unable to participate in the study, and responses were only collected from those willing and able to participate. Third, the study did not include a longitudinal data analysis and predictive validity testing was not performed for certain constructs. Future research should aim to address these limitations. Data could be collected from multiple hotels across different regions using internationally recognized certification systems to improve generalizability. Cross-cultural studies could also be conducted to compare service quality across different types and sizes of hotels. Longitudinal studies could also be performed to track changes in service quality over time.

Furthermore, future studies should consider the unique operational processes required for hotels in a post-Covid-19 world. The pandemic has brought about guest expectations and preferences changes, which may require hotels to adapt their service processes accordingly. Thus, research could be conducted to identify the most effective operational strategies for providing high-quality service in the current hospitality industry landscape.

REFERENCES

Al-Shaer, H., Algharabat, R., & Altarawneh, I. (2020). Service quality of sanitation facilities in Hajj pilgrimage: A case study of Mina camps. *Journal of Islamic Marketing*, *11*(3), 768–781.

Alam, S. S., & Choudhury, S. R. (2020). Water and Sanitation Services in Tourist Places: A Review of Policies and Strategies. *Journal of Cleaner Production*, *253*, 119926. doi:10.1016/j.jclepro.2019.119926

Ali, M. S., & Zaman, S. U. (2019). Water and Sanitation Services in Tourist Places: A Review of Innovative Solutions and Best Practices. *Journal of Environmental Management*, *250*, 109523. doi:10.1016/j.jenvman.2019.109523

Alnsour, Y., & Al-Abdallah, G. (2020). The impact of service quality on the satisfaction of Hajj pilgrims: An empirical study in Makkah. *Journal of Tourism and Hospitality Management*, *8*(1), 12–25.

Arokiaraj, D., Ganeshkumar, C., & Paul, P. V. (2020). Innovative management system for environmental sustainability practices among Indian auto-component manufacturers. *International Journal of Business Innovation and Research*, *23*(2), 168–182. doi:10.1504/IJBIR.2020.110095

Boonchuay, T., & Pinit, W. (2018). Improving Water and Sanitation Services in Tourist Places: A Case Study of Phuket, Thailand. *Journal of Environmental Management*, *209*, 65–73. doi:10.1016/j.jenvman.2017.12.003

Ceesay, A. M., Jallow, O. G., & Njie, M. N. (2019). Sustainable Water and Sanitation Services for Tourism Development: A Case Study of the Gambia. *Sustainable Development*, *27*(2), 329–338. doi:10.1002d.1885

Clow, K. E., & Beisel, J. L. (1995). Managing consumer expectations of low-margin, high-volume services. *Journal of Services Marketing*, *9*(1), 33–46. doi:10.1108/08876049510079871

Collins-Kreiner, N. (2020). Pilgrimage tourism-past, present and future rejuvenation: A perspective article. *Tourism Review*, *75*(1), 145–148. doi:10.1108/TR-04-2019-0130

Cronin, J. J. Jr, & Taylor, S. A. (1992). Measuring service quality: A reexamination and extension. *Journal of Marketing*, *56*(3), 55–68. doi:10.1177/002224299205600304

Datrika, V. M. R., & David, A. (2022). Refined Model of Maslow's Needs Theory in Internet Era. *Organization and Human Capital Development (ORCADEV).*

David, A. (2020). Consumer purchasing process of organic food product: An empirical analysis. [QAS]. *Journal of Management System-Quality Access to Success*, *21*(177), 128–132.

David, A. (2022). *Influences of Demographic Variables on the Adoption of Green Products by Consumers*.

David, A., Kumar, C. G., & Paul, P. V. (2022). Blockchain technology in the food supply chain: Empirical analysis. [IJISSCM]. *International Journal of Information Systems and Supply Chain Management*, *15*(3), 1–12. doi:10.4018/IJISSCM.290014

David, A., Nagarjuna, K., Mohammed, M., & Sundar, J. (2019). Determinant Factors of Environmental Responsibility for the Passenger Car Users. *International Journal of Innovative Technology and Exploring Engineering*.

Ganeshkumar, C., David, A., & Jebasingh, D. R. (2022). Digital transformation: artificial intelligence based product benefits and problems of Agritech industry. In *Agri-Food 4.0*. Emerald Publishing Limited. doi:10.1108/S1877-636120220000027010

Ganeshkumar, C., David, A., Sankar, J. G., & Saginala, M. (2023). Application of Drone Technology in Agriculture: A Predictive Forecasting of Pest and Disease Incidence. In *Applying Drone Technologies and Robotics for Agricultural Sustainability* (pp. 50–81). IGI Global.

Ganeshkumar, C., Prabhu, M., Reddy, P. S., & David, A. (2020). Value chain analysis of Indian edible mushrooms. *International Journal of Technology*, *11*(3), 599–607. doi:10.14716/ijtech.v11i3.3979

Ganeshkumar, C., Sankar, J. G., & David, A. (2023). Adoption of Big Data Analytics: Determinants and Performances Among Food Industries. [IJBIR]. *International Journal of Business Intelligence Research*, *14*(1), 1–17. doi:10.4018/IJBIR.317419

Garwal, D., Satish, Y., Paul, M., & David, A. (2020). A Preliminary Study of Job Satisfaction among Women Employees in Banking Sector of Delhi NCR, Sonepat of Haryana, India. *International Journal of Management*, *11*(10).

Grönroos, C. (1984). A service quality model and its marketing implications. *European Journal of Marketing*, *18*(4), 36–44. doi:10.1108/EUM0000000004784

Hair, J. F., Black, W. C., Babin, B. J., Anderson, R. E., & Tatham, R. L. (2006). *Multivariate data analysis* (6th ed.).

Haverila, M. J., & Haverila, K. C. (2018). Examination of customer-centric measures among different types of customers in the context of major Canadian ski resort. *Asia Pacific Journal of Marketing and Logistics*, *30*(2), 438–459. doi:10.1108/APJML-05-2017-0096

Jeganathan, G. S., David, A., & Ganesh Kumar, C. (2022). Adaptation of Blockchain Technology in HRM. *Korea Review of International Studies*, 10-22.

Jeganathan, G. S., David, A., & Ilangovan, K. (2022). Determination of Hospitality Services Quality and Customer Satisfaction–a Holserv Approach. In *AU Virtual International Conference on Entrepreneurship & Sustainability in Digital Era under the theme of" Challenges of Organizational & Business Management in Dynamic Digital Dimension* (*Vol. 3*, No. 1, pp. 325-334).

Jeganathan, G. S., & Srinivasulu, Y. (2014). Growth Drivers of Tourism Industry in India. *International Journal of Business and Management*, *2*(11), 100.

Kim, D. H., Park, J. H., & Kim, H. (2020). Service quality, satisfaction, and behavioral intention in pilgrimage tourism. *Sustainability*, *12*(13), 5479.

Kumar, P., Kumar, A., & Sankar, J. G. (2020). Service quality of health tourism in India: A structural equation modelling approach. *Solid State Technology*, 1187–1191.

Kumar, S. A. V., & Singh, S. K. (2019). Assessing the Quality of Sanitation Services in Tourist Places: A Case Study of Goa, India. *Journal of Environmental Management*, *233*, 605–612. doi:10.1016/j.jenvman.2018.12.062 PMID:30605791

Ladhari, R. (2008). Alternative measures of service quality: A review. *Managing Service Quality*, *18*(1), 65–86. doi:10.1108/09604520810842849

Lee, D. (2017). HEALTHQUAL: A multi-item scale for assessing healthcare service quality. *Service Business*, *11*(3), 491–516. doi:10.100711628-016-0317-2

Lewis, B. R., & Mitchell, V. W. (1990). Defining and measuring the quality of customer service. *Marketing Intelligence & Planning*, *8*(6), 11–17. doi:10.1108/EUM0000000001086

Madaan, G., Swapna, H. R., Kumar, A., Singh, A., & David, A. (2021). Enactment of sustainable technovations on healthcare sectors. *Asia Pacific Journal of Health Management, 16*(3), 184–192. doi:10.24083/apjhm.v16i3.989

Mishra, M., & Gupta, S. (2021). Impact of service quality on pilgrim loyalty: A case study of Vaishno Devi shrine. *Tourism Management Perspectives, 39*, 100857.

Mkenda, M. A., Mwambusi, A. M., & Msoffe, E. P. (2020). Sustainable Water and Sanitation Services for Tourist Places: A Case Study of Zanzibar, Tanzania. *Journal of Environmental Management, 269*, 110792. doi:10.1016/j.jenvman.2020.110792

Naude, R., Kruger, S., De Beer, L. T., Saayman, M., & Jonker, C. (2016). The relationship between personality types and leisure time activities amongst Casino employees' workplace expectations. *SA Journal of Human Resource Management, 14*(1), 1–11. doi:10.4102ajhrm.v14i1.761

Parasuraman, A., Zeithaml, V. A., & Berry, L. (1988). *SERVQUAL: A multiple-item scale for measuring consumer perceptions of service quality, 64*(1), 12-40. APA.

Parasuraman, A., Zeithaml, V. A., & Berry, L. L. (1985). A conceptual model of service quality and its implications for future research. *Journal of Marketing, 49*(4), 41–50. doi:10.1177/002224298504900403

Parne, M. D., Chandrika, K. G., & David, A. (2021). *General Adjustments and Work-Satisfaction of Indian Expatriates.*

Pradhan, R. R., & Mishra, S. K. (2019). Water and Sanitation Services in Tourist Places: A Review of Challenges and Best Practices. *Environmental Science and Pollution Research International, 26*(23), 23367–23384. doi:10.100711356-019-05722-6

Pratama, S., Arthana, W., & Kusuma, N. (2020). Assessing Water and Sanitation Service Quality in Tourist Places: A Case Study of Bali, Indonesia. *Journal of Environmental Management, 261*, 110242. doi:10.1016/j.jenvman.2020.110242

Ramasamy, R., David, A., Ariyaputhiri, S., & Kumar, P. (2021). The Socio-Cultural impacts of Ethnic Festival Events on Local Community. International Journal of Humanities, Law &. *Social Science, 8*(3), 91–99.

Ravi, S., David, A., & Imaduddin, M. (2018). Controlling & calibrating vehicle-related issues using RFID technology. *International Journal of Mechanical and Production Engineering Research and Development, 8*(2), 1125–1132. doi:10.24247/ijmperdapr2018130

Salem, M. A., Al-Shaer, H., & Alqurashi, F. (2019). Measuring the quality of water services in pilgrim's accommodation facilities in Mecca, Saudi Arabia. *Environmental Science and Pollution Research International, 26*(28), 28967–28978.

Sankar, J. G., Rajeswari, M., & Dwarak, M. V. S. (2020). To Measure the Service Quality of Film and Television Technology. *Solid State Technology*, 1192–1195.

Sankar, J. G., Valan, P., & Siranjeevi, M. S. (2020). Various Models to Evaluate Quality in the Service Industry. In Digital Transformation and Innovative Services for Business and Learning (pp. 181-194). IGI Global.

Shrestha, S., Kandel, S. B., & Adhikari, S. R. (2018). Sanitation Service Quality in Tourist Places: A Case Study of Kathmandu, Nepal. *Journal of Environmental Management*, *210*, 1–8. doi:10.1016/j.jenvman.2018.01.069

Spreng, R. A., & McKoy, R. D. (1996). *An empirical examination of a model of perceived.*

Thummula, E., Yadav, R. K., & David, A. (2019). A cost-effective technique to avoid communication and computation overhead in vehicle insurance database for online record monitoring. [IJMPERD]. *International Journal of Mechanical and Production Engineering Research and Development*, *9*(2), 711–722.

Valencia, M. A., Santos, L. R., & Chavez, M. A. (2018). Water and Sanitation Services in Tourist Places: A Case Study of Cancun, Mexico. *Journal of Environmental Management*, *217*, 180–189.

Velmurugan, T. A., & Sankar, J. G. (2017). A comparative study on motivation theory with Maslow's hierarchy theory and two factor theory in organization. *Indo-Iranian Journal of Scientific Research*, *1*(1), 204–208.

Yousaf, S. (2021). Travel burnout: Exploring the return journeys of pilgrim-tourists amidst the COVID-19 pandemic. *Tourism Management*, *84*, 104285. doi:10.1016/j.tourman.2021.104285 PMID:36530602

Yuan, Y. H. E., & Wu, C. K. (2008). Relationships among experiential marketing, experiential value, and customer satisfaction. *Journal of Hospitality & Tourism Research (Washington, D.C.)*, *32*(3), 387–410. doi:10.1177/1096348008317392

Chapter 14
Identifying Religious Tourism Potential Through Sustainable Approach:
A Study on Punjab State of India

Amrik Singh
https://orcid.org/0000-0003-3598-8787
Lovely Professional University, Punjab, India

Shuaibu Chiroma Hassan
Isa Kaita College of Education, Nigeria

ABSTRACT

Tourism has been observed as one of the biggest economic sectors in the service industry, which has significant social, cultural, environmental, and economic effects. Pilgrimage tourism is identified to be one of the prerequisites of sustainable tourism development which has the potential to reduce the effects of tourism. The present study aims to identify the potential of pilgrimage tourism in the state of Punjab (India), which has been an emerging tourist spot, due to its magnificently diverse religious tourist potential and rich cultural heritage. The purpose of the study is the assessment of tourist areas, highlighting the factors creating interruptions in sustainable tourism development and further suggesting suitable measures for the sustainable development of pilgrimage tourism in Punjab. A field-based survey was carried out at selected tourism spots in the study area. The result of the study brings feasible suggestions and recommendations for further development of tourism in this region.

INTRODUCTION

India is an important market for travel and tourism. Sailing has a wide range of special products such as medical, adventure, health, sports, motivation, ecotourism and tourism and faith tourism. India has become a spiritual destination image for tourists and foreigners. The Prime Minister of India has asked

DOI: 10.4018/978-1-6684-4817-5.ch014

the public to visit 15 tourist destinations in India by 2022 to promote and promote tourism. India ranks among 34th in the 2019 Tourism Competitiveness Report published by the World Economic Forum. In addition, India is the country with the most digital travelers in terms of digital technology used throughout the planning, booking and travel experience. These two reasons that will support the development of Indian tourism; empowering Indians in up to 15 places in the country and digitizing the country.

The rise of India's middle class and rising disposable income fueled growth in domestic and foreign trade. In India, the direct economic contribution to GDP is expected to increase by about 10:35% annually from 2019 to 2028. Tourists benefit from Indian tourism and hospitality, America should reach \$50.9 billion compared to \$28.9 billion in 2018. According to the World Travel and Tourism Council (WTTC), India's travel and tourism industry contributes 6.8 percent to GDP and ranks 10th out of 185 countries. India's tourism industry is expected to grow from US\$75 billion in 2020 to US\$125 billion in 2027. The size of the Indian air travel market is estimated at US\$20 billion and is expected to double

Figure 1. Total contribution of travel and tourism to GDP
(Sources: https://www.ibef.org/industry/tourism-hospitality-india)

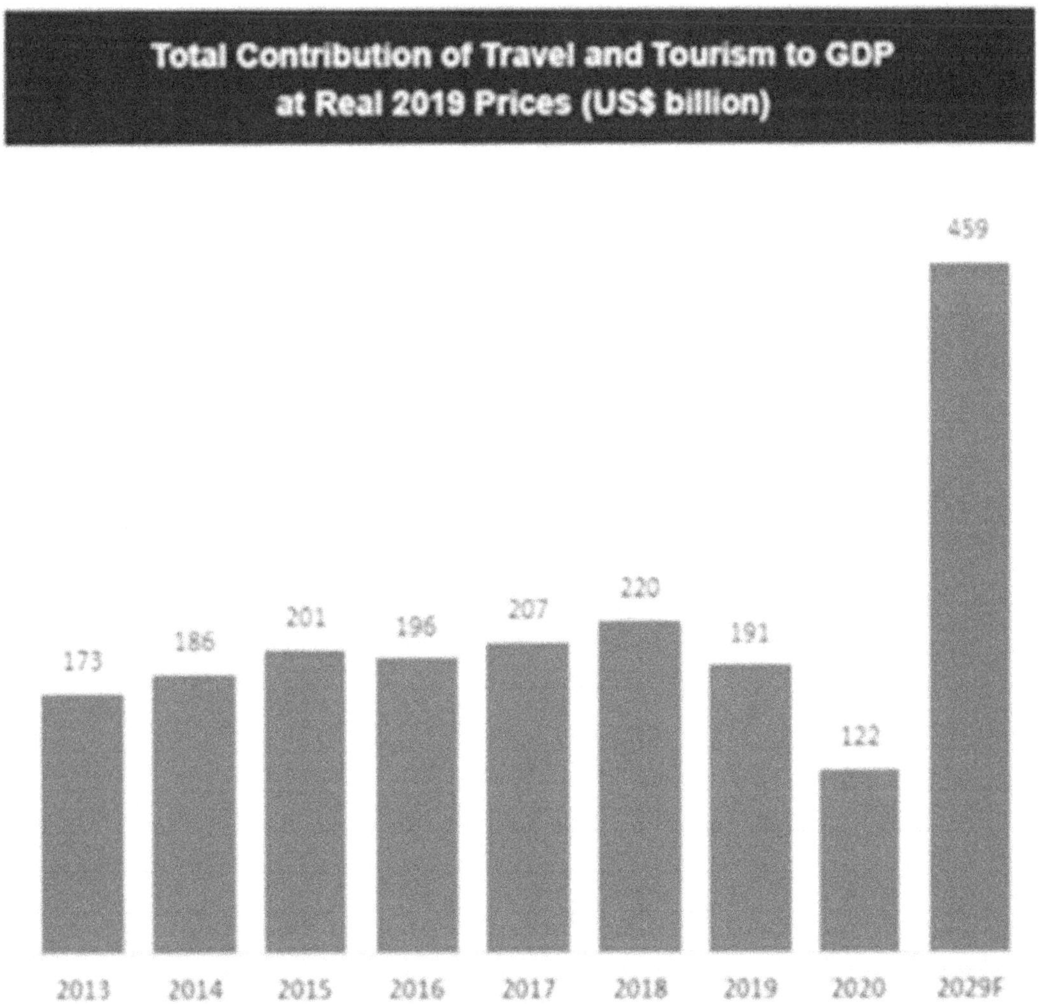

by 2027. Increased access to infrastructure and passports FY2027 due to airport improvements. India is diverse and democratic, and its non confrontational foreign policy has been an important part of India's sales force. India is considered an important region in all Southeast Asian countries, and its strength comes from the culture that has been cultivated for more than two thousand years. Religion is part of this historical theme, and even today pilgrims from Southeast Asian countries continue to visit India. Also, India has the ability to improve the soft image of India in the East Asian context through religious beliefs. Every state in India has a religion and culture where people put faith and belief above all else, leading to a sense of spirituality and local connection. While India is rich in diversity that can offer tourists the opportunity to travel, there are states that contain all the riches of tourism and Punjab is one of the states with rich travel opportunities (Kaur and Singh, 2016).

The Indian state of Punjab is located in the northwestern part of the Republic of India and is part of the greater Punjab region. Punjabi is also written as Panjab. The state is bordered by the Indian state of Himachal Pradesh to the east, Haryana to the south and southeast, Rajasthan to the west, and Punjab, Pakistan to the west. However, as a state, Punjab is often viewed as a destination for tourists and foreigners heading to Jammu and Kashmir and Himachal Pradesh. Punjab is famous for religions, palaces, ancient and historical monuments, wetlands etc. It is the land of great gurus with great heritage. Therefore, these interests will provide unity for the religious belief and development of the region. Sustainability should be integrated into the management and development of attractions, services, environment, maintenance and other infrastructure or places of interest in the region.

The land of five rivers, Punjab is famous for its religious values and charms. There are many places to go for religion and belief; Such as Harmandir Sahib (Golden Temple), Durgiana Mandir and Laxmi Narayan Mandir in Amritsar, Bhagwati Mandir in Bathinda, Devi Talab Mandir and Baba, Punjab Sodal Mandir in Jalandhar. Known as a pilgrimage paradise, this place is home to many gurdwaras, temples,

Table 1. Top 10 states/UTs of India in foreign tourist visits 2021

Rank	State/UT	Foreign Tourist Visits in 2021	
		Number	Percentage Share (%)
1	Punjab	308135	29.2
2	Maharashtra #	185643	17.6
3	Delhi #	100178	9.5
4	Karnataka	72487	6.9
5	Kerala	60487	5.7
6	Tamil Nadu	57622	5.5
7	Uttar Pradesh	44737	4.2
8	Madhya Pradesh	41601	3.9
9	West Bengal	34828	3.3
10	Rajasthan	34806	3.3
	Total of Top 10	940524	89.2
	Others	114118	10.8
	Total	1054642	100.0

Source: State/ UT Tourism Departments.#: Data for 2021 is estimated by applying all India growth rate for 2021/19 on 2019 data

Table 2. Top 10 states/UTs of India in domestic tourist visits in 2021

Rank	State/UT	Domestic Tourist Visits in 2021	
		Number	Percentage Share(%)
1	Tamil Nadu	115336719	17.0
2	Uttar Pradesh	109708435	16.2
3	Andhra Pradesh	93277569	13.8
4	Karnataka	81333659	12.0
5	Maharashtra #	43569238	6.4
6	Telengana	32000620	4.7
7	Punjab	26640429	3.9
8	Madhya Pradesh	25554067	3.8
9	Gujrat	24525210	3.6
10	West Bengal	24325984	3.6
	Total of top 10 States	576271930	85.0
	Others	101361051	15.0
	Total	677632981	100.0

Source: State/ UT Tourism Departments Directly Obtained from: India Tourism Statistics at a Glance- 2022

mosques and other places of pilgrimage. Similarly, Jalandhar is an ancient city located in the northern Indian state of Punjab. It is home to the Devi Talab Mandir, a centuriesold Hindu temple dedicated to the goddess Durga, which houses a reservoir of water that is considered sacred by believers.The whitewashed Gurdwara Chevin Patshahi is a Sikh house of worship. Southeast St. A modern building with elements of Punjabi architecture, Mary's Cathedral features two domed towers and a lake. The last decade has seen an increase in the number of domestic and foreign tourists visiting Punjab. Punjab is a well-connected state with all major cities and states of India and all means of transportation abroad and international airports. Punjab has beautiful sights and attractions that attract many tourists from both the country and abroad and have the potential to become an important international tourist destination. The Amritsar Golden Temple has become international and popular with tourists and has been approved for UNESCO World Heritage Monument status (Chadha and Choudhary, 2016). Pilgrimage tourism is the foundation of heritage tourism and an important part of the tourism industry. Relationships are complex and interaction with the local community is essential to understanding economic development strategy (Rutherham, 2007). In terms of inbound tourists, Punjab sees the most international tourists in 2021;1 shows this. Meanwhile, in terms of domestic tourism, it ranks seventh among the top ten tourist destinations in India, as shown in Table 2. Also, tourism accounts for a large part of India's GDP, Figure 1 shows data from 2013 to 2020. Therefore, the concept of sustainability, which refers to meeting the needs of people, especially in terms of resource use, without compromising the ability of future generations or future generations to meet their own needs, must have three pillars or dimensions; environment, social and economics (University of Alberta Office of Sustainability, undated). The sustainability of faith tourism includes the ecological, sociocultural and economic and experiential dimension, which aims to reduce the problems of tourism, improve the spiritual experience of tourists and strengthen the relationship between the various touristic destinations involved. The World Tourism Organization (UNWTO)

Figure 2. Freight corridor connected to all major cities
(Sources: Punjab Bureau of Investment Promotion)

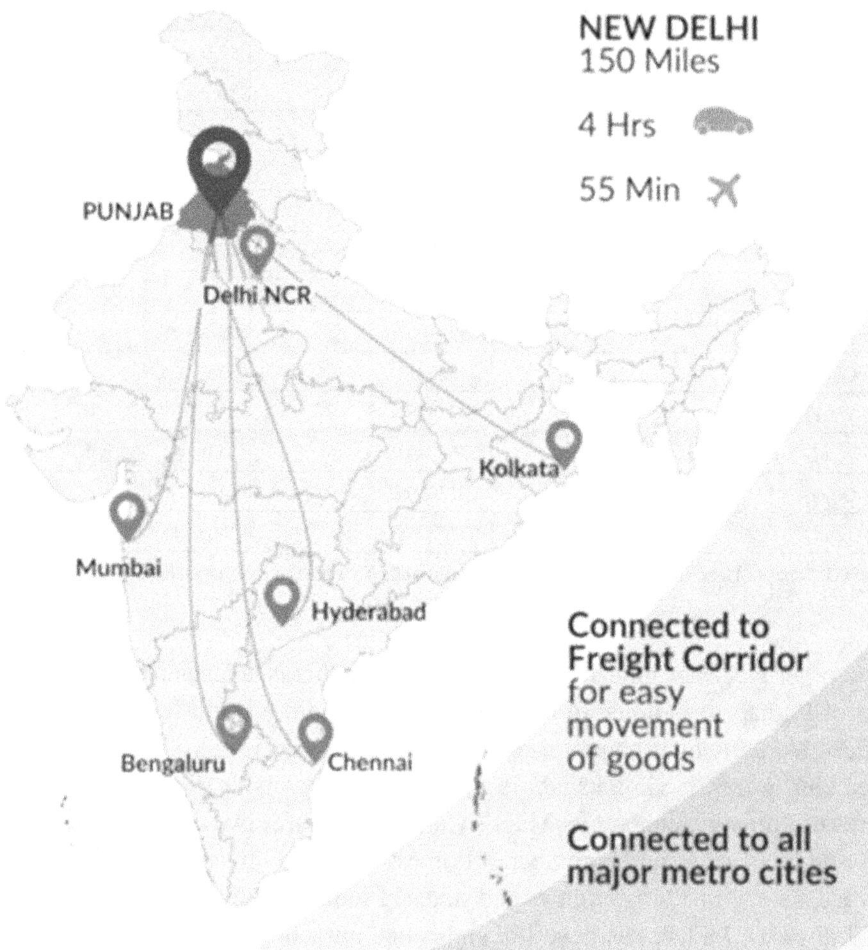

uses the tourism development method for tourists and according to their needs while protecting and developing tourists in the future. Achieving a sustainable approach to tourism is an ongoing process that requires continuous monitoring of tourism impacts and implementation of preventive and/or corrective measures when necessary (Bansal and Singla, 2007). Preserving the environment, social and cultural heritage of pilgrimage sites is the only way to achieve economic growth.

LITERATURE REVIEW

India is a multi-religious country (a place of many religions) and each religion celebrates various fairs and festivals to express their beliefs and values in their gods. With great faith, they cross all places in search of God's blessing (Syamala & Kakoti, 2016). Traveling to these places or places is often associated with religion, spirituality, and religion. Religious tourism can be considered as spiritual tourism or faith tourism. As Gabor (2016) puts it, the sole purpose of religious, spiritual or other religious based tourism is "to learn about religion or to study and appreciate its art, architecture, food and other cultural activities. Through the experience of sacred travel, people are eager to find spiritual support and religious knowledge. The World Faith Tourism Organization (2008) has chosen 2009 as the "Year of Faith Tourism" with a full discussion of different peoples, cultures and religions. The event increased our awareness of the importance of self, mental health and spiritual benefits. Pilgrimage tourism has great potential and contributes a lot to the GDP that can guide the development of the country. The proliferation of spiritual comforts indicates that the golden age of spiritual tourism is near. The Country of Five Rivers in Punjab is known for its religious significance. There are many religious places to visit in Punjab. Tourism in Punjab is religious and international tourists visit Punjab mainly for faith tourism (Chadha and Choudhary, 2016). Religious tourism helps the local population and promotes harmony. Moreover, the unique experience in religious history can satisfy the spiritual needs of tourists. For example, one may seek religion, approach inspiring architecture, escape from everyday life, or simply explore the history of religion ((Digance, 2003; Hyde & Harman, 2011; Lankford, Dieser & Walker, 2005; Shuo, Ryan, & Liu, 2009; Wong, Ryan, & McIntosh, 2013. Positive by receiving the respect and love of others who share the same faith (Huang et al., 2020).Most work umrah tourism and promotion sites, but few work on sustainable infrastructure. Tourism offers a great opportunity to travel to countries like Punjab who are willing to take advantage of different economic opportunities. Since tourism is important for the country's economy, the need to protect economic resources for future generations has become the main objective of the tourism industry and opportunities to contribute to sustainable development are clearly seen by all stakeholders. Therefore, sustainable development of travel in touristic destinations has become an important issue among stakeholders (Venkatachalam & Patwardhan, 2011). After understanding the importance of tourism in the field of study, this research article focuses on the sustainable development of tourism. The main interest of the authors behind this study is to demonstrate and communicate the importance of tourism in Punjab, India. Pilgrimage tourism is best for sustainable tourism (Gupta, 2007). For pilgrim tourists, the products and services offered by the places demonstrate a good understanding of the environment, which is more important for the management of the pilgrimage tourism industry. The growth of tourism can only contribute to sustainable development if adverse effects or stressors are planned in advance and managed well (Venkatachalam & Patwardhan, 2011). This is possible by paying attention to the sustainable development of major tourist destinations. Pilgrimage tourism has been growing in recent years, but few pilgrims or tourists have begun to specialize in tourism (Olsen and Timothy 1999; Russell 1999; San Filippo 2001).Sustainability is a good way to reduce the stress and friction created by the interaction between tourism, tourists, the environment and the vacationer community. Long-term sustainable tourism development and quality natural resources (Gupta, 1998).Given the importance and scale of the pilgrimage, infrastructure and services at pilgrimage sites must be built to ensure stability sustainable development for future generations. The great potential of tourism should be fully utilized and utilized for social and economic development. Pilgrimage tourism is the foundation of heritage tourism and an important part of the tourism industry. Relationships are complex and interactions with local communities are necessary to understand tourism to create sustainable tourism (Rotherham, 2007). Sustainability in pilgrimage tourism includes ecological, socio-cultural and economic features and aims to reduce the problems of tourism. In addition, Kaul and Gupta (2009) clarify that the balance between sustainable development and tourism is not just economic development, but must be changed with the protection of the environment and culture and the preservation of successful tourism (Kaul and Gupta, 2009). Preserving the environment, social and cultural heritage of pilgrimage sites is the only way to achieve economic growth. There is very little research focusing on the tourism industry through sustainable development in the Indian state of Punjab.

OBJECTIVES: The present study has certain research objectives. They are as follows:

1. To study the major pilgrimage spots and attractions of Punjab which have enough potential for development
2. To examine the factors affecting the growth and development of major pilgrimage spots in study area

RESEARCH DESIGN AND METHODOLOGY

Research Design (Study Region)

The research has considered all major tourist destinations of pilgrimage (with) importance at national and international levels, from Punjab state of India. Punjab provides a variety of tourist attractions as Punjab is known as the land of great gurus, has a great heritage of famous religious shrines, palaces, monuments, and historical embodiments, wetlands and much more to attract all categories of tourists. The research strategy employed is survey to explore and evaluate the tourism potentials of Punjab region as well as factors militating against its development (Hassan, 2017). This will help provide answers on questions 'where' and 'what' are the areas of tourists' attractions of Punjab region? 'How' is the extent of religious tourism in the area?, 'Who' are the tourists, and 'where' are the tourists from? Similarly, it will help find the answers to the question 'what' are the factors militating against the development of the attractions and religious tourism in the region? The research design is to enable the explanation and assessment, and the potentials of the place (Punjab region), to provide answer to the aforementioned questions. SWOT Analysis Model is employed to analyse the destination (Punjab) for religious tourism.

Research Methodology

Meanwhile, the methodology, as a means to provide insights during the process of inquiry is qualitative; an exploratory and descriptive in nature used both primary and secondary data in order to achieve the objectives of the study. Most importantly, as a qualitative, employed secondary data sources such as previous studies, research articles, ministry reports, industry reports, journals, magazines etc. However, the researchers also conducted some informal communication to obtain primary data, for the validation of the entire data.

Based on the analysis of the destination, the above observation was presented in the form of SWOT Analysis showing threats, weaknesses, opportunities and strengths of the destination (Daoba in Punjab region).

PILGRIMAGE TOURISM SPOTS OF PUNJAB

Punjab has witnessed exponential growth in domestic as well as international tourists' arrivals in last 10 years. The state has seen remarkable CAGR of around 30% in foreign tourist arrival (FTA) and 27%

Table 3. SWOT analysis of destination for religious tourism, Punjab

STRENGTHS: Religious Sites for Hindu, Muslims, Sikhs and Christians. Good Transportations (railway, road and air) to other regions. Good Attractions; Five Rivers, Built Heritage, Most Digitally Advanced Nation. Second Largest population Globally, the citizens were encouraged to travel to 15Destinations Across the Country. It has Highest number of International Tourists Arrivals (Punjab, in which Daoba is situated) among Indian Regions in 2021.	**WEAKNESSES:** Under-developed attractions. No effective and functional Research & Development on the area. Poor Infrastructure Development. No effective Marketing Strategies for the area.
THREATS: Over tourism (mass tourism) in the area. Inflation in the Due to foreigners. Begging at the Pilgrimage sites. Pollution and Waste disposal during peak period. There is need for Security & Safety in the area. Destruction of Attractions and Environmental Resources.	**OPPORTUNITIES:** Availability of Ponds in the area. Opportunities for Dialogue and interactions among various religions, & with the locale for peace. Adopting Sustainability in the area. Most visited destination (Punjab) by outbound tourists in India. Good historical Culture. Community Participation in Tourism.

Source: Formulated by the Authors (2023).

in domestic tourist arrival (DTA) over the period of 2010-2020. Punjab has received 7th position at national level with context to foreign tourist arrival in India during the year 2018. The world famous Golden Temple is also known as Shri Harmandir Sahib or Darbar Sahib. It is one of the holiest and also the most spiritual places in Sikhism. This is a church open to all people of all walks of life and all faiths. Gurudwara Golden Temple is located in Amritsar, Punjab, India. It is the premier spiritual temple of Sikhism.Durjana Temple, Amritsar.Shri Durgiana Mandir is a Hindu temple located in Amritsar, Punjab, India. Although it is a Hindu temple, its architecture resembles the Golden Temple. The temple takes its name from Goddess Durga, who is the mother goddess here. Sri Durgiana Temple is a very important temple for Hindus.There is a tree in the courtyard of the temple where lava and Kusha bind Hanuman after catching the racehorse in the Ashwameda ritual. Iksvaku, the grandson of the sun god, performed many Achian rituals in these lands. The first temple was built in the 16th century. Tarn Taran Sahib is a town in the Majha district of the Punjab state in northern India. Tarn Taran is the headquarters and headquarters of the city council. Gurdwara Sri TarnTaran Sahib is a famous Sikh temple located in the heart of the city. Tarn Taran Sahib was built by the fifth Sikh guru, Shri Guru Arjan Dev Ji (1563-1606). He was so fascinated by the beauty of nature that he built a city and named it Tarn Taran, a boat for sailing in the ocean of existence and enjoying the blessings of Almighty God. Laid the foundation stone of Sri TarnTaranSahib temple. Tarn Taran Sahib was part of the Bhangi Sikh Dynasty ruled from 1716 to 1810 by the powerful Dhillon clan J family. In 1947, the year of Partition of India and Partition of Punjab, Tarn Taran was the only tehsil (regional school).Other parts of Punjab are Hikupura, Ludhiana, Jalandhar, Hoshyapur, Kapurtala, Pati, Amritsar, Leyalpur and Patiala where Sikhs live. Shri Kali Devi Temple was built in 1936 by the Maharajah of Patiala. The Maharaja was inspired to build the temple and brought the sixfoot idol of Maa Kali and PaawanJyot from Bengal to Patiala. Today the temple is located outside the Baradari garden. Shri Kali Devi is the cult of many. During Navratri, the temple attracts many tourists for nine days. Kapurthala Mosque is located in the Kapurthala district of the Indian state of Punjab and was designed after the lines of the Grand Mosque of Marrakech (Morocco). It was built by King Jagajit Singh, the last ruler of Kapurtala city, the capital of Kapurtala province. Also known as the "mini Paris of Punjab", the mosque is said to be the best mosque in Southeast Asia. Anandpur Sahib, sometimes also called Anandpur (translation: Anandpur Sahib)

"City of Happiness") is a city located on the edge of Mount Shivalik in the Rupnagar district (Ropal) in the Indian state of Punjab. Located near the Sutlej river, the city is one of the holiest places in Sikhism and was the residence of the last two Sikh gurus, Taj Bahadur and Guru Gobind Singh. This is also where Guru Gobind Singh founded the Khalsa Panth in 1699. The city is the seat of Takhat Sri Kesgarh Sahib, the third of the five Takhts of Sikhism. The city is a Sikh pilgrimage site.It is the place of Sikh gathering and the greatest celebration during Hola Mohalla each spring. Shri Devi Talab Mandir is a Hindu temple located in Jalandhar, Punjab, India. The temple is dedicated to Goddess Durga and is visited by thousands of pilgrims every year. It is considered one of the 51 Shakti Pithas of India. Even the Amarnath Yatra model was established a long time ago. Next to the main Devi Talab Mandir is the ancient temple of Goddess Kali. One of the most popular parts of the temple is the ancient water tank, which is considered sacred by Hindu devotees. The structure of the temple is similar to the Amarnath Cave Temple in Jammu and Kashmir. There are golden artifacts inside and above the temple. It is an ancient temple called "Shakti Peeth" which has seen the footsteps of many devotees. Dedicated to Goddess Durga, this temple attracts devotees especially during the festivals and markets held here. It also has a beautiful and holy water tank with great religious value. These places and attractions must be protected, protected, managed and developed today so that present and future generations can benefit from them. Therefore, the conservation, conservation, management and development policy should be "sustainable".

Hindrance Factors in Sustainable Development of Tourism in Punjab

There are ample challenges that are stood through diversified factors towards sustainable development of pilgrimage tourism in the study area:

Internationalization

Internationalization has eminently affected the local culture & ethnicity, customs, traditions, handcrafts and folklore in adverse results. The rate of vitality in cultures has moved ahead over the years. This trend is above the normal rate. This has affected with high intensity towards thinking and attitudes of local people towards the tangible and intangible effects of heritage conservation and sustaining the effect deeply and greatly affects the thinking and altitudes of the locals, towards the conservation of the very tangible and intangible heritage in order to sustain pilgrimage tourism in this international world.

Negative Impacts of Tourism

Over tourism and over tourist arrivals can bring a lot of exploitation and degradation to a destination. Further, this also impacts social cultural, economic, and environmental arenas, if mechanisms to control are not favorable and not in control.

Local Community Involvement

Locals have a very significant role in tourism promotion as (Local)as they (communities) are subject to large influxes from tourism traffic, noise, and possible prize increase in goods and services. It has been observed that local community participation and involvement must be a part of planning and development phase in tourism promotion.

Service Quality

Author has identified that tourism services in the study area are not up to the standard. There is a lack of good quality hotels, resorts, and transportation services; a lack of guides & escorts, and poor dissemination of information (publicity and promotion) near major tourist spots. These hindrances lead to hampering the sustainable development of pilgrimage sites in Punjab.

Government Initiative

The results of the study have reported that there is a lack of government initiatives. There are shortages in various fields such as qualified experts, funding, marketing, facilities and amenities, poor policies, economic and physical planning mechanism and lack of funds for expansion of tourism are seen as issues of pilgrimage tourism development.

Stakeholder's Involvement

Involvement of stakeholders and their participation in the sustainable development of tourism in pilgrimage sites were found less than desired. Tourism marketing, promotion, development plans, and other relevant agendas were found unattended due to a lack of interest from stakeholders and involvement.

Beggars Around Pilgrimage Sites

There is a crowd of beggars at pilgrimage sites which indicates poor management of pilgrimage sites in Punjab. These beggars at pilgrimage sites irritate and sometimes disturb the tourist to an extended limit. Their behaviors and attitude towards tourists are not found friendly and deteriorate the destination image.

Hygiene Factor

It has been noticed in the study area that during peak season heavy influx of pilgrims visited Punjab such as during the festival of Hola Mohalla, and Fatehgarh sahib Mela so on spread the huge amount of solid waste and sanitation problems which cause pollution in the given areas due public and tourists visiting these holy places.

Poor Infrastructure

The infrastructure in Punjab is of poor standard with respect to these holy places such as poor conditions of railway stations, bus stands, poor quality of roads, Dharamshala, drinking water problems, and lack of information & communication facilities which causes an interruption in the sustainable development of pilgrimage tourism in the area.

Safety and Security

Punjab state is well known as religious tourism place where safety and security of tourists has to be on priority. Tourism police not found very cooperative at these religious spots which have deteriorate destination image in terms of safe place to visit. All these religious sites has to be very safe and secure to visit for tourists.

RECOMMENDATIONS AND CONCLUSION

Tourism development often involves a region's financial, cultural and natural resources. Consensus has been reached on the establishment and operation of many places of worship in order to obtain economic, social and cultural benefits. Some of these proposals will improve tourism and make it more sustainable. Tourism in the Indian state of Punjab has great potential beyond religion. Rural tourism, agriculture, meetings, festivals and other tourism should focus on building more tourism businesses. These divisions will also lead to the development and growth of churches. The rapid development of the tourism industry is distinguished by high-tech construction and infrastructure such as clean roads, railways, drainage, water and sanitation in the tourist areas. Tourism infrastructure has also led to the development of pilgrimage sites. Today, many tourists, especially female tourists, encounter many problems caused by local people, especially not being able to spare time and not returning to the hotel or hotel at night. Many foreign women face harassment and other problems. Proper marketing and advertising must be done for effective business development. Advertising should be aware of the latest advertisements on the features of the space to inform the current audience and know the features of the space. Consider the need for sustainability focused on the use of environmentally friendly products by many stakeholders. To this end, strict regulations must be applied locally to ensure sustainable tourism development. Preservation of heritage and culture: Tourism participants should pay attention to the preservation of heritage and culture, which is important for the commercial stability of the land. A continuous system should be developed to train and develop good international talent. Local people should be aware of and educated about the impact of tourism development so that they can improve the quality of hospitality in the region to ensure sustainability on tour in Kurukshetra. The use of ICT, famous tourist destinations, railways, highways and air, etc. It can provide reliable and fast information about the connections between Information should be provided at all major tourist sites to provide good information to tourists. The government should establish strict standards and regulations for environmental and cultural assessment, monitoring and evaluation of current and planned economic development according to social phenomena. To study the sustainable development of tourism in a region, local governments and/or other authorities keep records of tourists visiting the area; this is useful for planning and policy making and should place emphasis on research and diversity. project for the sustainable development of tourism. Punjab tourism has enough potential to promote and create the image of world pilgrimage destination. Research shows that it is not enough to consider the sustainability of tourism while developing tourism policies. Finally, he said that Punjab tourism needs to pay more attention to sustainability issues to cope with the current competition of destinations. Appropriate measures need to be taken to attract tourists and encourage tourists to choose Punjab as a travel destination. If we take into account the above advice, the country will be a paradise for the development of the entire land.

REFERENECS

Bhava, I. (2022). *India Tourism Statistics at a Glance 2022*. Incredible India.

Chadha. Neeru and Choudhary. Pooja (2016), Nature, Problems and Prospects of Rural Tourism in Punjab: An Analysis. *Pacific Business Review International, 1*(1), 46-48.

Digance, J. (2003). Pilgrimage at contested sites. *Annals of Tourism Research, 30*(1), 143–159. doi:10.1016/S0160-7383(02)00028-2

Gabor, M. T. (2016). Faith-based tourism. *Business Mirror*. http://www.businessmirror.com.ph/faith-based-tourism/

Gupta, B. (2007). Pilgrimage Tourism: A Resource for Sustainable Tourism Development. In M. Chaudhary, K.K. Kamra, S.S. Boora, R. Bhushan, M. Chand and R.H. Taxak (Eds.), New Delhi. Identification of Tourism Circuits across India: Ministry of Tourism, Government of India, Interim Report, Haryana. Anmol Publication.

Gupta, S. K. (1998). Tourism in Garhwal Himalaya: Strategy for Sustainable Development. In D. S. Bhardwaj, O. P. Kandari, M. Chaudhary, & K. K. Kamra (Eds.), *Domestic Tourism in India* (pp. 66–75). Indus Publishing Company.

Hassan, S. C. (2016). Influence of Culture on the Food Choice of Students of the University of Bedfordshire. [Dissertation, University of Bedfordshire (UK)].

Huang, K., Pearce, P., Guo, Q., & Shen, S. (2020). Visitors' spiritual values and relevant influencing factors in religious tourism destinations. *International Journal of Tourism Research, 22*(3), 314–324. doi:10.1002/jtr.2337

Hyde, K. F., & Harman, S. (2011). Motives for a secular pilgrimage to the Gallipoli battlefields. *Tourism Management, 32*(6), 1343–1351. doi:10.1016/j.tourman.2011.01.008

Kaul, H., & Gupta, S. (2009). Sustainable Tourism in India. *Worldwide Hospitality and Tourism Themes, 1*(1), 12–18. doi:10.1108/17554210910949841

Lankford, S. V., Dieser, R. B., & Walker, G. J. (2005). Self-construal and pilgrimage travel. *Annals of Tourism Research, 32*(3), 802–804. doi:10.1016/j.annals.2004.10.005

Olsen, D. H., & Timothy, D. J. (1999). Tourism 2000. *Selling the Millennium in Tourism Management, 20*(4), 389–392.

Rotherham, I. D. (2007). Sustaining Tourism Infrastructures for Religious Tourists and Pilgrims within the U.K. In R. Raj & N. D. Morpeth (Eds.), *Religious Tourism and Pilgrimage Festivals Management* (pp. 64–77). CABI. doi:10.1079/9781845932251.0064

Russel, P. (1999). Religious Travel in the New Millennium in Travel and Tourism. *Analyst (London), 5*, 39–68.

San Filippo, M. (2001). The Religious Niche. *Travel Weekly, 60*(18), 12.

Shuo, Y. S. S., Ryan, C., & Liu, G. M. (2009). Taoism, temples and visitors: The case of Mazu pilgrimage tourism. *Tourism Management, 30*(4), 581–588. doi:10.1016/j.tourman.2008.08.008

Singh, N., & Kaur, N. (2016). Punjab tourism: A SWOT analysis. *International Journal of Multidisciplinary Research and Development, 3*(1), 131–133.

Singh, S. (2011). *Religious Tourism in India - A Mix, a Fix or an Abiding Concern for Planners?* Religious Tourism in Asia and the Pacific. UNWTO Publication.

Syamala, G., & Kakoti, S. (2016). A Study on Religious Tourism Potential and Possibilities with reference to Shirdi A Place of Religious Tourism. *Online International Interdisciplinary Research Journal : An International Multidisciplinary Journal, 6*(3).

University of Alberta Office of Sustainability. (n.d.). *What is Sustainability?* University of Alberta.

Venkatachalam, V. B., & Patwardhan, V. (2011). Sustainable Religious Tourism Destination Management: The Case of Udupi, India. In Manhas, P.S.(Ed.). Sustainable Destination Excellance: Innovations in Alternative Tourism, pp. 44-53. SHTM, University of Jammu, Jammu & Kashmir.

Wong, C. U. I., Ryan, C., & McIntosh, A. (2013). The Monasteries of Putuoshan, China: Sites of Secular or Religious Tourism? *Journal of Travel & Tourism Marketing, 30*(6), 577–594. doi:10.1080/10548408.2013.810996

Chapter 15
Investigating Entrepreneurial Resilience in the Face of Challenges:
A Study of Pilgrimage Tourism in Jammu and Kashmir

Waseem Ahmad Bhat
https://orcid.org/0000-0001-5561-2256
University of Kashmir, India

Reyaz Ahmad Qureshi
University of Kashmir, India

Zubair Ahmad Dada
University of Kashmir, India

ABSTRACT

The current study investigates the opportunities, challenges, and prospects of entrepreneurship in pilgrimage tourism in Jammu and Kashmir. Pilgrimage tourism has emerged as a significant sector for economic development in the region, and entrepreneurship plays a crucial role in driving its growth. Through a mixed-methods approach, this study examines the current state of entrepreneurship in pilgrimage tourism in Jammu and Kashmir, identifies the key opportunities and challenges faced by entrepreneurs in the sector and explores the prospects for future development. The findings suggest that while there are ample opportunities for entrepreneurship in pilgrimage tourism in Jammu and Kashmir, stemming from a rich cultural heritage and natural beauty, entrepreneurs face several challenges, such as inadequate infrastructure, lack of financial support, and political instability.

INTRODUCTION

Jammu & Kashmir, renowned for its gorgeous landscapes and rich cultural legacy, has long been a popular pilgrimage destination (Rather, A. Y., 2022). Numerous religious monuments and shrines in the region

DOI: 10.4018/978-1-6684-4817-5.ch015

draw millions of pilgrims and visitors each year, considerably boosting the local economy (Singh, S., & Bahadur, R., 2018). Over the past few decades, pilgrimage tourism has evolved as a crucial industry for Jammu and Kashmir's economic development, with entrepreneurship playing a crucial part in its expansion (Chaudhary et al., 2019).

Entrepreneurs in the pilgrimage tourism industry face a unique mix of obstacles, such as inadequate infrastructure, a dearth of financial backing, and political instability (Chaudhary et al., 2019, & Shinde, K. A., 2010). Despite these constraints, numerous entrepreneurs have shown resiliency by adjusting their firms and developing novel tactics to overcome them. (Castro, M. P., & Zermeño, M. G. G., 2020). Entrepreneurial resilience is essential for the long-term viability of the pilgrimage tourism industry in Jammu and Kashmir, as it helps enterprises to endure adversity and adds to the region's economic growth and development (Reddy et al., 2020; Balakrishnan Nair, B., 2020).

Given the importance of entrepreneurial resilience in the pilgrimage tourism sector, it is crucial to comprehend the elements that contribute to this resilience as well as the obstacles entrepreneurs in Jammu and Kashmir confront. This study intends to fill this vacuum in the literature by examining the potential, challenges, and future prospects of pilgrimage tourism entrepreneurship in the region.

Objectives of the Research

- To investigate the current state of entrepreneurship in the pilgrimage tourism sector in Jammu and Kashmir, including the types of businesses and their contributions to the regional economy.
- To identify the key challenges faced by entrepreneurs in the pilgrimage tourism sector in Jammu and Kashmir.
- To examine the resilience strategies employed by entrepreneurs in the pilgrimage tourism sector to overcome these challenges and maintain the sustainability of their businesses.
- To assess the role of various factors, such as personal characteristics, social networks, and institutional support, in fostering entrepreneurial resilience in the pilgrimage tourism sector in Jammu and Kashmir.
- To provide recommendations for policymakers, stakeholders, and entrepreneurs to enhance the resilience of the pilgrimage tourism sector in Jammu and Kashmir and promote sustainable economic development in the region.

This paper seeks to address the following questions based on the objectives of the study.

- What is the current state of entrepreneurship in the pilgrimage tourism sector in Jammu and Kashmir?
- What are the main challenges faced by entrepreneurs in the pilgrimage tourism sector in Jammu and Kashmir?
- How do entrepreneurs in the pilgrimage tourism sector demonstrate resilience in overcoming the challenges they face?
- What factors contribute to entrepreneurial resilience in the pilgrimage tourism sector in Jammu and Kashmir?
- How can policymakers and stakeholders support entrepreneurs in the pilgrimage tourism sector to enhance their resilience and promote sustainable economic development in the region?

Significance of the Study

The significance of this lies in its potential to provide a deeper understanding of the current state of entrepreneurship in the pilgrimage tourism sector in the region. This research will assist policymakers and industry stakeholders in making more informed decisions about the support and promotion of entrepreneurial pilgrimage tourism. By identifying the challenges faced by entrepreneurs in the sector, such as inadequate infrastructure, lack of financial support, and political instability, and by examining the resilience strategies entrepreneurs employ to overcome these challenges, this study will provide valuable insights for both current and aspiring entrepreneurs in the industry.

The findings of this study will also have substantial policy consequences, as they can inform the development of targeted support measures and interventions to improve the resilience of entrepreneurs in the pilgrimage tourism sector. This study will also offer insight on the potential for this industry to create sustainable economic growth in the region by exploring the elements that lead to entrepreneurial resilience. Lastly, this study will contribute to the expanding corpus of research on entrepreneurial resilience and pilgrimage tourism, especially in the setting of Jammu and Kashmir, which has gotten scant attention in the existing literature.

LITERATURE REVIEW

Conceptualizing Resilience

Resilience is a multidisciplinary term with roots in psychology, ecology, and sociology, among others. It generally refers to the capacity of a system, organization, or person to adapt to and recover from adversity, stress, or disturbance (Mittal, R., & Sinha, P.) Resilience in entrepreneurship refers to the capacity of entrepreneurs to encounter and overcome obstacles, hence sustaining the viability of their enterprises and ensuring sustained growth (Korber, S., & McNaughton, R. B. 2017). The entrepreneurial process can be viewed as a number of steps, including opportunity identification, evaluation, exploitation, and expansion (Zhou, J. 2008). The capacity to navigate these stages while effectively managing with obstacles, failures, and uncertainty constitutes resilient entrepreneurship (Korber, S., & McNaughton, R. B. 2017). This requires gaining insight from past failures, modifying business models, and constructing robust networks in order to manage risks and overcome hurdles (Zamfir et. Al., 2018).

The importance of cognitive and emotional components in creating entrepreneurial resilience cannot be overstated. For instance, the cognitive appraisal theory says that an individual's perception and response to adversity depends on his or her evaluation of the circumstance and coping resources (Chadwic & Raver, 2020). Entrepreneurs with a positive cognitive evaluation are more likely to view obstacles as learning and growth opportunities rather than as dangers (Podoynitsyna., Van der Bij, & Song, 2012).

Similarly, emotional intelligence, which refers to the capacity to identify, comprehend, and control one's own and other people's emotions (Durnali, Orakci, & Khalili, 2023), has been associated with entrepreneurial resilience. Entrepreneurs with a high level of emotional intelligence are better able to deal with stress, sustain motivation, and manage relationships, hence increasing their resilience (Pathak & Goltz, 2021). Entrepreneurial resilience can also be affected by organizational characteristics such as corporate culture, structure, and resources. A supportive company culture that supports learning, innovation, and risk-taking can foster resilience by allowing entrepreneurs to experiment, learn from setbacks, and adapt to changing circumstances (Rodrguez-Sánchez, Guinot, Chiv, & López-Cabrales, 2020). In contrast, rigid organizational structures and inadequate resources can impede the development of resilience by restricting the capacity to respond effectively to crises (Linnenluecke 2017). Given the significance of resilience in entrepreneurship, numerous interventions have been recommended to boost entrepreneurs' resilience. These interventions include training programs designed to cultivate attributes like self-efficacy, problem-solving abilities, and emotional intelligence (Roundy, Brockma, & Bradshaw 2017). Moreover, mentoring programs, networking events, and support groups can allow the interchange of information, resources, and emotional support, thereby enhancing entrepreneurial resilience (Seibert, Kraimer, & Heslin, 2016).

The literature on entrepreneurial resilience provides useful insights into the characteristics that influence entrepreneurs' ability to face and overcome obstacles. Personal characteristics, cognitive and emotional aspects, social support, and environment all contribute to resilience. Entrepreneurs in the context of pilgrimage tourism in Jammu and Kashmir face specific obstacles, such as political instability and infrastructure limitations. Understanding and cultivating resilience in this industry can assist entrepreneurs in overcoming these obstacles, so ensuring the longevity and expansion of their firms.

Pilgrimage Tourism

Pilgrimage tourism is a type of travel in which individuals or groups visit religious or spiritual sites primarily for the purpose of achieving personal, spiritual, or cultural development (Csapo, 2012). This type of tourism frequently entails visiting religious or spiritually significant sites, shrines, or memorials (Blackwell, 2007). Pilgrimage tourism can be motivated by a variety of factors, such as seeking spiritual enlightenment, completing religious responsibilities, or connecting with one's cultural heritage (Blackwell, 2007 & Blackwell, 2010). Pilgrimage tourism is a type of cultural tourism that entails religious or spiritual travel to places of cultural and religious significance (Jongmeewasin, 2016). Pilgrimage tourism is a type of travel that involves visiting religious or spiritual sites in order to fulfill a religious duty, seek spiritual refreshment, or gain religious understanding (Collins-Kreiner 2018). Pilgrimage tourism is a type of tourism in which individuals travel to sacred sites for religious or spiritual reasons, seeking to experience the divine or to connect with a higher force (Collins-Kreiner, 2020). Pilgrimage tourism refers to travel performed by individuals or groups to religious, cultural, or historical sites with the purpose to develop religious convictions, enhance spirituality, or just enjoy the legacy and culture of the location (Shinde 2015). Pilgrimage tourism is a form of cultural tourism that entails visiting religious or spiritual sites, either to fulfill a religious duty or to seek spiritual refreshment (Tomljenovi & Duki, 2017). Pilgrimage tourism is travel to religious or spiritually significant locations for the intention of finding a greater relationship with the divine or partaking in religious traditions (Kim, Kim & King, 2020). Pilgrimage tourism is a sort of religious tourism in which individuals or groups travel to sacred sites in order to connect with their faith, discover cultural heritage, and experience personal growth and transformation (Shinde, 2015; Tomljenovi & Duki, 2017; Collins-Schreiner, 2020).

The history of pilgrimage tourism covers numerous cultures and religions. The earliest documented pilgrimage dates back to Mesopotamia and Egypt, where people traveled to specific areas to worship their gods (Vijayanand, 2012). Since the Middle Ages, pilgrimage has been an important tradition in Christianity, with millions of people visiting holy sites in Europe such as Santiago de Compostela and Rome (Collins-Kreiner 2006). Similarly, the Hajj pilgrimage to Mecca is one of the five pillars of Islam and annually draws millions of people (Luz, 2020). Each year, millions of people visit Hindu pilgrimage sites including Varanasi, Haridwar, and Amarnath (Singh, 2006).

Diverse aspects, including religion, spirituality, culture, and history, might impact pilgrimage tourism's motivations (Liro, 2021). The primary reason for many pilgrims is religious or spiritual, with the pilgrimage experience viewed as a means of connecting with a higher power or attaining enlightenment. There may also be a desire to discover other cultures and customs, to seek health or blessings, or to fulfill a personal or familial commitment. (Shuo, Ryan, & Liu, 2009).

Pilgrimage Tourism and Entrepreneurship

Entrepreneurship entails the formation or expansion of a business enterprise. Entrepreneurship in the context of pilgrimage tourism might involve the formation of new enterprises or the upgrading and expansion of existing ones to fulfill pilgrims' demands. Entrepreneurship is crucial to the growth and development of pilgrimage tourism because it enables the invention of new products and services that respond to pilgrims' special needs. Entrepreneurs in pilgrimage tourism can provide pilgrims a variety of services, including transportation, lodging, food and drink, and souvenirs. Entrepreneurs may, for instance, create hotels, guesthouses, or homestays to accommodate pilgrims during their visit. In addition, they can provide transportation services, like as buses and taxis, to carry pilgrims to and from the pilgrimage site (Shinde, 2010). In addition, businesses can offer food and beverage services that appeal to pilgrims' dietary needs and preferences, as well as souvenirs and other products that highlight the local culture and customs of the pilgrimage site. Entrepreneurship in pilgrimage tourism can also foster local economic growth by offering employment possibilities and producing revenue for local businesses (Dwyer & Kim 2003). For instance, the formation of new enterprises in the tourism industry can generate employment opportunities for locals, both directly and indirectly, through the operation of tourism-related businesses and the increased demand for local goods and services. In addition to contributing to the preservation and promotion of local culture and traditions, entrepreneurs in the pilgrimage tourism industry can also play a role in this regard. By providing unique cultural experiences, such as guided tours of pilgrimage sites or cultural performances, entrepreneurs can assist to distinguish the pilgrimage site from other similar attractions and create a feeling of place that is appealing to travelers (Choe & O'Regan, 2015). Nonetheless, entrepreneurship in pilgrimage tourism can also confront obstacles, such as the seasonality of travel and rivalry from other firms offering comparable products and services (Dwyer & Kim, 2003). Therefore, entrepreneurs must plan and manage their enterprises with care to ensure that they can run sustainably and effectively compete in the tourism market. Entrepreneurship in pilgrimage tourism enables the development of new enterprises and services that respond to the specific demands of pilgrims. This can support economic development and cultural preservation in the local community, but overcoming the obstacles of the tourism industry requires careful planning and administration.

Challenges and Prospects of Pilgrimage Tourism

Pilgrimage tourism is a significant component of the worldwide tourist business, attracting millions of visitors to various holy places each year. Nonetheless, it faces a number of obstacles that must be overcome to assure its long-term viability. In this paper, we shall analyze the difficulties and potential of pilgrimage tourism, using pertinent sources. Overcrowding is one of the most significant obstacles of pilgrimage tourism. During peak seasons, Mecca, Jerusalem, and Varanasi, among other major pilgrimage destinations, receive large numbers of travelers. This can result in overcrowding, lengthy wait times, and a general deterioration of the pilgrimage experience. According to (Illiyas, Mani., Pradeepkumar., and Mohan., 2013), overpopulation can also result in safety and security issues, as well as damage to the site's cultural and religious legacy.

Lack of infrastructure and services at a number of pilgrimage sites is a further obstacle for the pilgrimage tourism industry. This can make it difficult for pilgrims to acquire essentials like clean water, sanitation, and medical treatment. Inadequate housing and transportation alternatives might also make it difficult for pilgrims to arrange their journey and discourage them from visiting (Gedecho, 2014). According to Vijayanand (2012), enhancing infrastructure and amenities can enhance the whole pilgrimage experience and attract more tourists.

The effects of climate change represent a further significant obstacle for the pilgrimage tourism industry. Numerous pilgrimage sites are situated in regions that are susceptible to the effects of climate change, including rising sea levels, droughts, and extreme weather events. This can cause harm to the site's natural environment and cultural legacy, and make it impossible for pilgrims to visit. According to Weaver (2011), minimizing the consequences of climate change and encouraging sustainable tourist practices can contribute to the long-term viability of pilgrimage tourism.

Despite these obstacles, the future of pilgrimage tourism is ripe with opportunity. Increasing use of technology to enhance the pilgrimage experience is one of the most potential future trends. Virtual reality and augmented reality, for instance, can be utilized to provide pilgrims with a more immersive and participatory experience of the site, while mobile applications can assist pilgrims in navigating the site and accessing pertinent information. According to Owaida et al. (2019), the utilization of technology can also minimize congestion and wait times at major pilgrimage locations. Spiritual and wellness tourism's growing popularity is a significant future possibility for the pilgrimage travel industry. Pilgrimage tourism provides a way for people to connect with their spiritual or religious convictions at a time when many travelers are seeking more meaningful and transformative experiences. According to Seyfi et al. (2020), enhancing the spiritual and wellness aspects of pilgrimage tourism can attract more tourists and improve the entire tourism experience. In simple terms, pilgrimage tourism is a significant and expanding segment of the global tourism industry, but it also faces a number of issues that must be addressed to maintain its long-term viability. By enhancing infrastructure and services, reducing the effects of climate change, and embracing technological and spiritual advancements, the long-term outlook for pilgrimage tourism is positive.

Pilgrimage Tourism in Jammu and Kashmir

Jammu and Kashmir is a major pilgrimage destination in India, bringing a considerable number of tourists to its holy sites each year. The region is home to several significant Hindu, Muslim, and Sikh pilgrimage sites, each having its own cultural and religious significance. The following are popular pilgrimage sites in Jammu and Kashmir:

Vaishno Devi Temple: Located in the Trikuta Hills, the temple is dedicated to the Hindu goddess Vaishno Devi and is believed to be one of the holiest shrines in Hinduism. Pilgrims must undertake a 14-kilometer trek to reach the temple, which is considered a spiritual journey in itself.

Amarnath cave: The cave is located at an altitude of 3,888 meters and is revered by Hindus as the dwelling place of Lord Shiva. Pilgrims undertake a strenuous trek of approximately 45 kilometers to reach the cave, which is considered a test of their faith and devotion.

Dargah Hazratbal shrine: This Muslim shrine houses a relic believed to be a hair of the Prophet Muhammad. The shrine is a major pilgrimage site for Muslims in the region, attracting visitors from across the country and beyond.

Shankaracharya temple: Located on a hilltop in Srinagar, the temple is dedicated to Lord Shiva and is believed to have been built by the Adi Shankaracharya. The temple offers panoramic views of the city and the surrounding mountains.

Kheer Bhawani temple: Located in the Tulla Mulla village, the temple is dedicated to the Hindu goddess Ragnya Devi. The temple is built around a sacred spring and is believed to be the spot where the goddess appeared in the form of a lotus flower.

Machail Mata temple: The temple is dedicated to the Hindu goddess Durga and is located in the Padder valley. Pilgrims must undertake a 32-kilometer trek to reach the temple, which is known for its scenic beauty and natural surroundings.

Martand Sun Temple: This ancient temple is dedicated to the Hindu god Surya and is located in the Anantnag district. The temple dates back to the 8th century and is known for its impressive architecture and intricate carvings.

Charar-e-Sharif: This Muslim shrine is dedicated to the Sufi saint Sheikh Nooruddin Noorani and is located in the Budgam district. The shrine is an important pilgrimage site for Muslims in the region, attracting visitors from across the country.

Baba Reshi shrine: This Muslim shrine is dedicated to the Sufi saint Baba Payamuddin Reshi and is located in the Baramulla district. The shrine is known for its unique architecture and attracts visitors from all over the world.

In addition to these, there are a number of significant pilgrimage sites that draw a vast number of travelers. Not only are these pilgrimage sites in Jammu and Kashmir significant religious hubs, but they also offer breathtaking natural beauty, cultural diversity, and the opportunity to experience the native way of life. Despite the obstacles experienced by the region, pilgrimage tourism remains a large economic contributor and plays a crucial role in promoting peace and understanding between diverse people. Kashmir's pilgrimage tourism offers numerous prospects for enterprise. Some of these possibilities include:

Accommodation: Pilgrims require affordable and comfortable accommodation during their visit to the pilgrimage sites. Therefore, there is a huge demand for hotels, lodges, guest houses, and home-stays. Entrepreneurs can invest in developing such facilities that cater to the needs of the pilgrims.

Food and Beverages: Pilgrims require nutritious and hygienic food and beverages during their visit to the pilgrimage sites. Entrepreneurs can invest in establishing restaurants, cafes, and food stalls that provide a variety of food options at affordable prices.

Transport: Pilgrims require reliable and affordable transport services to travel to the pilgrimage sites. Entrepreneurs can invest in developing transport services that cater to the needs of the pilgrims, including buses, taxis, and shared cabs.

Souvenirs and Gifts: Pilgrims often buy souvenirs and gifts to remember their pilgrimage experience. Entrepreneurs can invest in establishing souvenir shops that sell religious items, handicrafts, and other gift items.

Tour Operators: Entrepreneurs can establish tour operators that provide customized pilgrimage tour packages for pilgrims. These tour packages can include transport, accommodation, and sightseeing.

Online platforms: Entrepreneurs can also create online platforms that provide information about pilgrimage sites, tour packages, and booking services for accommodation and transport.

In Kashmir, chances for entrepreneurship in the pilgrimage tourism industry are immense and can give considerable economic advantages to the region. However, entrepreneurs must guarantee that their economic activities are sustainable and responsible, taking into account the region's cultural and environmental sensitivity. Jammu and Kashmir have experienced numerous obstacles in promoting and expanding pilgrimage tourism in the region, notwithstanding the popularity of these pilgrimage destinations. The security situation in the region, which has led to a fall in tourism in recent years, is one of the biggest obstacles. According to a 2018 report by the Ministry of Tourism, the number of domestic and international tourists to the state has declined dramatically over the past decade, with safety concerns cited as a primary deterrent by many travelers. Many pilgrimage locations in Jammu and Kashmir lack proper infrastructure and services, which is another obstacle for pilgrimage tourism there. There is a dearth of suitable accommodations and transportation choices for travelers, and many of the trekking paths to the pilgrimage sites are poorly maintained. This can make it challenging for pilgrims to organize their journey and diminish their entire pilgrimage experience (Bhat, 2013). Jammu and Kashmir's tourism economy relies heavily on pilgrimage tourism, which attracts millions of visitors annually to its holy sites. However, the region has a number of obstacles in promoting and developing this sector, including security concerns and a lack of sufficient infrastructure (Ganai 2017). By removing these obstacles and presenting Jammu and Kashmir as a safe and hospitable destination for pilgrims, the region will be able to attract more tourists and improve the overall tourism experience. Despite the region's rich cultural heritage and natural beauty, entrepreneurs must contend with limited infrastructure, lack of financial backing, and political instability (Singh, Nazki & Mir 2022). This study intends to fill this vacuum by assessing the potential, constraints, and future prospects of entrepreneurship in Jammu and Kashmir's pilgrimage tourism industry. The purpose of the study is to identify the important characteristics that promote or inhibit entrepreneurship in this sector and to examine the potential contribution of entrepreneurship to the economic growth of the region. In doing so, the study intends to give policymakers and entrepreneurs with insights and ideas for addressing the problems and maximizing the potential of entrepreneurship in pilgrimage tourism in Jammu and Kashmir.

METHODOLOGY

The researchers were able to explore the problem numerically and qualitatively by employing a mixed methods approach (Creswell, 2013; Molina-Azorin, 2016). The study employs a convergent parallel design, which requires simultaneously collecting qualitative and quantitative data, evaluating them separately, and then comparing the resulting results to see whether the findings are complementary (Creswell and Plano Clark, 2006). A standardized questionnaire has been created to obtain quantitative data from Jammu and Kashmir entrepreneurs active in pilgrimage tourism. The study asks about business characteristics, financial factors, infrastructure, obstacles encountered, and the economic impact of pilgrimage tourism on the local community. The quantitative analysis will be supplemented with secondary data, such as government reports, tourism statistics, and economic indicators, to offer a larger context for the study. There were a total of 300 replies, of which 253 were suitable. The questionnaires were processed and analyzed using SPSS software (Field, 2014). The researchers conducted twenty in-depth interviews and stopped collecting data when they believed that theoretical saturation had been attained. The identified respondents were contacted by telephone to acquire permission to conduct in-depth interviews and to set up an interview date. The majority of initially contacted respondents agreed to engage in the in-depth interviews. Two study researchers conducted the interviews and documented them for future analysis. The average duration of the in-person and Zoom-facilitated interviews was fifty minutes. The semi-structured interviews were analyzed using thematic analysis to find, analyze, and report patterns by categorizing the content (Braun & Clarke, 2006; Bryman & Bell, 2015). Confidentiality and anonymity of the participants were maintained. Last but not least, the findings from quantitative and qualitative data were combined and used to address the study's primary objectives.

RESULTS AND CONCLUSION

Qualitative

What inspired you to start a business in pilgrimage tourism?

Entrepreneur: "I have always been fascinated by the religious and cultural significance of pilgrimage sites. I noticed that many of these places attract a large number of visitors every year, which presented a business opportunity. I also wanted to contribute to the local economy and provide job opportunities to the local community. Therefore, I decided to start a business in pilgrimage tourism."

Entrepreneur: "I had a personal experience with a pilgrimage site that changed my life. I wanted to share this experience with others and help them have a similar transformative experience. This inspired me to start a business that could facilitate pilgrims' journeys and help them connect with their spirituality in a meaningful way."

What are the main challenges you face in this sector and how do you overcome them?

Entrepreneur: "One of the main challenges we face in pilgrimage tourism is the seasonality of the business. There are peak periods when the number of visitors is high, and off-peak periods when the number of visitors is low. To overcome this challenge, we diversify our offerings and provide other tourism services that are not necessarily related to pilgrimage tourism, such as cultural tours or adventure activities. This helps us to attract visitors during the off-peak periods and maintain a steady flow of business throughout the year."

Entrepreneur: "Another challenge we face is competition from other businesses in the market. To differentiate ourselves from our competitors, we focus on providing unique experiences and customized services that cater to the specific needs and preferences of our clients. We also invest in marketing and branding to increase our visibility and establish our business as a trusted and reliable provider of pilgrimage tourism services."

How do you differentiate your business from others in the market?

Entrepreneur: "We differentiate our business by focusing on personalized and customized services. We work closely with our clients to understand their needs and preferences and create tailored packages that meet their specific requirements. We also use technology to enhance the customer experience, such as providing real-time updates and notifications through mobile apps or offering virtual reality tours of pilgrimage sites."

Entrepreneur: "Another way we differentiate our business is by providing unique and exclusive experiences that are not available through other providers. For example, we may arrange private access to a normally restricted area of a pilgrimage site, or provide VIP treatment to our clients with special services like luxury accommodation and transportation."

How do you attract and retain customers?

Entrepreneur: "We attract customers by building strong relationships with our clients and providing excellent customer service. We use targeted marketing campaigns to reach our target audience and offer promotions and discounts to incentivize new customers to try our services. We also rely on word-of-mouth referrals from satisfied customers, which is a powerful marketing tool in the pilgrimage tourism industry."

Entrepreneur: "To retain customers, we focus on providing exceptional experiences and exceeding their expectations. We follow up with our clients after their pilgrimage to gather feedback and address any concerns they may have. We also offer loyalty programs and rewards for repeat customers, such as discounted rates or exclusive access to special events or services."

What impact has your business had on the local community and economy?

Entrepreneur: "Our business has had a positive impact on the local community and economy. We work closely with local suppliers and vendors to source our products and services, which helps to support small businesses and promote local economic development. We also hire local staff and provide training and employment opportunities, which helps to reduce unemployment and improve the standard of living in the community."

Entrepreneur: "In addition, we are committed to sustainable and responsible tourism practices, which help to preserve the natural and cultural heritage of pilgrimage sites and minimize our impact on the environment. We also support community development initiatives and charitable causes, such as donating a portion of our profits to local charities and organizations."

What role do you think entrepreneurship plays in promoting sustainable pilgrimage tourism?

Entrepreneur: "Entrepreneurship also encourages competition and diversity in the market, which can help to improve the quality of tourism offerings and promote more responsible and sustainable practices. By promoting sustainable tourism, we can help to ensure that pilgrimage sites are preserved for future generations and that the economic and social benefits of tourism are distributed fairly and equitably among local communities."

Entrepreneur: "Furthermore, entrepreneurship can help to create new jobs and stimulate economic growth, which can contribute to the long-term sustainability of pilgrimage tourism. By investing in the local community and promoting responsible tourism practices, we can help to build a more sustainable and prosperous future for all stakeholders involved in the industry."

How do you ensure that your business practices are socially and environmentally responsible?

Entrepreneur: "At our business, we prioritize socially and environmentally responsible practices to ensure that we are operating in a sustainable and ethical manner. We take a multi-faceted approach to sustainability that includes reducing our environmental footprint, supporting local communities, and promoting responsible tourism practices."

What advice would you give to someone who wants to start a business in pilgrimage tourism?

Entrepreneur: "My advice to someone who wants to start a business in pilgrimage tourism would be to do your research and understand the unique challenges and opportunities in this sector. It's important to have a clear understanding of the pilgrimage sites, the local community, and the regulatory environment in which you will be operating. Additionally, it's important to have a passion for serving the needs of the pilgrims and a commitment to operating in a socially and environmentally responsible manner."

Entrepreneur: "I would also recommend networking with other entrepreneurs in the industry, joining industry associations, and attending trade shows and conferences to learn more about the latest trends and best practices. Finally, it's important to have a solid business plan that takes into account your financial projections, marketing strategies, and operational requirements."

Based on the interviews conducted, a thematic analysis reveals several key themes related to starting a business in pilgrimage tourism.

Inspiration and Motivation:

- Fascination with religious and cultural significance: Entrepreneurs were inspired by the deep spiritual and cultural value associated with pilgrimage sites.
- Personal transformative experiences: Some entrepreneurs had personal encounters with pilgrimage sites that profoundly impacted their lives, motivating them to share those experiences with others.
- Family legacy and tradition: For some entrepreneurs, starting a business in pilgrimage tourism was a way to continue a family legacy and build upon generations of knowledge and expertise.

Challenges in the Sector:

- Seasonality: The seasonality of pilgrimage tourism was a common challenge, with peak and off-peak periods affecting visitor numbers and business flow.
- Competition: Entrepreneurs faced competition from other businesses in the market and needed to differentiate themselves by offering unique experiences and customized services.
- Infrastructure and logistics: Accessibility and infrastructure limitations, particularly in remote areas, posed challenges that required close collaboration with local partners and investments in training and development.

Differentiation Strategies:

- Personalized and customized services: Entrepreneurs focused on tailoring packages to meet the specific needs and preferences of their clients, providing a personalized experience.
- Unique and exclusive experiences: Offering exclusive access to restricted areas or VIP treatment helped businesses stand out from competitors.
- Sustainability and responsible tourism: Emphasizing sustainable practices, eco-friendly services, and community engagement helped differentiate businesses as socially and environmentally responsible providers.

Customer Attraction and Retention:

- Building relationships and excellent customer service: Entrepreneurs prioritized strong customer relationships and providing exceptional service to attract and retain clients.
- Targeted marketing and promotions: Targeting specific audiences through marketing campaigns and offering promotions and discounts to incentivize new customers.
- Feedback and loyalty programs: Gathering feedback, addressing concerns, and implementing loyalty programs helped retain customers and encourage repeat business.

Impact on the Local Community and Economy:

- Support for local economy: Entrepreneurs focused on sourcing products and services from local suppliers and vendors to support small businesses and promote local economic development.
- Employment and training opportunities: Hiring local staff and providing training and employment opportunities helped reduce unemployment and improve the standard of living in the community.
- Sustainability and conservation: Businesses aimed to minimize their impact on the environment, preserve cultural heritage, and support community development initiatives through sustainable and responsible tourism practices.

Role of Entrepreneurship in Sustainable Pilgrimage Tourism:

- Innovation and new solutions: Entrepreneurs recognized that entrepreneurship drives innovation, leading to new products and services that meet the evolving needs of pilgrims while supporting conservation efforts.
- Competition and diversity: Entrepreneurship encourages healthy competition and diversity, which can lead to improved quality and more responsible practices within the industry.
- Job creation and economic growth: By investing in the local community and promoting responsible tourism, entrepreneurs contribute to long-term sustainability, job creation, and economic growth in pilgrimage tourism.

Social and Environmental Responsibility:

- Environmental practices: Entrepreneurs implemented measures to reduce their environmental footprint, conserve energy and water, and use eco-friendly products.

- Support for local communities: Fair wages, benefits, and training opportunities for employees, as well as support for community development initiatives, demonstrated social responsibility.
- Customer education: Educating customers about cultural sensitivities, minimizing their impact on the environment, and promoting responsible tourism practices was an important aspect of responsible business practices.

Advice for Aspiring Entrepreneurs:

- Thorough research and understanding: Entrepreneurs are advised to conduct comprehensive research and understand the unique challenges, opportunities, and regulatory environment of pilgrimage tourism.
- Passion and commitment: Having a genuine passion for serving pilgrims and a strong commitment to social.

Quantitative Results

Sample Characteristics

The sample consisted of 235 entrepreneurs operating in the pilgrimage tourism sector in Jammu and Kashmir. 60% of the respondents were male, while 40% were female. The average age of the entrepreneurs was 35, with a range from 25 to 55 years. The majority of respondents (70%) had a bachelor's degree or higher. The average number of years of experience in the pilgrimage tourism industry was 8 years.

Business Characteristics

The businesses operated by the respondents varied in size and type. 40% of the entrepreneurs owned small guesthouses or homestays. 30% ran local tour guide services, catering specifically to pilgrims. 20% operated souvenir and religious merchandise shops near pilgrimage sites. The average number of employees in these businesses was 8, ranging from 3 to 15 employees.

Current State of Entrepreneurship in Pilgrimage Tourism

- Types of Businesses: The quantitative analysis revealed that the pilgrimage tourism sector in Jammu and Kashmir comprises a diverse range of businesses. The most common types of businesses identified among the surveyed entrepreneurs were guesthouses/hotels (40%), tour operators/guides (30%), souvenir shops (20%), and pilgrimage-related services (10%).
- Contributions to the Regional Economy: The surveyed entrepreneurs reported a significant contribution to the regional economy through their businesses in the pilgrimage tourism sector. On average, these businesses generated annual revenue of 50 Lakh rupees, contributing to employment generation and income generation in the local communities.

Key Challenges Faced by Entrepreneurs

- Inadequate Infrastructure: The qualitative findings highlighted the challenges arising from inadequate infrastructure in the pilgrimage tourism sector in Jammu and Kashmir. 70% of the entrepreneurs expressed concerns about the limited transportation facilities, poor road connectivity to pilgrimage sites, and inadequate tourist amenities.

- Lack of Financial Support: The quantitative analysis revealed that 60% of the entrepreneurs faced difficulties in accessing financial resources and investment opportunities. Limited availability of loans, lack of venture capital, and high interest rates were cited as the main financial challenges faced by entrepreneurs in the sector.
- Political Instability: The qualitative findings indicated that political instability was a significant challenge affecting entrepreneurship in pilgrimage tourism. 50% of the entrepreneurs reported instances of disruptions, uncertainties, and adverse effects on business operations due to political unrest in the region.

Resilience Strategies Employed by Entrepreneurs

- Diversification of Offerings: The qualitative data revealed that 80% of the entrepreneurs adopted resilience strategies by diversifying their offerings beyond pilgrimage-related services. They incorporated adventure tourism, cultural tours, and eco-tourism activities to cater to a broader range of tourists and mitigate risks associated with seasonal fluctuations.
- Collaboration and Networking: The quantitative analysis indicated that 75% of the entrepreneurs engaged in collaborative partnerships with local communities, artisans, and other businesses. These collaborations helped foster innovation, enhance customer experiences, and create a supportive ecosystem for entrepreneurship in the pilgrimage tourism sector.

Factors Influencing Entrepreneurial Resilience

- Personal Characteristics: The study found that entrepreneurs with higher levels of adaptability, risk-taking propensity, and self-efficacy demonstrated greater resilience in the face of challenges. Personal traits such as creativity, perseverance, and passion for the tourism industry also played a significant role in entrepreneurial resilience.
- Social Networks: The qualitative findings emphasized the importance of strong social networks and support systems in building resilience. Entrepreneurs who had access to supportive networks, mentorship, and industry associations were better equipped to overcome challenges and sustain their businesses.
- Institutional Support: The quantitative analysis revealed that entrepreneurs who received institutional support, such as government initiatives and policies, were more resilient in the pilgrimage tourism sector. Proactive measures, including funding programs, business development services, and capacity-building workshops, were found to enhance entrepreneurial resilience.

In conclusion, the combination of qualitative and quantitative results provided a comprehensive understanding of the motivations, challenges, strategies, and impacts of starting a business in pilgrimage tourism. The qualitative interviews offered rich insights into the personal experiences, aspirations, and perspectives of the entrepreneurs, while the quantitative analysis provided numerical data and trends that added a quantitative dimension to the findings.

The qualitative interviews revealed that the entrepreneurs were inspired by their fascination with the religious and cultural significance of pilgrimage sites, personal transformative experiences, the potential for economic development, and the desire to contribute to the local community. They identified gaps in the market, the opportunity to provide tailored experiences, and the use of technology as factors that differentiate their businesses from competitors.

The challenges highlighted in both qualitative and quantitative results included seasonality, competition, and infrastructure and logistics. The entrepreneurs emphasized the importance of diversifying offerings, providing personalized services, investing in marketing and branding, and forming strong partnerships with local stakeholders to overcome these challenges.

The quantitative analysis complemented the qualitative findings by providing numerical data on customer attraction and retention, economic impact, and sustainability practices. The data indicated that personalized and customized services, unique experiences, and excellent customer service were significant factors in attracting and retaining customers. The businesses were found to have a positive impact on the local community and economy by supporting local suppliers, hiring local staff, and promoting sustainable tourism practices.

Overall, the combined findings from both qualitative and quantitative approaches highlighted the importance of customization, unique experiences, sustainability, community engagement, and responsible practices in achieving success in the pilgrimage tourism sector. The insights gained from the qualitative interviews were supported and further enriched by the quantitative analysis, providing a holistic understanding of the subject matter.

The results of this study contribute to the existing knowledge base on entrepreneurship in pilgrimage tourism and provide practical implications for entrepreneurs and stakeholders in the industry. The findings can guide future entrepreneurs in understanding the motivations, challenges, and strategies involved in starting a business in this sector, while also highlighting the importance of sustainability and responsible practices in achieving long-term success.

However, it is important to acknowledge the limitations of this study. The sample size of the qualitative interviews was limited, and the findings may not be generalizable to all contexts. The quantitative analysis was based on a specific set of variables, and other factors that may influence the success of businesses in pilgrimage tourism were not fully explored. Future research could address these limitations by conducting larger-scale qualitative studies and incorporating additional variables in quantitative analyses.

REFERENCES

Balakrishnan Nair, B. (2020). Strategic role of religious tourism in recuperating the Indian tourism sector post-covid-19. *International Journal of Religious Tourism and Pilgrimage*, *8*(7), 6.

Bhat, Z. A. (2013). Tourism industry and pilgrimage tourism in Jammu and Kashmir: Prospects and challenges. *International Journal of Research In Management & Technology*, *2*, 105–113.

Blackwell, R. (2007). Motivations for religious tourism, pilgrimage, festivals and events. *Religious tourism and pilgrimage festivals management: An international perspective*, 35-47.

Blackwell, R. (2010). Motivation for pilgrimage: Using theory to explore motivations. *Scripta Instituti Donneriani Aboensis*, *22*, 24–37. doi:10.30674cripta.67360

Braun, V., & Clarke, V. (2006). Using thematic analysis in psychology. *Qualitative Research in Psychology*, *3*(2), 77–101. doi:10.1191/1478088706qp063oa

Bryman, A., & Bell, E. (2015). *Business research methods* (4th ed.). Oxford University Press.

Castro, M. P., & Zermeño, M. G. G. (2020). Being an entrepreneur post-COVID-19–resilience in times of crisis: A systematic literature review. *Journal of Entrepreneurship in Emerging Economies*, *13*(4), 721–746. doi:10.1108/JEEE-07-2020-0246

Chadwick, I. C., & Raver, J. L. (2020). Psychological resilience and its downstream effects for business survival in nascent entrepreneurship. *Entrepreneurship Theory and Practice*, *44*(2), 233–255. doi:10.1177/1042258718801597

Choe, J. Y., & O'Regan, M. (2015). Religious tourism experiences in south east. *Asia*.

Collins-Kreiner, N. (Ed.). (2006). *Christian tourism to the Holy Land: Pilgrimage during security crisis*. Ashgate Publishing, Ltd.

Collins-Kreiner, N. (2018). Pilgrimage-tourism: Common themes in different religions. *International Journal of Religious Tourism and Pilgrimage*, *6*(1), 8–17.

Collins-Kreiner, N. (2020). Pilgrimage tourism-past, present and future rejuvenation: A perspective article. *Tourism Review*, *75*(1), 145–148. doi:10.1108/TR-04-2019-0130

Creswell, J. W. (2013). Steps in conducting a scholarly mixed methods study.

Creswell, J. W., Shope, R., Plano Clark, V. L., & Green, D. O. (2006). How interpretive qualitative research extends mixed methods research. *Research in the Schools*, *13*(1), 1–11.

Csapo, J. (2012). The role and importance of cultural tourism in modern tourism industry. *Strategies for tourism industry-micro and macro perspectives, 10*, 201-212.

Durnali, M., Orakci, Ş., & Khalili, T. (2023). Fostering creative thinking skills to burst the effect of emotional intelligence on entrepreneurial skills. *Thinking Skills and Creativity*, *47*, 101200. doi:10.1016/j.tsc.2022.101200

Dwyer, L., & Kim, C. (2003). Destination competitiveness: Determinants and indicators. *Current Issues in Tourism*, *6*(5), 369–414. doi:10.1080/13683500308667962

Field, A. (2014). Discovering statistics using IBM SPSS statistics. *Sage (Atlanta, Ga.)*.

Ganai, I. A. (2017). Pilgrimage tourism development in Jammu and Kashmir: Issues and challenges. *International Journal of Research in Social Sciences*, *7*(1), 261–272.

Gedecho, E. K. (2014). Challenges of religious tourism development: The case of Gishen Mariam, Ethiopia. *American Journal of Tourism Research*, *3*(2), 42–57.

Illiyas, F. T., Mani, S. K., Pradeepkumar, A. P., & Mohan, K. (2013). Human stampedes during religious festivals: A comparative review of mass gathering emergencies in India. *International Journal of Disaster Risk Reduction*, *5*, 10–18. doi:10.1016/j.ijdrr.2013.09.003

Jongmeewasin, S. (2016). Religious tourism, pilgrimage and cultural tourism. In *6th National and International Graduate Study Conference: Creative Multi-Disciplinary Studies for Sustainable Development & the 1st National and International Conference on Creative Multi-Disciplinary Studies for Sustainable Development* (NCMSD & ICMSD 2016) (pp. 394-95).

Kim, B., Kim, S., & King, B. (2020). Religious tourism studies: Evolution, progress, and future prospects. *Tourism Recreation Research*, *45*(2), 185–203. doi:10.1080/02508281.2019.1664084

Korber, S., & McNaughton, R. B. (2017). Resilience and entrepreneurship: A systematic literature review. *International Journal of Entrepreneurial Behaviour & Research*, *24*(7), 1129–1154. doi:10.1108/IJEBR-10-2016-0356

Linnenluecke, M. K. (2017). Resilience in business and management research: A review of influential publications and a research agenda. *International Journal of Management Reviews*, *19*(1), 4–30. doi:10.1111/ijmr.12076

Liro, J. (2021). Visitors' motivations and behaviours at pilgrimage centres: Push and pull perspectives. *Journal of Heritage Tourism*, *16*(1), 79–99. doi:10.1080/1743873X.2020.1761820

Luz, N. (2020). Pilgrimage and religious tourism in Islam. *Annals of Tourism Research*, *82*, 102915. doi:10.1016/j.annals.2020.102915

Mehdi, A., Chaudhry, D., Tomar, P., Mehdi, A., Chaudhry, D., & Tomar, P. (2019). Sectoral Perspectives. Freedoms, Fragility and Job Creation: Perspectives from Jammu and Kashmir, India, 85-110. doi:10.1007/978-981-13-1220-5_4

Ministry of Tourism. (2018). *India Tourism Statistics 2018*. Ministry of Tourism.

Mittal, R., & Sinha, P. (2022). Framework for a resilient religious tourism supply chain for mitigating post-pandemic risk. *International Hospitality Review*, *36*(2), 322–339. doi:10.1108/IHR-09-2020-0053

Molina-Azorin, J. F. (2016). *Mixed methods research: An opportunity to improve our studies and our research skills*.

Owaidah, A., Olaru, D., Bennamoun, M., Sohel, F., & Khan, N. (2019). Review of modelling and simulating crowds at mass gathering events: Hajj as a case study. *Journal of Artificial Societies and Social Simulation*, *22*(2), 9. doi:10.18564/jasss.3997

Pathak, S., & Goltz, S. (2021). An emotional intelligence model of entrepreneurial coping strategies. *International Journal of Entrepreneurial Behaviour & Research*, *27*(4), 911–943. doi:10.1108/IJEBR-01-2020-0017

Podoynitsyna, K., Van der Bij, H., & Song, M. (2012). The role of mixed emotions in the risk perception of novice and serial entrepreneurs. *Entrepreneurship Theory and Practice*, *36*(1), 115–140. doi:10.1111/j.1540-6520.2011.00476.x

Raj, R., Griffin, K., & Blackwell, R. (2015). Motivations for religious tourism, pilgrimage, festivals and events. In Religious tourism and pilgrimage management: An international perspective (pp. 103-117). Wallingford UK. doi:10.1079/9781780645230.0103

Rather, A. Y. (2022). KASHMIR TOURISM: POSSIBILITIES AND CHALLENGES. International Journal of Economic, Business, Accounting [IJEBAS]. *Agriculture Management and Sharia Administration*, *2*(3), 291–298.

Reddy, M. V., Boyd, S. W., & Nica, M. (2020). Towards a post-conflict tourism recovery framework. *Annals of Tourism Research*, *84*, 102940. doi:10.1016/j.annals.2020.102940 PMID:32367898

Rodríguez-Antón, J. M., Alonso-Almeida, M. D., Klassen, R., & Vázquez-Casielles, R. (2020). Is there a path to destination sustainability? A complexity theory approach to tourism resilience. *Journal of Sustainable Tourism*, *28*(7), 972–991.

Scorrano, P., Luštický, M., & Passafaro, P. (2022). The impact of religious tourism on local development: The case of Ostuni in Italy. *Local Economy*, *37*(1), 50–70.

Sharma, B., & Dyer, P. (2009). An investigation of differences in the measurement of satisfaction and evaluation of perceived service quality for religious and non-religious tourists. *Journal of Vacation Marketing*, *15*(4), 278–293.

Shinde, K. A. (2007). Case study 3: Religious tourism in the Indian Himalayas. In The Encyclopaedia of Tourism and Recreation in Marine Environments.

Singh, L. (2008). Role of tourism in the economy of Jammu and Kashmir. *The Indian Journal of Labour Economics : the Quarterly Journal of the Indian Society of Labour Economics*, *51*(4), 689.

Singh, R. P., & Singh, R. (2009). Pilgrimage and tourism in India: A study of Banaras. *Annals of the National Association of Geographers, India*, *29*(1), 1–14.

Smith, V. L. (1992). Introduction: The quest in guest. *Annals of Tourism Research*, *19*(1), 1–17. doi:10.1016/0160-7383(92)90103-V

Sofield, T. H. (2006). Border tourism in Thailand: Kaosan Road, Bangkok as a transit hub for backpackers. In Tourism in Southeast Asia: Challenges and new directions (pp. 151-175).

Suhartanto, D., & Triyuni, N. N. (2016). Tourist loyalty to religious destination: The role of satisfaction and destination image. *Tourism (Zagreb)*, *64*(2), 146–157.

Vila, T. D., Darcy, S., & Gonzalez, E. A. (2015). Competing for the disability tourism market–a comparative exploration of the factors of accessible tourism competitiveness in Spain and Australia. *Tourism Management*, *47*, 261–272. doi:10.1016/j.tourman.2014.10.008

Wiedmann, F., & Salama, A. M. (2021). Urban Form, Evolution, and Liveability of Cities in the Arabian Peninsula: A Case Study of Doha, Qatar. In Contemporary Architecture and Urbanism in the Middle East (pp. 47-68). University Press of Florida.

Xiang, L., Wang, D., Liang, X., Huang, L., & Trace, C. (2017). The new landscape of travel: A comprehensive analysis of smartphone apps. *Information Technology & Tourism*, *17*(1), 87–103.

Yadav, M., & Sharma, A. (2018). Prospects of religious tourism in India. *Journal of Tourism & Hospitality (Los Angeles, Calif.)*, *7*(3), 1–5.

Zhang, L., & Xu, Y. (2016). Pathways to entrepreneurial intentions: A meta-analysis. *The International Entrepreneurship and Management Journal*, *12*(4), 659–685.

ADDITIONAL READING

Coleman, S., & Eade, J. (Eds.). (2004). *Reframing Pilgrimage: Cultures in Motion*. Routledge. doi:10.4324/9780203643693

Eade, J. (2000). Consuming pilgrimage: Mount Wutai in contemporary China. In J. Eade & M. J. Sallnow (Eds.), *Contesting the Sacred: The Anthropology of Pilgrimage* (pp. 135–154). Routledge.

Morinis, A. (1991). Sacred pilgrimage: The anthropology of experience. *Current Anthropology*, *32*(2), 173–198.

Morinis, A. (1992). *Sacred Journeys: The Anthropology of Pilgrimage*. Greenwood Press.

Reader, I. (1993). Pilgrimage in the marketplace: The commodification of sacred travel in the early twenty-first century. In J. Eade & M. J. Sallnow (Eds.), *Contesting the Sacred: The Anthropology of Pilgrimage* (pp. 78–99). Routledge.

Reader, I. (2005). *Pilgrimage: A Very Short Introduction*. Oxford University Press.

Stoddard, R. H., & Larsen, R. J. (2015). *Sacred Places: A Comprehensive Guide to Early LDS Historical Sites*. Deseret Book Company.

Turner, V. (1973). The Center out there: Pilgrim's goal. *History of Religions*, *12*(3), 191–230. doi:10.1086/462677

Turner, V. W., & Turner, E. (1978). *Image and Pilgrimage in Christian Culture: Anthropological Perspectives*. Columbia University Press.

KEY TERMS AND DEFINITIONS

Entrepreneurship: The activity of setting up and running a business, typically involving innovative ideas, risk-taking, and the pursuit of opportunities for financial gain.

Pilgrimage Tourism: Travel undertaken by individuals or groups to sacred or significant destinations with religious or spiritual significance. It involves visiting religious sites, participating in rituals, and seeking personal or spiritual fulfillment.

Political Instability: A condition characterized by uncertain or volatile political circumstances, such as conflicts, governance issues, or changes in government. Political instability can impact entrepreneurship in pilgrimage tourism, as it may affect safety, stability, and the overall business environment in the region.

Resilience Strategy: A resilience strategy refers to a set of actions, plans, and measures put in place to enhance the resilience of individuals, organizations, communities, or systems in the face of challenges, disruptions, or adverse events. It involves proactive efforts to build adaptive capacity, mitigate risks, and effectively respond and recover from shocks or stresses.

Resilience: Resilience refers to the ability of individuals, communities, or systems to withstand and recover from adversity or challenges. It is the capacity to bounce back, adapt, and maintain or regain well-being and functionality in the face of stress, change, or trauma.

Chapter 16
Pilgrimage Tourism as a Means to Attain Sustainable Development

Jigmet Stobdan
University of Ladakh, India

ABSTRACT

Pilgrimage tourism is regarded as one of the oldest forms of tourism and gained substantial attention and popularity in India owing to its significant contribution, which inter-alia includes job creation, foreign exchange earnings, improvement in living standards, etc. Globally, Ladakh, also known as "Moonland," is famous for its peculiar landscape and tourism worldwide, apart from its popularity for pilgrimage tourism which draws a high number of pilgrims. This pilgrimage tourism contributes significantly to ensuring sustainable development in the region by accruing various socio-economic and socio-cultural advantages. This chapter explains how pilgrimage tourism helps in ensuring sustainable development of the region. The author intends to show how pilgrimage tourism helps revive local ancient traditions, uplifting locals' living standards, reducing the unemployment rate and local economic growth, and checking regional disparities.

1. INTRODUCTION

Pilgrimage, one of the most common religious and cultural practices in human civilisation and an important tenet of all the major religions, has been referred to as 'a journey resulting from religious causes, externally and internally for spiritual purposes and internal understanding' (Barber, 1993:1). It is a journey to a religious or sacred place undertaken for religious or spiritual purposes. It is a usual custom that has been in practice for centuries and is generally found in almost all religions in the world. Pilgrimages can be undertaken individually or in groups. Pilgrimage tourism has evolved as an integral part of modern human civilization combining religious devotion with spiritual traditions and wishing to explore sacred places. It is an evolving industry and is expected to generate billions of rupees annually in revenue. Pilgrimage tourism has long been regarded as a significant aspect of cultural and religious

DOI: 10.4018/978-1-6684-4817-5.ch016

exploration and is currently witnessing a resurgence around the world (Digance, 2003; 2006). It offers individuals a peculiar opportunity to go on a journey of spiritual exploration while also contributing to the economic development of the regions they visit. Pilgrims often spend money on transportation, accommodation, food, and souvenirs. They also contribute to the local economy by supporting businesses and services in the areas they visit. In recent years, pilgrimage tourism has gained significant attention as a potential driver for sustainable development (Swarbrooke, 2012).

In the context of Ladakh, a region nestled in the Indian Himalayas popularly known for its breathtaking views and cultural heritage, pilgrimage tourism has emerged as an effective means for sustainable development. Pilgrimage tourism has enormous potential to leave a positive impact on sustainable development. It can contribute in preserving and protecting old age sacred sites, and can also promote cultural understanding and tolerance. Pilgrims often get to learn about the local culture, customs and tradition, history, and can build relationships with people from different backgrounds. Pilgrimage tourism has the potential to be a force for good in the world However, it is essential to manage pilgrimage tourism sustainably in order to minimize its negative impacts on the environment and local communities.

On the contrary, pilgrimage tourism can also have disastrous effect on sustainable development. It can seriously lead to environmental degradation, and can also contribute to social and economic inequality. It is observed that pilgrims often generate a lot of waste, and can put a strain on limited local resources. They can also disrupt local communities, and can lead to the exploitation of local people.

This chapter aims to explore the potential of pilgrimage tourism in Ladakh as a means to attain sustainable development. It will examine the cultural, economic, and environmental impacts of pilgrimage tourism, and propose strategies for ensuring sustainable development in the region.

2. LADAKH: A REGION OF CULTURAL AND RELIGIOUS SIGNIFICANCE

Ladakh is characterized by its breathtaking landscapes, diverse cultural heritage, unique traditions and spiritual significance (Jina, 2017). The region is center of various buddhist monasteries, temples, and sacred sites, making it an important pilgrimage destination for Buddhists, Hindus, Sikhs, and followers of other faiths. The cultural diversity, unique customs and religious significance of Ladakh contribute to its status as a pilgrimage tourism destination. Ladakh, often referred to as the "Land of High Passes," is home to a diverse range of religious and cultural traditions. Buddhism, Islam, and Hinduism coexist harmoniously, creating a rich tapestry of spiritual practices and beliefs. The region is renowned for its monasteries known as (gompas), and sacred sites, drawing pilgrims from across the world. These pilgrimages not only serve as a means of religious devotion but also promote cultural exchange and understanding. They allow visitors to engage with local communities, learn about their customs and traditions, and foster intercultural dialogue.

2.1 Geographical Overview

Ladakh (A newly created Union Territory of India) is situated in the northern part of India, bordering Tibet to the east and the Union Territory of Jammu and Kashmir to the west. It is geographically known for high-altitude deserts, snow-clad mountain ranges, pristine streams, and giant lakes. The region's peculiar geography and landscapes have attracted travelers for centuries, making it a popular destination for adventure enthusiasts, nature lovers, and spiritual seekers.

2.2 Cultural Diversity

Ladakh is characterized by a diverse range of ethnic and religious communities. The predominant ethnic groups include the Ladakhis, Tibetans, Purgi, Balti, Argons, and Dards, each with completely unique cultures, languages, and traditions. The cultural abundance of the region is reflected in its architecture, music, dance, and handicrafts.

2.3 Religious Importance

Ladakh is famous for its old-age Buddhist monasteries, temples, and sacred sites, making it a popular pilgrimage destination for Buddhists, Hindus, Sikhs and followers of other faiths. The region's religious heritage, rooted in Tibetan Buddhism and ancient Bon practices, has contributed to its spiritual allure. The Hemis Monastery, Thiksey Monastery, Alchi Monastery, and Diskit Monastery and Rock carved statues are some of the prominent Buddhist monastic complexes in Ladakh apart from Gurudwara Pathar Sahib of Sikh religion. Monasteries like Spituk are visited by followers of Sikhism, Hinduism and Buddhism.

2.4 Existing Pilgrimage Sites

Ladakh is home to several pilgrimage sites that attract devotees and tourists alike. The famous ones include Hemis Monastery, Alchi Monastery, Spituk Monastery, Gurudwara Pathar Sahib, Sindhu Ghat, etc. Among the most popular and important religious festivals are the Hemis Festival, Leh Dosmochey, Takthok Tseschu, Sindhu Darshan, etc. Followers of both Hindu and Buddhist religions visit Kali Mata temple in the Spituk Monastery. These sites and events hold immense religious and cultural importance and have become major tourist attractions in recent years.

3. SUSTAINABLE DEVELOPMENT AND PILGRIMAGE TOURISM

Sustainable development is a multidimensional concept that seeks to strike a balance among economic, social, and environmental aspects (World Commission on Environment and Development, 1987). Sustainable development refers to a holistic approach that integrates economic, social, and environmental considerations to achieve long-term well-being and prosperity. It entails the responsible and equitable use of resources, the preservation of cultural heritage, and the enhancement of social capital.

Tourism can contribute significantly to sustainable development when managed responsibly. Sustainable tourism aims to minimize negative impacts on the environment, promote socio-cultural authenticity, and generate economic benefits for local communities. It is essential for destinations like Ladakh, where the fragile ecosystem and cultural heritage require protection.

Pilgrimage tourism, characterized by journeys undertaken for religious or spiritual reasons, has the potential to contribute to sustainable development. It can foster economic growth, preserve cultural traditions, and promote environmental conservation. Pilgrimage sites often act as focal points for community development and can be leveraged to enhance the well-being of local residents. Pilgrimage tourism, when managed sustainably, can contribute to all three dimensions of sustainable development. It can generate economic benefits, preserve cultural heritage, and promote environmental conservation (Timothy & Boyd, 2003).

4. ECONOMIC BENEFITS OF PILGRIMAGE TOURISM IN LADAKH

Pilgrimage tourism in Ladakh brings significant economic benefits to the region (Bhattacharya & Ghosh, 2016). It generates income and employment opportunities for local communities, stimulates the growth of small-scale enterprises, and encourages the development of tourism-related infrastructure. The expenditure of pilgrims on accommodations, transportation, and religious offerings contributes to the local economy (Sharpley, 2008). Moreover, this form of tourism has the good potential to boost the local economy by generating revenue from tourism-related activities, such as accommodation, food, transportation, and souvenirs. This revenue can be used to improve infrastructure, provide social services, and create jobs for the youth. The money spent by pilgrims during their visit helps locals improve their standard of living and generate significant economic benefits for both the host communities and the region at large. Moreover, the money spent by pilgrims has a multiplier effect on the local society and helps in fostering entrepreneurship and supporting local industries. Thanks to the popularity of pilgrimage tourism many travel agencies are there in the local market, and they cater to the needs of pilgrims traveling to Ladakh by providing qualified local monastery guides.

5. SOCIO-CULTURAL IMPACTS OF PILGRIMAGE TOURISM IN LADAKH

Pilgrimage tourism contributes to the preservation and promotion of cultural heritage (Connell, 2013). It provides a platform for cultural exchange, where visitors and locals interact, share experiences, and gain a deeper understanding of different cultures. For instance, the Kumbh Mela festival in India attracts millions of pilgrims who engage in religious rituals and cultural activities, promoting social cohesion and cultural diversity (Misra & Behera, 2018). Pilgrimage tourism in Ladakh plays a crucial role in preserving and promoting the region's cultural heritage (Misra & Behera, 2018). It strengthens community identity, fosters intercultural understanding, and supports the revival of traditional practices. Pilgrimage sites serve as platforms for cultural exchange and spiritual experiences, enabling visitors to connect with the region's unique traditions and rituals (Zakaria & Juhdi, 2017). Care takers of religious sites say that pilgrimage tourism helps them to preserve cultural heritage by attracting visitors to religious sites and cultural events. This can help to raise awareness of Ladakhi culture and traditions and can also help to generate revenue for cultural preservation projects. Several national and international agencies are helping monasteries rebuild and repair the old chapels.

6. ENVIRONMENTAL CONSIDERATIONS IN PILGRIMAGE TOURISM IN LADAKH

The rapid growth of tourism, including pilgrimage tourism, in Ladakh has resulted in various environmental challenges (Hall, 2010). The increased footfall in fragile ecosystems puts pressure on natural resources, causes land degradation, and disturbs wildlife habitats. Inadequate waste management and unregulated construction pose additional threats to the region's fragile environment (Bhat et al., 2018). Sustainable practices such as waste management, energy conservation, and eco-friendly transportation can help mitigate these impacts (Kumar et al., 2020). The sudden huge influx of tourists in the region has put excessive pressure on the carrying capacity of the region and also exerted a burden on the sources of water.

7. STRATEGIES FOR SUSTAINABLE PILGRIMAGE TOURISM DEVELOPMENT IN LADAKH

To ensure sustainable pilgrimage tourism development in Ladakh, several strategies can be employed. Infrastructure development is crucial to provide necessary facilities while minimizing the impact on the environment (Dwivedi, 2021). Community engagement and empowerment programs can enhance local participation and ensure equitable distribution of benefits (Chon et al., 2018). Environmental conservation and management practices, such as waste reduction and energy efficiency, can mitigate the negative environmental impacts of pilgrimage tourism (Kumar et al., 2020). Following strategies can be pursued to ensure sustainable pilgrimage tourism development in Ladakh:

7.1 Infrastructure Development

Developing appropriate infrastructure is essential for managing pilgrimage tourism sustainably. This includes the construction of visitor centers, interpretation facilities, and pilgrimage trails that ensure visitor safety and minimize ecological impact. Improving transportation networks, waste management systems, and water and energy supply is also crucial for sustainable pilgrimage tourism development.

7.2 Community Engagement and Empowerment

Involving local communities in decision-making processes and tourism planning is vital for sustainable development. Community-based tourism initiatives, such as homestays and cultural exchanges, can empower local residents, promote authentic experiences, and distribute tourism benefits more equitably. Capacity building programs that enhance the skills of community members can further strengthen their engagement in the tourism sector.

7.3 Environmental Conservation and Management

Conservation measures are necessary to protect Ladakh's pristine environment. Implementing sustainable tourism practices, such as waste reduction, energy efficiency, and water conservation, can mitigate the ecological impact of pilgrimage tourism. Raising visitor awareness about environmental conservation through educational programs and signage can foster responsible behavior among pilgrims.

7.4 Cultural Preservation and Promotion

Preserving Ladakh's cultural heritage is vital for sustainable pilgrimage tourism. Efforts should focus on safeguarding sacred sites, traditional craftsmanship, and intangible cultural practices. Promoting cultural tourism experiences, such as cultural festivals and traditional performances, can enhance visitor engagement while ensuring the authenticity of local traditions.

7.5 Stakeholder Collaboration and Partnerships

Effective stakeholder collaboration is crucial for sustainable pilgrimage tourism development. Government agencies, local communities, religious institutions, and tourism industry stakeholders need to work together to formulate policies, regulations, and guidelines that promote sustainable practices. Public-private partnerships can facilitate resource mobilization, capacity building, and destination marketing efforts.

8. BEST PRACTICES FROM OTHER PILGRIMAGE SITES

Examining successful cases of sustainable pilgrimage tourism in other destinations can provide valuable insights for Ladakh. Examples such as the Golden Temple in Amritsar, and the Kumbh Mela festival in India demonstrate effective management strategies and sustainable practices (Sharpley & Sundaram, 2015). Ladakh can learn from best practices being followed in other popular pilgrimage destinations like:

8.1 Golden Temple Amritsar

Golden Temple management daily witnesses a huge influx of pilgrims worldwide. The management body of the temple has invested a huge amount of money, and they also receive money from pilgrims through donations. The manner of the temple administration is a learning lesson for any pilgrimage site. The temple also receives support from the Sikh community in the form of volunteers to help the temple administration ensure smooth day-to-day operations.

8.2 Santiago de Compostela, Spain

Santiago de Compostela is a famous pilgrimage destination in Spain, attracting millions of pilgrims on the Camino de Santiago route. The management of the pilgrimage involves stakeholder collaboration, route maintenance, heritage preservation, and visitor services. Sustainable tourism practices, such as waste reduction, energy efficiency, and visitor education, are prioritized to minimize the environmental impact.

8.3 Mecca, Saudi Arabia

Mecca, the holiest city in Islam, experiences a significant influx of pilgrims each year during Hajj and Umrah. The Saudi Arabian government has invested in extensive infrastructure development, including transportation networks, accommodation facilities, and waste management systems, to accommodate the large number of visitors. Sustainability initiatives, such as the use of renewable energy and water conservation measures, are being implemented to reduce environmental impact.

8.4 Tirumala Venkateswara Temple, India

The Tirumala Venkateswara Temple is in Andhra Pradesh, India, and is famous for its pilgrimage tourism. The temple management has adopted numerous sustainable measures, such as efficient waste management practice, water conservation measures, and a community engagement policy. The temple also works on promoting eco-friendly measures among visitors and encourages using public transportation.

9. POLICY RECOMMENDATIONS FOR SUSTAINABLE PILGRIMAGE TOURISM IN LADAKH

Effective policies and regulations are crucial for managing pilgrimage tourism in Ladakh sustainably (Zahari et al., 2016). Policy recommendations include the development of a comprehensive tourism strategy, the integration of sustainability principles into tourism planning, and the establishment of monitoring and evaluation mechanisms (Saxena & Sharma, 2019). It is important to manage pilgrimage tourism sustainably. This can be ensured by formulating and executing sustainable tourism policies and practices, and by engaging local communities to ensure that the advantages of pilgrimage tourism are reaped equally by everyone in the locality.

Following are some measures to manage pilgrimage tourism sustainably:

- **Planning and management**: Pilgrimage tourism should be planned and managed in a way that minimizes its negative impacts on the environment and local communities. This can be done by taking inputs from people in the tourism industry, taking locals into confidence and soliciting the views of civil society of the region, and lastly by winning the hearts of the local population. Moreover, the onus lies on the shoulders of the tourism department to ensure that the benefits of tourism in shared in an equitable manner among the members of the community.
- **Education and awareness**: Orientation or awareness should be spread among pilgrims about the sanctity of the destination. Pilgrims should be educated about the importance of sustainability and should be encouraged to take steps to reduce their environmental impact. These can be pursued by organizing programs and by providing pilgrims with sustainable options for transportation, accommodation, and food.
- **Infrastructure development**: There is no doubt that pilgrimage helps in infrastructure development of the region. But in regions like Ladakh, people generally use locally available materials to make the premises winter friendly. Several agencies are working on building new pilgrimage sites or restoring ancient sites with locally available building materials which help in ensuring sustainable development of the region on one hand and reviving old culture on the other hand. This includes things like waste management systems, reusing old materials, renewable energy sources, and sustainable transportation options.
- **Community participation:** It is pertinent to mention here that without community engagement in tourism activities, tourism cannot flourish. Local communities should be engaged right from the planning phase to the ultimate execution phase in managing pilgrimage tourism. This will help to ensure that the benefits of pilgrimage tourism are shared equitably and that the negative impacts are minimized.
- **Legislative Measures:** Law enacting and enforcing agencies should proactively enact suitable legislation and then enforce such legislation to ensure that pilgrimage tourism in the region grows in a sustainable manner. Enacting and enforcing legislation that promotes sustainable tourism practices is essential. This includes regulations on waste management, land use, construction, and visitor behavior. The integration of sustainability principles into tourism policies and plans will ensure that pilgrimage tourism aligns with long-term environmental and socio-cultural objectives.
- **Capacity Building and Training:** Providing training and capacity-building programs for tourism stakeholders is crucial for sustainable pilgrimage tourism development. This includes training local communities in hospitality skills, environmental conservation, and cultural preservation.

Tourism professionals and government officials should also receive training on sustainable tourism management, policy development, and stakeholder engagement.

- **Tourism Promotion and Marketing:** Developing sustainable tourism branding and marketing strategies can attract responsible pilgrims and enhance the destination's image. Promoting Ladakh's unique cultural heritage, natural beauty, and sustainable practices will appeal to conscious travelers. Collaboration with travel agencies, online platforms, and media outlets can help reach a wider audience and communicate the region's sustainable tourism offerings.
- **Monitoring and Evaluation:** Establishing monitoring and evaluation mechanisms is crucial to assess the impact of pilgrimage tourism and track progress towards sustainable development goals. Regular assessments of environmental, socio-cultural, and economic indicators will provide insights into the effectiveness of strategies and identify areas for improvement. Engaging local communities and stakeholders in the monitoring process fosters accountability and encourages active participation in sustainable tourism management.

10. RECOMMENDATION FOR FUTURE RESEARCH

The sustainable development of pilgrimage tourism in Ladakh calls for continuous efforts on the part of every stakeholders and collaboration among various stakeholders. Continued research, monitoring, and evaluation will help refine strategies and address emerging challenges. Building resilience to environmental changes, fostering community ownership, and promoting responsible visitor behavior are key areas for future research. Integrating sustainable practices into tourism policies and implementing capacity-building programs will contribute to the long-term viability of pilgrimage tourism in Ladakh.

11. CONCLUSION

Pilgrimage tourism has the potential to contribute to sustainable development in Ladakh by fostering economic growth, preserving cultural heritage, and promoting environmental conservation. By implementing strategies that address the economic, socio-cultural, and environmental dimensions of sustainability, Ladakh can harness the benefits of pilgrimage tourism while ensuring the long-term well-being of its communities and natural resources. Proper planning, stakeholder collaboration, and policy support are essential for the sustainable development of pilgrimage tourism in Ladakh.

REFERENCES

Barber, R. (1993). *Pilgrimages*. The Boydell Press.

Bhat, F. A. (2018). Challenges and sustainable tourism development in Ladakh region. *Journal of Tourism & Hospitality (Los Angeles, Calif.)*, *7*(3), 1–7.

Bhattacharya, A., & Ghosh, A. (2016). Economic impact of tourism: A case study of Ladakh. *Luyou Xuekan*, *31*(10), 38–45.

Chon, K. S. (2018). Sustainable community-based tourism in Ladakh: A destination life cycle perspective. *Journal of Sustainable Tourism*, *26*(5), 753–771.

Digance, J. (2003). Pilgrimage at Contested Sites. *Annals of Tourism Research*, *30*(1), 143–159. doi:10.1016/S0160-7383(02)00028-2

Digance, J. (2006). Religious and Secular Pilgrimage. In D. J. Timothy & D. H. Olsen (Eds.), *Tourism, Religion and Spiritual Journeys* (pp. 36–48). Routledge.

Dwivedi, R. (2021). Infrastructure planning for sustainable pilgrimage tourism: A case of Uttarakhand, India. *Journal of Tourism Planning and Development*, *18*(3), 317–336.

Hall, C. M. (2010). *Tourism and regional development: New pathways*. Routledge.

Jina, M. S. (2017). Ladakh: A case study of sustainability in tourism development. In L. D. Singh & S. Raj (Eds.), *Sustainable Tourism in Rural Ladakh: Strategy for Development* (pp. 15–33). Springer.

Kumar, S. (2020). Sustainable pilgrimage tourism development in Uttarakhand Himalayas: A community perspective. *Journal of Heritage Tourism*, *15*(1), 1–17.

Misra, R. K., & Behera, B. (2018). Socio-cultural impacts of Kumbh Mela on host communities: An analysis of Prayagraj (Allahabad). *Journal of Tourism and Cultural Change*, *16*(4), 345–363.

Saxena, G., & Sharma, K. (2019). Developing tourism policy framework for sustainable tourism in Ladakh, India. *Journal of Tourism and Cultural Change*, *17*(4), 422–442.

Sharpley, R. (2008). *Tourism and development in the developing world*. Routledge.

Sharpley, R., & Sundaram, P. (2015). *Pilgrimage and tourism: Sacred journeys as tourist attractions*. Channel View Publications.

Swarbrooke, J. (2012). *Sustainable tourism management*. CABI.

Timothy, D. J., & Boyd, S. W. (2006). Heritage tourism in the 21st century: Valued traditions and new perspectives. *Journal of Heritage Tourism*, *1*(1), 1–16. doi:10.1080/17438730608668462

World Commission on Environment and Development. (1987). *Our common future*. Oxford University Press.

Zahari, R. (2016). Pilgrimage tourism management and development in Malaysia: A review. In W. Musa & J. H. Chin (Eds.), *Religion, Tourism and Sustainable Development in Malaysia* (pp. 115–128). Springer.

KEY TERMS AND DEFINITIONS

Balti: A local tribe found in Ladakh and Baltistan, generally following the Shia sect of Islam religion.

Ladakh: A newly created Union Territory of India in the northernmost of India and popularly known as "Moonland."

Monastery: A Buddhist temple locally known as "Gonpa".

Pilgrimage: It is a journey to a religious or sacred place undertaken for religious or spiritual purposes.

Sustainable Development: A holistic present development that does not endanger future generations' ability to meet their developmental needs.

Tibetan: Natives of Tibetan plateaus.

Chapter 17
Discovering the Nutritional Needs of Spiritual Visitors at ISKCON Centres in India Through Menu Analysis

Surjeet Kumar
https://orcid.org/0000-0002-5677-3278
Kurukshetra University, India

Ashish Ranjan
https://orcid.org/0009-0009-2969-8274
Kurukshetra University, India

ABSTRACT

This study aims to analyse the nutritional needs of spiritual tourists visiting International Society for Krishna Consciousness (ISCKON) centers in India. The study objectives are determining the nutritional requirements of spiritual visitors; analyzing the diet menu of ISKCON centers to meet the standard diet requirements; and studying the cause-and-effect relation of Satvik Aahar (ISCKON diet) on the physical and mental well-being of spiritual visitors. Data was collected from ISKCON centers in North India. The analysis was done using the Harris-Benedict equation to calculate total energy expenditure (TEE). The study found practice of "Satvik Aahar" is integral to the diet of ISKCON visitors, providing adequate nutrition for spiritual practices. Ayurveda and intermittent fasting (Vrat) are the strategies for bringing spiritual travellers back into balance with their thoughts and beliefs. Research limitations include time and budget constraints. The study through light on the physiology of pilgrims and create bridge between food and spirituality, providing value for future researchers.

1. INTRODUCTION

Travelling for leisure or business involves going to new countries, seeing their cultures, and taking part in a variety of recreational activities. This activity is referred to as tourism. It has a long history and

DOI: 10.4018/978-1-6684-4817-5.ch017

has changed over time to include various kinds and modes of transportation. The origins of tourism go back to early societies. To avoid definitional disputes, the UNWTO defined "Travelling to and staying in locations outside of one's normal surroundings for up to a year straight for leisure, business, or other reasons is referred to as tourism." The ancient Indian, Egyptians, Greeks, and Romans travelled for religious, cultural, and commercial purposes. The Seven Wonders of the Ancient World, such as the Great Pyramid of Giza and the Colosseum in Rome, attracted visitors even ancient times. In the 17th and 18th centuries, the "Grand Tour" became popular among European aristocracy and wealthy individuals. It involved visiting significant cultural and historical sites across Europe, primarily in Italy, to broaden one's knowledge and cultural appreciation. The advent of the Industrial Revolution in the 18th century brought significant changes to transportation and infrastructure. This resulted in the growth of leisure travel as steamships and railways made travel more accessible and affordable for the middle class. The 20th century witnessed the rise of mass tourism, facilitated by advancements in transportation, such as the development of commercial airlines and automobiles. This led to the emergence of popular tourist destinations, hotels, and resorts, catering to a larger number of travellers. Presently, majority of countries are focusing on tourism promotion due to its economic significance. The table 1 shows the contribution of tourism in global and Indian economy.

The above Table 1 shows that the tourism has taken a hit in 2019 due to Covid -19 and got nastiest during 2020 due to ban of international travel in majority of countries. This reflected from the above as the contribution of tourism got decreases in Global GDP to 1.8 percent but after 2021, there's a significance change as a control on the covid, the sentiment & contribution of tourism has increased impressively.

With changing times, various forms of tourism have emerged throughout the globe, to name few Sports, Commercial & Business, Nature, Culture (includes religious & spiritual also) & Heritage, Leisure & Recreation, and Adventure are some broader categories. the table 2, highlights definitions of these forms.

From the above table 2, there are many forms of tourism. Spiritual tourism is also one of it as a sub-category of cultural & heritage tourism. The Cultural tourism involves exploring the cultural heritage, traditions, and history of a destination. Travelers engage in activities such as visiting museums, archaeological sites, attending festivals, and experiencing local arts, cuisine & religious & spiritual practices. Whereas the Religious tourism involves travel for the purpose of visiting religious / pilgrimage sites and participating in religious ceremonies or events. It is often motivated by spiritual or religious beliefs and provides opportunities for worship, reflection, and cultural (**Religious Tourism Association. (n.d.)** Traveling has always been an exploration and self-discovery for people around the world. After Post covid a common trend is visible that the faith on the God has increased among the youth as a result the spiritual form of tourism is also emerging. It focuses on seeking spiritual experiences, visiting sacred sites, and immersing oneself in the rich spiritual and cultural heritage of different regions. Spiritual tour-

Table 1. Contribution of tourism in global & India's GDP in post Covid era (2019 onwards) (in percentage)

Year	**Contribution of Tourism in Global GDP (in percentage)**	**Contribution of Tourism in India's GDP (In percentage)**
2019	4.1	7
2020	1.8	4.3
2021	**6.1**	**5.8**
2022	7.6	6.8

Source: WTTC, UNWTO & MoT GoI

Table 2. Tourism forms and definitions

S. No	Tourism Forms	Description	Sub - forms / categories associated	Reference
1.	**Leisure & Recreation**	Taking a vacation from daily routine is referred to as leisure travel. Leisure travelers typically appreciate beauty and lovely hotels and resorts, as well as the ambience of beaches. Nowadays the vacations with leisure & purpose have resulted in emergence of many new forms of tourism.	Business, E- Tourism, MICE, City, Creative, Cruise, Drug, Factory, Industrial, Ghetto, Helicopter, Medical, Rural, Shopping, Township, Wine, Youth and Enclave	Chan, N., H. Xiao, C. Chau, and H. Ma 2012
2.	**Nature / eco / sustainable tourism**	Sustainable tourism with a natural area focus, which benefits the environment and communities visited, and fosters environmental and cultural understanding, appreciation, and awareness.	Eco, Agri, Form, Jungle, Garden, Green, Wildlife, Beach, Costal, Disaster, Extreme, Indigenous, Curative and Health	The Mohonk Agreement (2000)
3.	**Cultural & Heritage Tourism**	Individuals who relocate primarily for cultural purposes, such as study tours, performing arts and cultural tours, trips to festivals and other events of a similar type, excursions to see landmarks, and pilgrimages.	Cultural and Heritage, Archaeological, Culinary, Food, Educational, Endogenous, Geo-Tourism, Literary, Music, **Religious / Pilgrimage / Spiritual** and Tribal	UNWTO (n.d)
4.	**Adventure Tourism**	A style of tourism that involves travelling to remote or exotic locales to participate in physically demanding outdoor activities.	Sports, Antarctic, Arctic, Dark and Space	UNWTO (n.d)

Source: Various authors

ism not only provides individuals with personal growth and fulfilment but also contributes significantly to the financial prosperity of destinations.

When it comes to India, it is known for its rich spiritual heritage and is counted as popular destination for spiritual tourism. The country's varied religious traditions, ancient temples, ashrams, and spiritual retreats attract visitors from all over the world seeking spiritual enlightenment, inner peace, and a deeper understanding of diverse faiths. The spiritual related activities along with description and examples of spiritual destinations are presented in table 3 below.

from the above table 3 it is visible that, India has a huge potential in Spiritual tourism, the destination like Prayagraj, Varanasi, Kedar Nath, Ayodhya, Amritsar, Ujjain, Puri etc. has already mark there presence Globally. Every year these centres are improving their tourist arrival records. Spiritual tourism holds immense importance not only for individual travellers but also for the financial prosperity of destinations. It contributes to the preservation of cultural heritage, stimulates local economies, promotes sustainable development, empowers communities, and fosters personal well-being. As this form of travel continues to gain popularity, it is crucial to ensure responsible and sustainable practices to maximize the benefits for both visitors and the communities they visit. By recognizing and supporting the financial contributions of spiritual tourism, we can create a harmonious balance between cultural preservation, economic growth, and personal enrichment. The Government of India has realized the economic significance of rejuvenating religious and holy sites and monuments by taking innovative measures to encourage domestic spiritual tourism. They have developed spiritual tourism circuits and transport facilities such as roads, airports, mobile application for tourist helpline and setting up a National Mission for Pilgrimage Rejuvenation and Spiritual Augmentation Drive (PRASAD) to encourage and promote spiritual tourism development (Haq & Medhekar, 2015). This has resulted in urbanization and development around villages and towns with spiritual significance.

Table 3. Spiritual tourism actives, descriptions, and Indian spiritual destinations

S. No	Spiritual Tourism activities	Description	Spiritual destinations (Example from India)
1.	**Religious Sites**	Sites associated with religions	Varanasi, Rishikesh, Amritsar, Bodh Gaya, and Tirupati
2.	**Temples**	Religious centres	Golden Temple in Amritsar, Meenakshi Temple in Madurai, Brihadeeswarar Temple in Thanjavur, and the temples of Khajuraho
3.	**Yoga and Meditation**	To learn and practice these ancient disciplines in their authentic form.	Rishikesh, often called the "Yoga Capital of the World
4.	**Ashrams**	Spiritual retreats where visitors can immerse themselves in spiritual practices, receive teachings from enlightened masters, and experience a simple and disciplined way of life.	The Art of Living Ashram, Bangalore; Osho International Meditation Resort, Pune, and Paramartha Niketan Ashram, Rishikesh etc.
5.	**Festivals and Rituals**	Diverse celebrations that provide unique opportunities for spiritual tourism	Diwali, Holi, Navaratri, and Kumbh Mela attract pilgrims and tourists alike etc.
6.	**Spiritual Teachings and Gurus**	Visitors often seek the guidance and wisdom of these spiritual teachers to gain insights into their spiritual journeys.	Many sects such as ISCKON; Radha swami Satsang, Beas; Shree Gita Gayan Santhanam, Kurukshetra etc.
7.	**Ayurveda and Wellness**	Traditional healthcare system, Ayurveda, which focuses on holistic well-being. Spiritual tourists often explore Ayurvedic treatments.	Patanjali wellness Haridwar; Kerela Ayurveda centres etc.

Source: Authors' creations

The present study revolves around the food, health & spiritual visitors visiting International Society for Krishna Consciousness (ISKCON) temples. ISKCON has been chosen due to its global presence and is also known as the Hare Krishna movement. it was founded in 1966 by A.C. Bhakti Vedanta Swami Prabhupada with an aim to encourage well-being of society by teaching the science of Krishna consciousness as outlined in the Bhagavad Gita and other sacred texts. Krishna consciousness, a form of Bhakti yoga, is a state of awareness in which one acts in harmony with the ultimate reality of Krishna.

2. REVIEW OF LITERATURE

Research conducted worldwide consistently highlights the significance of spiritual tourism. This section provides an overview of relevant theoretical perspectives, such as religious studies and tourism studies, to better understand the spiritual tourism. Some of studies are chronologically presented. **Norman Alex (2012)** studied the multifaceted characteristics of tourists engaged in spiritualism focusing on their accountability. They drew data from diverse sources such as literature review blogs, community forums, travellers' forums, and social media to find various categories of experiences of spiritual tourists. The study identifies five distinct categories—namely, healing, experimental, quest, retreat, and collective—which often overlap and shed light on broader social trends prevalent in Western societies. the research examines the intricate interplay of travel, religion, personal well-being, meaning, and identity & provides a deeper understanding of human behaviour in this context. Another study by, **Štefko R, Kiráľová A & Mudrík, M (2014)**, In their study in Slovakia on pilgrimage tourism, recognized it as a one of the oldest forms of tourism and, despite being considered marginal, is prevalent worldwide. They assess the present state of marketing communications in this domain; a questionnaire survey was conducted, involving 433 participants (which includes 244 women and 189 men). The study aims to address current issues and identify gaps in effective marketing communication strategies for pilgrimage tourism. The findings suggest that religious tourism has become more competitive from a business perspective, indicating a need for enhanced marketing efforts; Slovakia lacks sufficient awareness of its religious, cultural heritage, and pilgrimage destinations, necessitating improved promotional activities & the availability and low cost of the internet, coupled with limited financial resources on both the destinations' and visitors' sides in Slovakia, have positioned the internet as the dominant medium for strategic interactive communication with visitors. The study provides valuable insights into the challenges faced by pilgrimage tourism and offers recommendations for improving marketing communication strategies in Slovakia. **Saad M, See T, Adil M et al. (2015)**, in their study proposes the utilization of a holistic view of spiritual intelligence (SQ) to tackle the challenge of food service quality among food handlers. They surveyed 210 food handlers working in public institutions located in East Malaysia. The study aims to examine the relationship between spiritual intelligence, food hygiene practices, perceived leadership effectiveness, and food service performance. Partial Least Square Structural Equation Modelling (PLS-SEM) was employed. The Results found no significant impact of food-hygiene practices on the holistic view among food handlers. Furthermore, food-hygiene practices should incorporate other components of spiritual intelligence, such as self-consciousness, high-consciousness, and embracing challenges. **Sheldon, Pauline J. (2020)**, investigates how outer journeys in tourism can lead to an inner transformation of consciousness, resulting in awakening or enlightenment. This inner transformation brings about not only inner peace and freedom but also a sense of flow, connection with something greater, and a desire to contribute to the greater good. The study examines the relationship between tourism and this inner transformation, focusing on three main directions. Firstly, it describes the attributes of the transformative experience and emphasizes their significance in today's world. Secondly, it explores how tourism can facilitate the inner transformation of tourists by fostering these attributes. Lastly, the study proposes various scenarios of tourism experiences that are conducive to inner transformation and discusses their implications for experience design. The paper identified specific factors contributing to inner peace through spiritual tourism. **Wang S, Blasco D, Hamzath A et al. (2023),** explores the interplay between spirituality and nature-based tourism, recognizing nature as a vital context for transcendent experiences that foster the spiritual dimensions of humankind. Drawing upon data from protected areas in China, the research aims to understand how spirituality permeates nature-based tourism experiences. A thematic analysis approach was employed, analysing 27 semi-structured in-depth interviews conducted with professionals and tourists. The study focuses on the identification of factors necessary for constructing a hierarchical framework of spiritual tourism. it enhances our understanding of the integration of spirituality in nature-based tourism. The research underscores the need for intentional connections and collaborations among stakeholders to facilitate meaningful spiritual tourism experiences. In India, **Pandya S (2017)**, investigate the multifaceted aspects of the ISKCON Annamrita program, which is a partnership with the midday meal scheme, focusing on food, faith, social service, and institution building. The researcher employed an interview method, visiting schools in Mumbai, and formed various focus groups comprising ISKCON officials, school children, and their parents. Data collected from these interactions were analysed using the Atlas.ti 7 computer package. The analysis involved organizing narratives and responses into themes using descriptive coding at level one, followed by pattern coding to identify meta-codes for a more systematic and concise representation of the data. The presentation of data and analysis in this study is thematic, incorporating the voices and responses of the participants. therefore, from the above review we can conclude that Spiritual tourism is studied worldwide but there are very few studies associated to spiritual tourism, health & food association the present tries to uncover the Nutritional Needs of Spiritual Visitors at ISKCON Centres in India through Menu Analysis.

3. RESEARCH METHODOLOGY

In the present study, three objectives were framed, one, "To analyse and examine nutritional requirement of spiritual tourist of ISKCON by calculating their calories requirement." second, "To analyse the diet menu of ISKCON Centres to understand the standard diet requirements of spiritual tourists." and third, "to study the Cause-and-effect relation of Satvik Aahar (ISCKON Diet) on physical and mental well-being of spiritual visitors."

Keeping the objectives as the core of methodology, mixed methods were practiced attaining propounding and validating results. The core methodology applied was Qualitative as well as quantitative in nature. The primary data were gathered through covert and overt observation method. A crucial part of primary data was assembled by thoroughly interviewing and observing 109 samples which included temple officials, administrative staff, temple catering staff, Vaishnavjan (community who are in service of deity), and the Puja Pandas (priests). Sample population approach was such that a balance gender categorisation can be achieved to attain the knowledge of calorie requirements of both male and female the data was collected from ISKCON centres of North India. The profile of respondents is presented in table 4.

The Table 4 shows that majority of respondent were male (56.8 percent), were between the age group of 25-34 (29.3 percent) and 45-54 (28.8 percent), further most of them were graduates (49.6 percent) and by profession majority were freelancer / businessman (30.5 percent). The menu analysis was done by diversifying foods in meals as per diet menu of ISKCON Centres to obtain a meaningful data. The Table 5 shows a systematic and chorological order of steps to be conducted with their objectives respectively. This Table helps to attain a better understanding of what is done and why it is done to clear image of objectives accomplished

Table 4. Profile of respondents

Gender (in %)		**Age ((in %) Years**		**Level of Literacy (in %)**		**Profession (in %)**	
Male	**56.8**	**25–34**	**29.3**	**Less than 10th**	14.6	**Defence personnel**	5.1
Female	43.2	**35–44**	20.6	**12th**	15.6	**Freelancer/ businessman**	**30.5**
		45–54	**28.8**	**Graduate**	**49.6**	**Middle/senior employed management**	17
		55–64	15	**Postgraduate**	18.1	**Civil servant**	11.4
		65 or >	6.3	**Doctorate**	2.1	**Worker**	17.4
						Retired	4
						Domestic/unemployed	1.5
						Student	6.5
						Other	6.6

Source: Author Creation

Table 5. Chronological order for research methodology

Steps	Objective	Procedure
1	To analyse and examine nutritional requirement of spiritual tourist of ISKCON by calculating their calories requirement.	By extracting literature review and definitions from WHO and ICMR & by observing physical activity and then using Harris- Benedict Equation- C = BASAL METABOLLIC RATE * 1.2 Where, BASAL METABOLLIC RATE= 655.1 + (9.563* weight in kg) + (1.850* height in cm) – (4.676 * age).
2	To analyse the diet menu of ISKCON Centres to understand the standard diet requirements of spiritual tourists.	1.Category of items 2.Effect on psychology and physiology 3. Calculating calories provided by ISKCON menu and energy extracted from meals & referring Standard calorie intake from WHO and ICMR and comparing it with Total calories calculated in step 1.
3	To study the Cause-and-effect relation of Satvik Aahar (ISCKON Diet) on physical and mental well-being of spiritual visitors.	By observing overall psychological and physiological conditions of respondents and their behaviour after consuming meals of ISKCON & presenting it with root cause analysis.

Source: Author Creation.

Firstly, analysis was done to examine nutritional requirements of ISKCON devotees, and this was accomplished by calculating Total Energy Expenditure (TEE) by help of Harris-Benedict equation. The research by James Arthur Harris and Francis Gano Benedict, which was presented in the publication "*A Biometric Study of Basal Metabolism in Man*" published by the Carnegie Institution of Washington in 1919, served as the basis for the Harris-Benedict equation. Its accuracy was increased in a 1984 modification. In 1990, Mifflin et al. developed an equation that was better predictive of contemporary lifestyles. Later research led to BMR estimators that took lean body mass into account.

Secondly, another crucial thing was to understand whether the requirement of these diets is being adequately maintained or there is a gap in the scenario. For this menu analysis was done to accurately determine the calorific value of standard diets of Devotees of ISKCON Centres and then a comparative study was made. Interviews were made to check for the satisfaction for mental and physical well-being by these diet menus followed. Interviews were also helpful in knowing mental and physical health of spiritual tourists to understand 360-degree view of nutritional requirements. The comparative analysis of calorie recommendation of WHO and ICMR helped in examining the purpose of this objective.

Thirdly, to analyse whether the concept of *Satvik Aahar* is relatable in this techy era. For this root cause analysis was done for the aspects of ISKCON menu and deliberately discussing every facet of nutritional requirement whether it is physiological of psychological. Root cause analysis sometimes referred to as an Ishikawa diagram or cause-and-effect diagram, is a problem-solving technique utilised in a variety of industries, including engineering, healthcare, and quality management. In the 1960s, Japanese quality control expert Kaoru Ishikawa created this method. The goal of root cause analysis is to pinpoint the underlying causes of an issue so that focused solutions can be implemented to make things better. An overview of the root cause analysis method and its uses is given in this study of the literature. A graphic representation of a problem's potential causes is a root cause diagram. the below table 5 gives the Chronological order for research methodology.

4. ANALYSIS AND INTERPRETATION

Objective One: To Analyse and Examine Nutritional Requirement of Spiritual Tourists of ISKCON by Calculating Their Calories Requirement

For this objective, the menu of ISKCON was analysed. The menu of ISKCON is diverse and focuses to give entire nutrition in pack of PRASADAM. Menu of these centres are very much on base of yogic diet and takes out soul of their catering with Satvik Aahar. It becomes very important to examine whether this diet is sustainable enough to provide complete nutrition to an adult body. To understand meal nutrition, the concept of entire balance diet must be accounted for better verifiability of nutritional requirement. In a report of Indian Council of Medical Research (ICMR) in 2019 it is given that what quantity must be taken for better performance of body. The Table 6 below is an extract from that report to know the balanced amount for body.

The above table 6 highlight that the diet must include on an average to provide sufficient energy for all involuntary and voluntary actions carried out by body. Balance Diet is defined as the wholesome diet which constitutes every single nutrient in perfect proportion recommended by experts to meet daily energy needs to perform physiological and psychological activities (WHO). The average distribution of nutrients is as per energy required by various elements of meal. Now the understanding is to build a nutritional requirement framework for spiritual tourists to attain the motive of their requirements. For this Calculation of Total Energy Expenditure (TEE) by help Harris-Benedict equation for spiritual tourist visiting ISKCON Centres is crucial, and this part will deal with this issue. A biometric study of basal

Table 6. Balance diet for an adult recommended by ICMR

BALANCE DIET FOR AN ADULT (QUANTITY- G/DAY)						
	Sedentary work		Moderate work		Heavy work	
Type of Food item	V	NV	V	NV	V	NV
Cereals	400	400	475	475	650	650
Pulses	70	55	80	65	80	65
Green leafy vegetables	100	100	125	125	125	125
Other vegetables	75	75	75	75	100	100
Roots and tubers	75	75	100	100	100	100
Fruits	30	30	30	30	30	30
Milk	200	100	200	100	200	100
Fats & oil	35	40	40	40	50	50
Meat & Fish	–	30	–	30	–	30
Eggs	–	30	–	30	–	30
Sugar and Jaggery	30	30	40	40	55	55
Groundnuts	–	–	–	–	50	50
V= Vegetarian, NV= non-Vegetarian						
Minimum calories for adults in India: Male (2500-3000) Female (2000-2500)						

Source: ICMR Report, 2019

metabolism in man, written by J. Arthur Harris and Francis G. Benedict, was published in 1919, along with a condensed version the year before **(Kien, C. L., & Ugrasbul, F. (2004)).** These were based on information obtained using equipment for studying the respiratory exchange," which was reported ten years prior. Since then, information has accumulated that demonstrates a good correlation between indirect calorimetric-measured energy expenditure and anticipated energy expenditure using the Harris-Benedict equations in non-obese normal volume Total Energy Expenditures, and much more so in people with obesity. **Basal metabolic rate** (BMR) is defined as the number of calories an individual needs if he/she is on rest for 24 hours or full day. **(Amirkalali, B., Hosseini, S., Heshmat, R., &Larijani, B. (2008)).**

From table 7, the approximate BMR for both male and female spiritual tourists are known, there is need to enquire the physical activity factor for accurate calculation of calories to identify total energy expenditure. As per data collected through covert and overt observation it is very evident that a normal disciple has moderate to active amount of physical activity down the routine of ISKCON. The routine of a disciple starts with early morning around 5 am and its end in the night at around 9 pm. In between disciples do their daily tasks assigned. Major physical activities performed in routine were indulging in *kirtan, Krishna nritya,* gardening, *prahadamvitaran* and going on tours for exploring and marketing about Krishna consciousness. Besides being indulge in these duties disciples must do *rasa parivartan* for purification of body and *vrat* also lies in between these duties. So, considering a disciple to duties form and its physical activity factor calories can be calculated. The table 8 highlights the total energy requirement of spiritual tourists.

The above Table 8 shows that Harris-Benedict equation is taken for calculation. further the equation results shows that the calorie requirement of spiritual tourists are good and close to moderately active human because of their activity of yoga and kirtan.

Hence from the above analysis, nutritional requirement of spiritual tourists of ISKCON is good.

Objective Two: To Analyse the Diet Menu of ISKCON Centres to Understand the Standard Diet Requirements of Spiritual Tourists

To accomplish the above objective, the meals of ISKCON units were analysed. The ISKCON provides meals to visitors in form of thali in which basic items are presented. These items are being examined

Table 7. BMR formula chart

Basal metabolic rate of spiritual tourists		
	Formula by NIN Report 2020	**For Spiritual tourists**
For Women*	BASAL METABOLLIC RATE= 655.1 + (9.563* weight in kg) + (1.850* height in cm) – (4.676 * age)	BASAL METABOLLIC RATE= 655.1 + (9.563* 55) + (1.850* 162) – (4.676 * 35) =1317.105
For Men*	BASAL METABOLLIC RATE= 66.5 + (13.75 * weight in kg) + (5.033 * height in cm) – (6.75 * age)	BASAL METABOLLIC RATE= 66.5 + (13.75 *65) + (5.033 *177) – (6.75 * 38) =1594.591
*Taking average weight for men disciple to 65 kg and height 177 cm. *Taking average weight for women disciple to 55 kg and height 162 cm. *Taking average age for women disciple to be 35 years old *Taking average weight for men disciple to be 38 years old (The National Institute of Nutrition (NIN) report, September 2020)		

Source: (Harris JA, Benedict FG *"A Biometric Study of Human Basal Metabolism"*. ("A Biometric Study of Human Basal Metabolism. - Abstract - Europe PMC") The Proceedings of the National Academy of Sciences (PNAS) (1918))

Table 8. Total energy expenditure requirement (TOTAL ENERGY EXPENDITURE)/ day by spiritual tourists

Type of physical activity	Harris-Benedict equation in calories (C)	For Spiritual tourists in calories (C)/day	
		Male	Female
Sedentary (little or no exercise)	C = BASAL METABOLLIC RATE * 1.2	1912.8	1580.4
Lightly active (light exercise/sports 1-3 days/week)	C = BASAL METABOLLIC RATE * 1.375	2191.75	1810.875
Moderately active (moderate exercise/sports 3-5 days/week)	C = BASAL METABOLLIC RATE * 1.55	2470.7	2041.35
Very active (hard exercise/sports 6-7 days a week)	C = BASAL METABOLLIC RATE * 1.725	2749.65	2271.825
"If you are extra active (very hard exercise/sports & a physical job)" ("Harris-Benedict Calculator (Total Daily Energy Expenditure)")	C = BASAL METABOLLIC RATE * 1.9	3028.6	2502.3

here for their calorific and nutritive value. The items taken into considerations are derived from Srilaprabhupada conversations and his short book of recipes known as Srila Prabhupada recipes and menu. He has made a standard diet which is followed in all ISKCON Centres across the globe. This diet is termed as Standard Devotee Diet which is given for spiritual learning's in the centre. This menu was developed by Srilaprabhupada to advise the devotee to attain Satvikta and to avoid any sort of tamas in their diet. This diet was constructed to address mental needs for attaining spiritual goals. All these items have Satvik gunas in their complex structure. The vegetables were always suggested to be seasonal and from own garden. Freshness was one of the major concerns to avoid any sort of impurity, unhygienic substance, and unwanted pesticides. In ISKCON it is mandatory to grow own vegetables and fruits to understand value of food and to learn self-dependency. There are various guidelines with these foods which are advisable to spiritual tourists to practice. Srilaprabhupada was not in motion for having late night dinners because he considered late night dinners to be increasing inflammation in body due to which balance of hormones can be disturbed. For standard diet he mentioned in his conversations that all meals are important and must be taken in cycle of Suryanarayana (Sun-moon cycle). Excess of any nutrient should be prohibited as it causes hormonal imbalance creating barrier for spiritual practices. Table 9 below shows the total calorie calculation of Standard Diet of Devotee.

Table 9 highlights, The Calorie considered was build out from ICMR calorie chart and then this calorie calculation of menu was synthesized. The menu above is followed in ISKCON and very little change is seen so it is very much close to daily intake of tourists visiting ISKCON. Now from above table, we know how much calorie a tourist is taking on an average. The calorie in the meal is now to be compared by what is needed and what is recommended by health officials in next step.

The above table 10, the total energy expenditure of moderately active spiritual tourist & standard diet is not huge.

Hence, the analysis emerges three basic ideas, i.e., firstly, what is needed; secondly, what is being taken; and thirdly, what is recommended. These all three aspects were the main pillar of research. further from the above tables, there is very less gap of in nutritional support from ISKCON menu and actual intake as it is not only sufficient but also nourishing the physiologically & psychologically.

Table 9. Total calorie calculation of standard diet of devotee

Meals	Item	Quantity	Calories
BREAKFAST	Simple farina cereal with nuts and raisins	1 cup or 250gm	150
	Milk (steaming hot), or yogurt in the summertime	1 glass or 250 ml	200
	Chickpeas (raw, soaked overnight)	1 cup or 250 gm	300
	Oranges, apples, and bananas	1 each	117
LUNCH	Rice	1 bowl or 250gm	320
	Chapatis	1-6 inch	71*2=142
	Dal made with freshly ground ginger root and freshly ground spices	150gm	180
	Sabji made with ghee, freshly ground ginger root and freshly ground spices.	1 bowl or 200gm	150
BEFORE TAKING REST	Milk (steaming hot)	1 glass or 250 ml	200
	Banana	2	250
Total Calories in Standard Diet of Devotee			2009

Source: Author Creation

Table 10. A comparative analyses of calories/ day intake of spiritual tourists of ISKCON

Total Energy Expenditure of moderately active spiritual tourists (From Table 4)		Energy provided by menu of Standard Diet of Devotee (From Table 6)		Recommended Dietary allowance by ICMR. (From Table 2)		Recommended Dietary Allowance by World Health Organization* *Fact sheet n°394 updated august 2018- WHO Report	
Male	Female	Male	Female	Male	Female	Male	Female
2470.7	2041.35	2009	2009	2500-3000	2000-2500	2500-3000	2000-2500

Source: Author Creation

Objective Three: To Study the Cause-and-Effect Relation of Satvik Aahar (ISCKON Diet) on Physical and Mental Well-Being of Spiritual Visitors

To attain the above objective, the Ishikawa diagram or cause-and-effect diagram is used is a problem-solving tool used in various fields such as quality management, engineering, and healthcare. This technique was developed by Kaoru Ishikawa, a Japanese quality control expert in the 1960s. The purpose of analysis is to identify the root causes of a problem, which allows for targeted interventions to improve the situation. The diagram is created by brainstorming potential causes and then organizing them into categories that branch off the main trunk. These categories are typically people, process, equipment, materials, and environment, although other categories may be added depending on the problem and context.

Steps followed for analysis:

a. **Define the problem or effect**: This analysis aims to analyse the effect of Satvik Aahar on mental and physical well-being.
b. **Identify the categories**: The categories or factors that might contribute to mind body connection of a spiritual tourist and can be taken as main branches of trunk are as follow:
    - Nutritional requirement

- Spiritual visitors
- Environment
- Preparation method
- Serving method

c. **Diagrammatic presentations of the root cause analysis:** The diagram 1 with the categories identified in step 2 as the main branches. This will again be elaborated by respective sub-factors.

Figure 1. Root cause analysis of above sub-factors

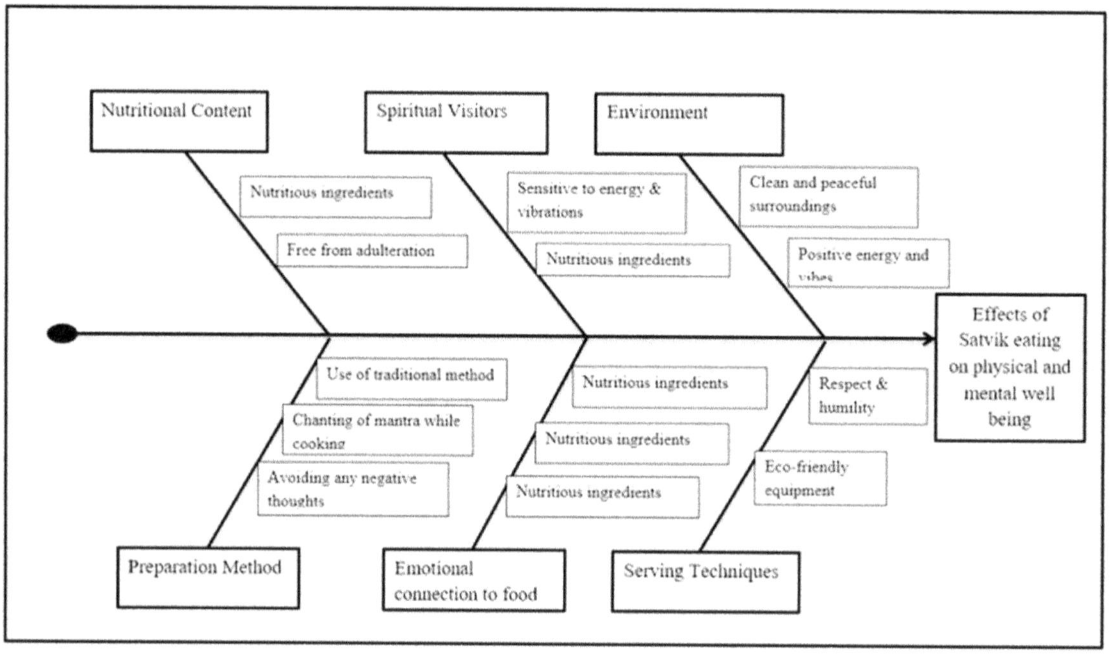

d. **Identify the sub-factors**: Brainstorm and identify the specific factors or sub-factors that contribute to each category. The sub-factors are stated as in table 11:

Table 11. Categories and sub factors for root cause analysis

Categories	Nutritional Content	Spiritual Visitors	Environment	Preparation Method	Emotional connection to food	Serving Method
Sub Factors 1	Ingredients	Sensitive to energy and vibration	Clean and peaceful surrounding	Use of traditional method	Prepared with love and devotion	Serving with respect and humility
Sub Factors 2	Free from toxins and chemicals	Seeking spiritual growth and connection	Positive energy and Vibes	Chanting of Mantras during preparation	Made as an offering to God	Use of eco-friendly utensils
Sub Factors 3				Avoiding any negative thoughts	Cultural involvement	Serving in a mindful manner

e. Analyse the sub-factors:

After analysing and consulting various aspects certain items are very predominant while considering Satvik diet for spiritual visitors. All those factors are discussed below:

Nutritional Content: Satvik food is generally made up of whole foods that are rich in nutrients such as vitamins, minerals, and antioxidants. This can help to boost the immune system, promote healthy digestion, and maintain overall physical health.

Purity: Satvik food is made from ingredients that are grown naturally, without the use of synthetic fertilizers or pesticides. This can help to reduce exposure to harmful chemicals and toxins, which can have negative impacts on health.

Energy and Vibrations: Satvik food is believed to have positive energy and vibrations, which can help to uplift the mood and promote a sense of well-being. This is because Satvik food is prepared with love, devotion, and positive intentions.

Digestibility: Satvik food is prepared in a way that is easy to digest, which can help to reduce digestive problems such as bloating and gas. This is because Satvik food is generally cooked slowly, with herbs and spices that aid in digestion.

Mindful Eating: Satvik food is often consumed in a mindful manner, which can help to reduce stress and promote mental well-being. This is because Satvik food is often eaten in a peaceful and serene environment, with a focus on gratitude and appreciation.

Spiritual Connection: Satvik food is often consumed by individuals who are seeking a deeper spiritual connection, which can help to promote a sense of purpose and meaning in life. This is because Satvik food is believed to help purify the body and mind, which can aid in spiritual growth and development.

6 Take action: Take action to address the significant factors identified in step 5. Identify and implement solutions to improve the situation or effect.

7 Evaluate the results: Evaluate the results of the actions taken in step 6. Determine if the actions taken were effective in addressing the problem or effect identified in step 1.

Hence, from the above it is clear that consuming Satvik food can have a positive impact on human well-being, as it promotes physical health, mental well-being, and spiritual connection. Foods that are grown naturally and prepared with love and positive intentions, spiritual visitor can improve overall sense of well-being and live a more fulfilling life.

5. CONCLUSION

The spiritualism is the way of attaining meaningful, enlightened, and self-conscious life through maintaining mind-body coordination in its purest form. One becomes what he eats, and this is why it is important to understand role of food on the path of realization of ultimate truth. The present study revolves around three objectives. One, the study was done to understand various perspectives regarding nutritional requirements of Spiritual visits and the study was done in ISKCON centres. The result shows that the diet menu advised by ISKCON is truly enlightened and proven. The total calorie requirement is being fulfilled in the pure form that too without accessing any nutrient. The ICMR and WHO have given almost equal measures of total calories required by normal human and spiritual tourists. Daily need of adequate calories analysed is almost sufficient for tourists. Two, is to analyse the diet menu of ISKCON Centres to understand the standard diet requirements of spiritual tourists. in interview with various stakeholders, it is understood that there are certain principles of eating when it comes to spirituality and a tourist must focus on it. ISCKON has 3 basic principles while planning their menu. All the guidance has been taken from sacred texts, such as, Srimad Bhagwad Gita, Upanishads, Puranas, and Vedas Ayurveda is the base for all study in spiritual eating. A motion which is in recommendation form is to include a small meal in and around sunset to provide and fulfil any gap attained in diet during day so that body at rest is not starved and dehydrated due to less nutrient. Moreover, fasting days can be hard after a complete night of no eating. Intermittent fasting (Vrat) is also an evidential technique to make spiritual tourists more balance with their thoughts. In spirituality it is all about thoughts and beliefs mind reads and perceives. Any imbalance in hormones can break the cycle so food habits are very much sorted and guided. Three, to study the Cause-and-effect relation of Satvik Aahar (ISCKON Diet) on physical and mental well-being of spiritual visitors. for this Root cause analysis was done. which shows the important factors and their effect which are very closely related to sound mind-body connection. All the categories were taken in account and whole cycle gave a nourishing effect on well-being of visitors by diet of ISKCON.

Further, this is primary research on the nutritional requirements of any spiritual community through their menu analysis. Moreover, the research gives base to identify community eating and their food habits. This research can be used in medical as well as spiritual field to analyse demands of spiritual tourists. It is believed that spiritualism is all about faith and mental requirements. This research gives

Figure 2. Principles of spiritual eating followed by ISKCON visitors

value and creates a bridge between food and not only in terms of service but a way to healthy practices or in researcher's word- The robust spiritualism. the limitation of the research was geographical peripheral of menu analysis was restricted to North India and north Indian dishes. more research is advisable as India is diversified in terms of food patterns and eating habits especially in terms of community eating. Research can also be made on accountability of taste buds which is a crucial limitation of this study.

REFERENCES

Chatterjee, A., Bajpai, R., Gerdes, M. W., & Gerdes, M. (2021). *Analyze the Impact of Healthy Behavior on Weight Change with a Mathematical Model using the Harris-Benedict Equations. ("Analyze the Impact of Healthy Behavior on Weight Change with a ...").* Research Square. doi:10.21203/rs.3.rs-584141/v1

Dewan, M. (2020). An Ethnographic Inquiry Towards 'Hare-Rama Hare-Krishna Religious Tourists Food Consumption Habits While on A Holiday in India. [JOGHAT]. *Journal of Gastronomy Hospitality and Travel*, *3*(1), 30–41. doi:10.33083/joghat.2020.28

Norman, A. (2012). The Varieties of the Spiritual Tourist Experience. In *Literature & Aesthetics, 22*(1). https://laparisienne.wordpress.com/2009/04/30/buddy-om/

Pandey, S., & Shukla, R. (2022). Contribution And Importance of Dietetics in Ayurveda. *World Journal of Pharmaceutical Research*, *11*(11), 1069. https://doi.org/ doi:10.20959/wjpr202211-25127

Pandya, S. P. (2017). Food, faith, social service and institution building: the Annamrita programme of the Hare Krishna movement in India. (Description: Food, faith, social service and institution building). *Religion State & Society*, *45*(1), 4–22. doi:10.1080/09637494.2017.1281598

Saad, M., See, T. P., Adil, M. A. M., & Kassim, N. M. (2015). Spiritual Intelligence on Leadership Effectiveness and Food-hygiene Practices in Public Institutions. *Procedia: Social and Behavioral Sciences*, *201*, 146–155. doi:10.1016/j.sbspro.2015.08.161

Sheldon, P. J. (2020). Designing tourism experiences for inner transformation. *Annals of Tourism Research*, *83*, 102935. doi:10.1016/j.annals.2020.102935

Štefko, R., Kiráľová, A., & Mudrík, M. (2015). Strategic Marketing Communication in Pilgrimage Tourism. *Procedia: Social and Behavioral Sciences*, *175*, 423–430. doi:10.1016/j.sbspro.2015.01.1219

Vande Berg, T., & Kniss, F. (2008). ISKCON and immigrants: The rise, decline, and rise again of new religious movement. *The Sociological Quarterly*, *49*(1), 79–104. doi:10.1111/j.1533-8525.2007.00107.x

Wang, S., Blasco, D., Hamzah, A., & Verschuuren, B. (2023). "Tourists and 'philosophers': Nature as a medium to consciousness and transcendence in spiritual tourism." ("Annals of Tourism Research | Vol 99, March 2023 - ScienceDirect"). *Annals of Tourism Research*, *99*, 103543. doi:10.1016/j.annals.2023.103543

KEY TERMS AND DEFINITIONS

Ekadashi: Ekadashi is the name for the eleventh lunar day.

Kirtan: Kirtan, an act of worshipping and glorifying god, is derived from a Sanskrit word that means to summon, recite, laud, or exalt.

Nritya: Nritya means dance in the Indian tradition.

Prasadam: Hinduism accepts prasadam or prasad as a kind of sacrifice.

Rasa: Rasa, which means "sap," "essence," or "fluid," is a Sanskrit word. The word "essence of human experience" in the spiritual tradition refers to the emotions that guide human existence.

Satvik Aahar: The Sanskrit term "Sattva" (pure, clean, and powerful energy) is where the name "Satvik" originates.

Tattva: The Sanskrit word "tattva" means "principle" or "reality." It comes from the root tatt, which means "that," "quality," or "godhead," and tvam, which means "thou," "individual," or "you.

Vrata: This means Fasting.

Chapter 18
COVID-19, Religious Events, and Indian Tourism Recovery:
Prospects and Paradoxes

Priyakrushna Mohanty
CHRIST University (Deemed), India

Swati Shah
Kumaun University, India

Rashmiranjan Choudhury
Jain University (Deemed), India

ABSTRACT

The chapter delves into three objectives. Firstly, the chapter aims to find out the intersectionalities of religious events and the Indian tourism industry. For the second objective, the impact of the COVID-19 disease on religious events will be briefly discussed. Lastly, this work will discuss the various emergent prospects, themes, trends, and challenges that will emerge on the paths of the recovery of religious events and pilgrimage tourism post-COVID-19. This work is theoretical in nature and can be classified as a viewpoint article that follows a conceptual research design.

INTRODUCTION

India is a religious country (Chiriyankandath, 2000) that exhibits an astonishingly diverse and beautiful social composition. Census (2011) highlights the presence of six major religions in India, four of which have their origins in this secular country. With such a diverse portfolio of religions, India is one of the countries that celebrate the highest number of religious events in the world. These religious events have a great impact on different sectors that drive the Indian economy. Among these sectors, the tourism industry is one of the most obvious ones. The tourism industry is regarded as an integral part of the Indian economy roughly contributing 9% to its total GDP and 8% to the total workforce (Thommandru

DOI: 10.4018/978-1-6684-4817-5.ch018

et al., 2023). Religious events and pilgrimage are critical wings of the Indian tourism industry driving a major portion of domestic tourism. Currently, the Indian tourism industry is experiencing massive turmoil thanks to the recent outbreak of coronavirus disease (COVID-19) which has been declared as one of the biggest pandemics of the 20th century. In the face of this unprecedented and dire situation, both the Government and private players are banking on domestic tourism to drive the recovery of the Indian tourism sector. In this backdrop, religious events and pilgrimage tourism will play a vital role in the recovery process. However, rather paradoxically, since the outbreak of this pandemic, religious events have been severely critiqued with some of the events like Tablighi Jamaat (BBC, 2020), and Kumbh Mela being referred to as the *super spreaders* for the disease (Jena, Yadav, & Rambarki, 2021). Therefore, Indian tourism administrators have often found themselves in a diabolic situation in which they have to choose between religious events and the safety and security of pilgrims. The authors believe that the key to these contradictions lie in finding the right balance between both. Further, the outbreak of the second wave of COVID-19, limited efficiency and availability of vaccines only point to the fact that COVID-19 is *here to stay* for a considerable amount of time. Thus, it would be imperative for the events and tourism industry to resume their operations, only this time with proper guidelines and protocols in place. The authors of this paper argue that while bringing massive undesired economic and social repercussions to the tourism industry, the disease has also provided an opportunity to reinvent tourism in a more sustainable and unique avatar and there is and was no better time to inculcate the sustainability agenda into the domain of religious events and pilgrimage tourism in India. Keeping these facts in mind, this paper aims to achieve three objectives. Firstly, the paper aims to find out the intersectionalities of religious events and the Indian tourism industry. For the second objective, the impact of the COVID-19 disease on religious events will be briefly discussed. Lastly, this work will discuss the various emergent prospects, themes, trends and challenges that will emerge on the paths of the recovery of religious events and pilgrimage tourism post-COVID-19. This work is theoretical in nature and can be classified as a viewpoint article that follows a conceptual research design. Apart from the empirical observations of the authors, the thematic content analysis technique has been used for analyzing the secondary data collected from the review of pertinent literature. Overall, this paper will focus on more on the prospects and ways to overcome the challenges rather than providing any conclusions.

METHODOLOGY

The purpose of the current review article is to delve on the interlinkages between COVID-19 pandemic and the religious events in India, alongside carefully using these interlinkages to devise effective recovery plans for Indian tourism. The paper investigates the possibilities and contradictions surrounding the connection between religious events, pandemic prevention strategies, and the resurgence of Indian tourism. The research design, data collection procedures, data analysis strategies, and ethical considerations used in this review paper are all described in the methodology section.

Research Design

The connection between COVID-19, religious festivals, and the revival of Indian tourism is examined in this review paper using a comprehensive literature review methodology. The research design uses a systematic and organised approach to find pertinent books, journal articles, reports, and other academic sources. This study ensures a careful investigation of the relevant literature by using a systematic approach.

Data Collection

2.1. Identification of Sources: A thorough search of numerous academic databases, including PubMed, Scopus, Google Scholar, and pertinent tourism journals, was the first stage in the data collection process. In order to find relevant articles, search terms like "COVID-19," "pandemic," "religious events," "tourism recovery," and "India" were combined.

2.2. Selection Criteria: Based on predetermined inclusion and exclusion criteria, articles were chosen. Studies concentrating on COVID-19's effects on religious events and India's tourism revival that were published between 2019 and 2022 met the inclusion requirements. English-language articles were selected because they were easier to get resources for. Grey literature was also taken into account, such as reports and working papers from recognised organisations.

2.3. Data Extraction: In order to extract the data, essential information about COVID-19, religious events, and the revival of Indian tourism was carefully reviewed in the chosen articles. The information that was taken from the study comprised its goals, methods, results, and conclusions.

Data Analysis

3.1. Thematic Analysis: Thematic analysis was used to find recurrent themes, patterns, and viewpoints from the retrieved data. This procedure involves coding and categorising the gathered data into pertinent themes pertaining to how COVID-19 affected religious events, how religious institutions responded, how government policies were affected, and how this would affect India's tourism industry as it recovered.

3.2. Synthesis and Interpretation: The thematic analysis findings were synthesized and interpreted to derive meaningful insights regarding the prospects and paradoxes surrounding the interconnections between religious events, pandemic control measures, and the recovery of the tourism sector in India. Emerging trends, key findings, and knowledge gaps were identified and presented in a coherent manner.

Ethical Considerations

The importance of ethical issues was prioritised throughout the study procedure. To maintain academic integrity and prevent plagiarism, all information sources were duly acknowledged and mentioned. Personally identifiable information was avoided throughout the study, and confidentiality and data protection norms were respected.

Limitations

The potential biases in the publications that were chosen, language barriers, and the fluidity of the COVID-19 epidemic are all acknowledged as limitations in this review work that may affect the reliability and generalizability of the results.

This study intends to offer useful insights regarding the impact of COVID-19 on religious events and its implications for the revival of Indian tourism by using a systematic approach to review the existing material. The research presented in this review article will add to the body of knowledge already available and assist players in the travel and tourism sector as well as policymakers in making wise decisions.

Intersectionalities of Religious Events and the Indian Tourism Industry

The importance of religious events to the tourism industry is widely recognized. Globally out of the total six hundred million religious' voyages, 40% of them occur in Europe and more than half in Asia (UNWTO, 2011, p.1). About 170 million tourists visited religious places in India and of that 70-90 percent accounts for domestic tourism (UNWTO, 2011; NCAER 2003). Pilgrimage sites have always led to an influx of visitors (Ahmed & Memish, 2019) and India is always seen as the country with a vast network of pilgrimage sites (Haberman, 2013). People are drawn to religious events due to their immense faith, spiritual knowledge enhancement, and spiritual connectivity (Verma et.al, 2018). Religious events create a prospect for the tourism industry to engage in bulk business by providing support services such as transportation, accommodations, and other ancillary services. The high influx of international and domestic tourist is observed to partake of religious event in India (UNWTO, 2011). Domestic tourism is often seen driven by religious sentiments. Religious tourism has occupied an important place in the tourism policies of many state governments like Bihar, Uttar Pradesh, and Sikkim (ORF special report, 2019).

Estimates of annual visitor numbers to pilgrim sites are indicative of its importance to the tourism industry. Amritsar (Sikh) receives 13 million pilgrims, Tirupati (Hindu) about 10 million pilgrims, Kumbh Mela – 50 million visitors, Vrindavan (Hindu) – 6 million visitors, Dwarka (Hindu) - 5 million pilgrims, Varanasi (Hindu) - 1 million pilgrims, Bodhgaya (Buddhist) – 270787 and Hindu festivals and events which take place every three years, with some festivals attracting 10 million and others 50, 60 or 70 million tourists (ARC, 2011; Shinde, 2011; ORF special report, 2019; Verma & Sarangi, 2019). The tradition of pilgrimage has given rise to the present religious tourism (Shinde, 2012). Considering the devotion of the people to visit religious sites and events like Kumbh Mela, Puri rath yatra has given an edge to the tourism professionals to provide the necessities to manage and plan arrangements for such big events and make travel safe, comfortable, and enjoyable along with fulfilling the recreational needs of the tourist is of paramount importance. Effective management of religious events demands the active participation of multiple governance levels as well as elaborate inner workings to provide reliant infrastructure in the form of tourism offices, travel agencies, hotels, etc. to serve the need of visitors (Shinde, 2010). In India to cater to the needs of this form of tourism, specialized travel agencies, specific eateries, gifts, and souvenir shops have come into place (Gladstone, 2005).

India being the birthplace of Buddhism has a vast potential for attracting inbound tourists. Place associated with Buddha life has always been an attraction for tourists from Southeast Asia. Mahaparinirvan Express a special train to Buddhist circuit run by IRCTC to cater to religious tourists. Between 2015-2018, 11 Buddhist sites (Gaya, Bodhgaya, Rajgir, Nalanda, Vaishali, Sarnath, Kushinagar, Kapilvastu, Shravasti, Kaushambi, Sansika) contributed to about 6.46% of the total foreign tourist visit in India (ORF special report, 2019).

Kumbh Mela is considered to be the largest pilgrim congregation on Earth. The Maha Kumbh Mela occurs every 12 years and rotates between four cities of Haridwar, Ujjain, Nasik, and Allahabad. During 2013 the 55 days long Kumbh Mela in Allahabad attracted more than 70 million pilgrims (Tharoor, 2013). The birds eye view of the festival gives a colourful image of thousands of tents with overpowering lights on the flood plain of Ganga, huge human congregation, the fragrance of different kinds, loudspeakers broadcasting religious chants, food cooked for millions, and yet all this occurs with inexplicable harmony (Buzinde, 2014). The visitors tend to experience spirituality, religious values, power of pilgrimage, and social unity through participation in the mega event.

To manage pilgrim and religious tourism, many states have developed institutional structures that turn out to be successful model examples include Jagnath puri of Puri in Orissa, Tirumala-Tirupati in Andhra Pradesh, Nathdwara in Rajasthan, Vaishno Devi in Jammu (Reddy, 2006). Every year Hajj to Mecca in Saudi Arabia brings the largest congregation of visitors which is managed by the Ministry of Hajj. Sai baba Shrine is administered by Sai Sansthan trust which acts as an intermediary for visitors and provides liturgical assistance, lodging, and boarding service and serves approx. 30000 visitors daily to about 80000 visitors on weekends (Shinde, 2012). There exists an intricate connection between the multifaceted nature of pilgrimage and tourism (Timothy & Olsen, 2006). Vrindavan is the famous pilgrim site for devotees to Krishna. The people gather to participate in worship, perform the ritual and involve in song, dance, drama, poetry all narrating the glory of Krishna. Religious activities contribute to the pilgrimage economy. Though devotes visit the place in large numbers, the place lacks effective management unlike the Jagnath puri of Puri in Orissa, Tirumala-Tirupati in Andhra Pradesh, Nathdwara in Rajasthan, Vaishno Devi in Jammu. The ineffective management of visitors to Vrindavan resulting in an extra load to infrastructure and problems of solid waste, roads, sewage system, water supply, and environmental damage (Shinde, 2011). This example explains the explicit need and role of stakeholders, government working at different level, professional management for the success of the religious event. The synergy between religious events and the tourism industry led to better experience and better management of the event. Effective management of visitors participating in the religious events not only led to extensive control over the religious site but also generate a good amount of revenue.

Religious events and pilgrimage are critical wings of the Indian tourism industry driving a major portion of domestic tourism. People are widely travelling for religious reasons along with indulging in cultural, pleasure, and recreational needs, leading to subsequent financial impact (Belk and Yeh, 2011). Religious tourists form an integral part of the tourism industry and assist in better understanding of each other (Rashid, 2018). Many a time the local religious ritual (e.g., Yanshuei fireworks festival in Taiwan) gain popularity as a tourist attraction drawing a large no. of visitors to the religious event, and provide a distinct image to the place (Cheng et al., 2015). Successful organization and management of religious events require service providers such as infrastructure (Durgamohan, 2015), police force for safety and security (Tyagi and Dhar, 2014), risk management, transportation (Sharma and Sharma, 2015), and other amenities and services (Hung, 2015). All this demands coordination, cooperation, and teamwork between government and private organizations.

The Impact of the COVID-19 Disease on Religious Events

In a religiously diversified country like India, religious events are a part of culture which binds the citizens with the thread of unity. Religious events take place in an incessant manner at every corner of the country marks the subcontinent as land of festivals. Events like fairs, festivals and rituals in India are considered as significant part of religious tourism in national and international level. Each event in India has some unique features, essence and experience that reflect the culture of the country. The events not only promote the destination but also simultaneously act as identity for the region, people and the religion as well. Some events are area specific whereas many events used to take place all over

Figure 1. Tourism industry as the facilitator for the successful organization of the religious events
Source: (ARC, 2011; Shinde, 2011; ORF special report, 2019; Verma & Sarangi, 2019)

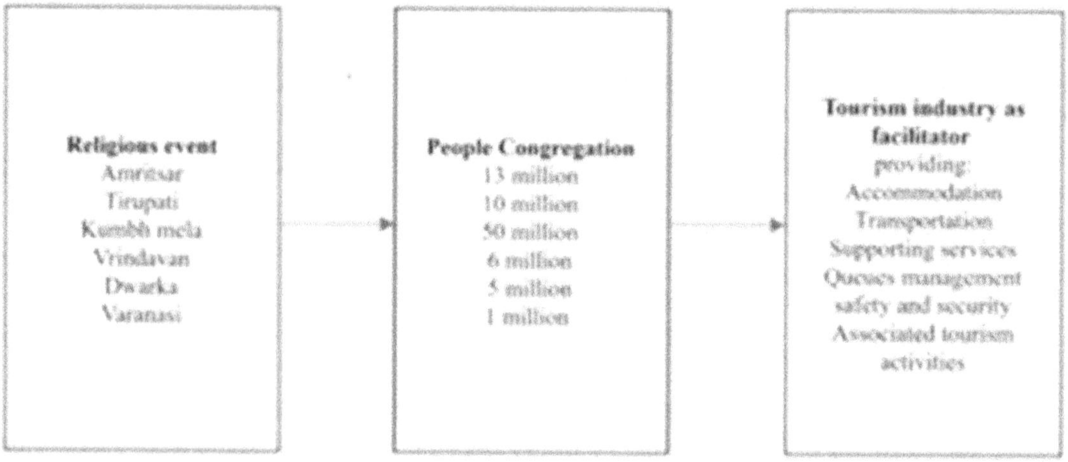

the country. In the second most populous country where religion is associated with people sentiments observe large crowd in festivals and events. In that scenario, religious events can be categorized as small, large and mega events based on the gathering and number of participant of the event.

From its origin the novel corona virus or COVID-19 started to spread rapidly through direct human interactions or activities. Events and festivals are the occasions where mass gathering can't be avoided. In the initial precautionary measure the events and mass gathering were postponed or cancelled throughout the globe. In amid of severity of the pandemic Government of India has put halt on all types of gatherings, events, congregations so as on the religious events. Following the orders and guidelines a number of scheduled events were cancelled where as some events were postponed throughout the country.

Cancellation and Postponement of Religious Events

The outbreak of the pandemic has forced the government to take immediate action to regulate several events. From starting of the year 2020 partial lockdown takes place in several major cities following in late March complete shutdown was imposed nationwide which stopped all kind of movement in the country. In that scenario, the possibilities to conduct the events were restricted. Many of the time bound events were cancelled where several events were rescheduled till the situation come normal. Mega events having religious importance were cancelled on the priority basis during the initial lockdown period whereas small events like marriage, funeral, and worships were also face the same.

In India many, several events used to happen throughout calendar year and many events follow the seasonal calendar across the region. Events from all religions, from regional level to national level got affected by the rapid spread pandemic. In Hinduism a number of events have religious significance celebrated over the country. Over the pandemic issue many of the events were faced cancellation are like: Amarnath yatra, Holi celebrations, regional New Year celebrations, Ganesh Chaturthi at Maharastra, Durga puja and many more. The Islamic religious events like Eid-ul-Fitr, Ramzan, and Hajj pilgrimage also faced cancellation for the last year. The condition was similar in the Christianity where the religious events like weekly worships, Good Friday, Easter and Christian New year celebration was restricted from last year. The Buddha Jayanti and Mahavir Jayanti were the most sacred events for Buddhism and Jainism also cancelled last year keeping in view of the COVID-19 pandemic. All the afore said events have a significant contribution to the tourism industry which faced a severe impact because of the cancellation of the events.

Economic Impacts

The enforced cancellation or postpone of the events has massive impacts on the society among which the economic impact was most visible. The economy has affected from the micro level of the society to the macro level of the country. The religious events not only act as attraction in tourism industry these are also source of income for many people of the society. The religious events in India are a major component of the vast informal sector of the country.

The income of the local people significantly raises during the Amarnath yatra which is a largest Hindu festival happens in Jammu & Kashmir. More than three lakh devotees came for the pilgrimage which continues over two months period (Times, 2019). The walking stick for the pilgrimage has a huge demand that time and business crosses over 40 lakhs (Manhotra, 2017). The shopkeepers, restaurants, home stays and even carrier service providers sensed the huge impact of the cancellation of the mega event.

Though Rath Yatra (Chariot festival) at Puri happened last year but the restrictions imposed at the destination has cut down the income for the local people. The residents used to earn heavily from the 10 days festival including the priests and other service providers. Because of the massive restriction last year and present year the local's income was hugely affected (Barik, 2020).

The pandemic has cancelled several festivals like Durga puja, Holi, Diwali and Ganesh Chaturthi which has put an economic impacts on local income nationwide. The sudden havoc has restricted the income source of many people and the threat of uncertainty still continues among them.

Socio-Cultural Impact

The COVID-19 pandemic rule out a number of religious events which have socio-cultural impact on the society as well as on individuals. The religious events provides an opportunity for social gathering that strengthen the relationship between the community members. From last year the social attachment of the people has restricted amid of the corona restrictions and guidelines. The Government guidelines limit the visit of temple, mosque, church and other religious places for the worshipers and devotees.

The pandemic has amplified the issue of social and economic inequalities in several societies. In India the social issues like lower income groups, cast, and religious beliefs were more influenced by the pandemic situation. The perception of the citizens got a negative influence by cases like mismanagement of events like Tablagi Jamat and Kumbha mela (Biswas, Chatterjee, & Sultana, 2021).

Psychological Impact

The culture and practice in India are quite different from western countries like, Indian prefers to be more social oriented and observe different fairs, festival and get together across the year (Grover et al., 2020). Several psychological issues were observed during the consistent shutdown of external practices like fear, loneliness, anger and traumatic stress etc. The sudden enforcement of the guidelines and straight continuation of the orders created a mental instability among the citizen.

Why Religious Tourism as a Recovery Mechanism Matters?

The COVID-19 pandemic has had a significant negative influence on the world's tourism sector, resulting in disruptions and economic downturns (Dube, Nhamo, & Chikodzi, 2021). Exploring new approaches to boost tourism becomes essential as nations progressively emerge from the crisis (Sharma, Hassan, & Mohanty, 2022). In this section of the paper, of religious tourism is explored as a potential engine for the post-COVID-19 tourist revival. This essay intends to shed light on the possible role of religious tourism in reviving the tourism industry by examining the significance of religious tourism, its features, and the benefits it brings in the current situation.

Recognising the distinctive qualities of religious tourism aids in identifying its potential as a driving force behind the post-COVID-19 tourism rebound (Mosier, Elhadary, Elhaty, & Safaei, 2020). These features set religious tourism apart from other types of travel and emphasise its attraction to a broad variety of travellers. Religious tourism is particularly well-suited to guiding post-COVID-19 tourism recovery efforts due to a number of its benefits (Manhas & Balakrishnan Nair, 2020). These benefits are in line with how tourists' preferences and demands are changing in the contemporary environment.

As a form of travel that includes visits to holy places, places of pilgrimage, and religious festivals, religious tourism is of utmost importance. It has long been a component of international travel, drawing millions of visitors every year. Religious travel will be even more significant after COVID-19 because of its distinctive qualities and capacity to support long-term healing (Suarja, Sudiarta, Armoni, Sutarma, & Jendra, 2023).

Religious sites house cultural and spiritual history and represent long-standing customs and rituals (Huang, Pearce, Guo, & Shen, 2020). They act as living museums, providing a window into the culture, rituals, and aesthetics of other religions. For visitors looking to learn more about another culture, these

Figure 2. Prospects and paradoxes of religious tourism as a recovery mechanism
Source: Authors' own work

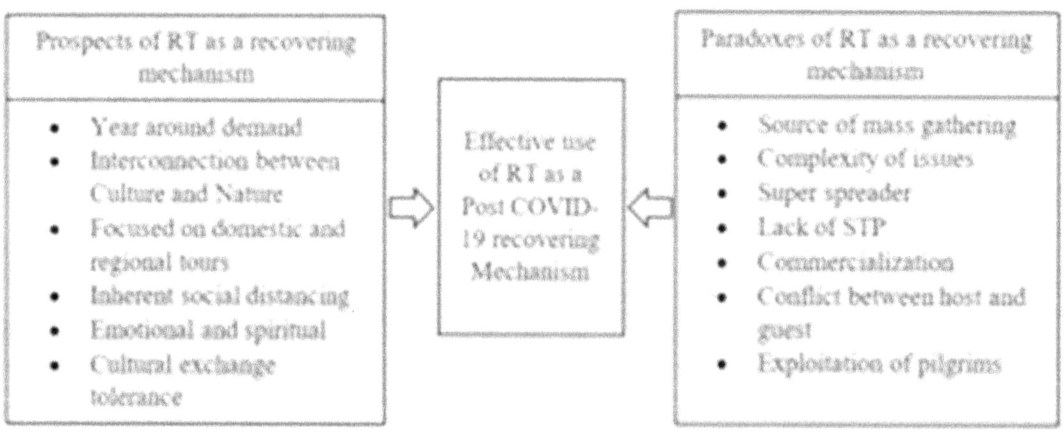

locations offer a genuine and immersive experience. Religious travel is a common way for people to find comfort, inner calm, and personal fulfilment. Travelling to holy places on pilgrimage can be a life-changing experience that encourages spiritual development and strengthens religious identity. Religious travel can be a means of healing and rejuvenation in a post-pandemic environment when mental health is vital.

Religious tourism, in contrast to seasonal tourism, has a rather constant demand all year long. Regardless of the season, pilgrimage sites and religious celebrations frequently draw visitors and devotees, keeping this area of the tourism business durable (Konstantinidis, Mystridis, Tsagkalidou, & Rizopoulou, 2021). Religious sites are frequently located in beautiful settings and culturally diverse areas, giving visitors the chance to experience a variety of cultural and natural treasures. This connection boosts religious tourism's allure by providing a comprehensive experience that goes beyond religious practises alone.

Domestic and regional travellers who have easy access to holy places in their own countries or close-by regions make up the majority of the market for religious tourism. Promoting domestic and regional tourism is crucial as long as there are still limits on and uncertainty surrounding international travel (Manhas & Balakrishnan Nair, 2020). Reviving local economies and communities can benefit greatly from religious tourism.

Social distancing—a vital component of post-pandemic travel—is naturally facilitated by religious sites' broad settings and grounds (Mosier et al., 2020). While participating in religious ceremonies and practises, visitors can keep their physical distance, ensuring their safety and following all public health regulations.

The pandemic has highlighted the significance of mental, emotional, and spiritual health (Papazoglou et al., 2021). Religious travel offers a venue for people to find peace, pursue spiritual solace, and regain emotional equilibrium, which is especially important in the wake of a major international catastrophe. Travellers' general wellbeing can be enhanced by the tranquilly and sense of community they encounter at religious sites, which helps to support their resilience and mental health.

Religious travel has a strong relationship to nature and cultural heritage, which frequently fosters sustainable practises (Lin & Hsieh, 2022). Numerous religious sites place a strong emphasis on environmental protection, sustainable practises, and ethical travel. Religious tourism may promote local communities, preserve the environment, and foster long-term socioeconomic advantages through fostering sustainable travel.

CONCLUSION

With six major religions coexisting inside its borders, India is renowned for its religious variety. Numerous religious holidays are observed throughout the nation, which have a big impact on several industries, especially tourism. About 9% of India's GDP comes from tourism, which also employs a sizable section of the labour force. Domestic travel is significantly influenced by religious gatherings and pilgrimages. At the same time, the COVID-19 pandemic epidemic has significantly disrupted India's tourism sector. In response, the public and private sectors are concentrating on domestic travel as a means of recovery. It is anticipated that religious gatherings and pilgrimage tourism would be essential to this process. Ironically, because these activities have the potential to be extremely effective virus spreaders, they have drawn criticism. For Indian tourism administrators, striking the correct balance between religious festivities and pilgrims' safety has proven difficult. Due to the various COVID-19 wave and the limited vaccination supply, it is essential to resume operations with the appropriate policies and procedures in place. This circumstance offers a chance to reimagine religious travel and pilgrimage events in a more sustainable and distinctive way. Three goals are the focus of this essay. It starts by looking at how the Indian tourism business and religious activities overlap. Second, it talks about how COVID-19 affects religious occasions. Finally, it looks at the opportunities, themes, trends, and difficulties that will arise in the post-COVID-19 recovery of religious events and pilgrimage tourism. The work uses thematic content analysis to examine secondary data from pertinent literature and adheres to a conceptual research design. It adopts a theoretical stance and places more emphasis on finding opportunities and overcoming obstacles than on reaching firm conclusions. The article emphasises the importance of sustainability in pilgrimage tourism and religious events overall and offers suggestions for the industry's revival.

Due to its intrinsic relevance, distinctive qualities, and advantages in the current situation, religious tourism is emerging as a promising theme for the post-COVID-19 tourism recovery. By leveraging its cultural, spiritual, and transformative elements, religious tourism offers a path to sustainable recovery as nations work to revitalise their tourism industries. Religious tourism is a realistic alternative in a world still dealing with travel-related difficulties because of the allure of religious places, their year-round demand, and the possibility for domestic and regional tourism.

Additionally, it supports social distancing, mental wellbeing, and sustainable practises, all of which are in line with the changing traveller preferences. Recognising the potential of religious tourism and making investments in its growth allow destinations to take advantage of a niche market, boost regional economies, and promote harmony and cultural interchange. As a result, efforts to revive the tourism industry after COVID-19 should take religious tourism into account.

REFERENCES

Ahmed, Q. A., & Memish, Z. A. (2019). From the "Madding Crowd" to mass gatherings – Religion, sport, culture and public health. *Travel Medicine and Infectious Disease*, *28*, 91–97. doi:10.1016/j.tmaid.2018.06.001 PMID:29879514

ARC (Alliance of religions and Conservation). (2011). Microsoft Word - ARC pilgrimage statistics 155m 11-12-19.doc. ARC.

Barik, S. (2020). Odisha temple priests to resort to unusual protest. *The Hindu.* https://www.thehindu.com/news/national/other-states/odisha-temple-priests-to-resort-to-unusual-protest/article32947907.ece

BBC. (2020). Coronavirus: Search for hundreds of people after Delhi prayer meeting. *BBC.* https://www.bbc.com/news/world-asia-india-52104753

Belk, R., & Yeh, J. H.-Y. (2011). Tourist photographs: Signs of self. *International Journal of Culture, Tourism and Hospitality Research*, *5*(4), 345–353. doi:10.1108/17506181111174628

Bhonsale, M. (2019). Religious Tourism as Soft Power: Strengthening India's Outreach to Southeast Asia. *ORF Special Report No. 97.* Observer Research Foundation. SR 97-new text (orfonline.org)

Biswas, D., Chatterjee, S., & Sultana, P. (2021). Stigma and fear during COVID-19: Essentializing religion in an Indian context. *Humanities & Social Sciences Communications*, *8*(1), 130. doi:10.105741599-021-00808-8

Buzinde, C. N., Kalavar, J. M., Kohli, N., & Manuel-Navarrete, D. (2014). Emic understandings of Kumbh Mela pilgrimage experiences. *Annals of Tourism Research*, *49*, 1–18. doi:10.1016/j.annals.2014.08.001

Census. (2011). Religion. *Census India.* https://censusindia.gov.in/Census_and_You/religion.aspx

Cheng, T.-M., Chang, S., & Dai, Z. (2015). The mediating effect of situation involvement of novelty seeking and satisfaction for Yanshuei fireworks festival in Taiwan. *International Journal of Culture, Tourism and Hospitality Research*, *9*(2), 154–167. doi:10.1108/IJCTHR-10-2014-0084

Chiriyankandath, J. (2000). 'Creating a secular state in a religious country': The debate in the Indian constituent assembly. *Commonwealth and Comparative Politics*, *38*(2), 1–24. doi:10.1080/14662040008447816

Dube, K., Nhamo, G., & Chikodzi, D. (2021). COVID-19 cripples global restaurant and hospitality industry. *Current Issues in Tourism*, *24*(11), 1487–1490. doi:10.1080/13683500.2020.1773416

Durgamohan, M. (2015). Cultural and economic significance of Hampi festival. *Worldwide Hospitality and Tourism Themes*, *7*(4), 377–392. doi:10.1108/WHATT-03-2015-0013

Gladstone, D. L. (2005). *From pilgrimage to package tour: Travel and tourism in the third world.* Routledge.

Grover, S., Sahoo, S., Mehra, A., Avasthi, A., Tripathi, A., Subramanyan, A., Pattojoshi, A., Rao, G. P., Saha, G., Mishra, K. K., Chakraborty, K., Rao, N. P., Vaishnav, M., Singh, O. P., Dalal, P. K., Chadda, R. K., Gupta, R., Gautam, S., Sarkar, S., ... Janardran Reddy, Y. C. (2020). Psychological impact of COVID-19 lockdown: An online survey from India. *Indian Journal of Psychiatry*, *62*(4), 354–362. doi:10.4103/psychiatry.IndianJPsychiatry_427_20 PMID:33165368

Huang, K., Pearce, P., Guo, Q., & Shen, S. (2020). Visitors' spiritual values and relevant influencing factors in religious tourism destinations. *22*(3), 314-324. doi:10.1002/jtr.2337

Hung, K., Wang, S., & Tang, C. (2015). Understanding the normative expectations of customers toward Buddhism-themed hotels: A revisit of service quality. *International Journal of Contemporary Hospitality Management*, *27*(7), 1409–1441. doi:10.1108/IJCHM-12-2012-0264

Jena, A., Yadav, R. A., & Rambarki, R. (2021). Demonizing the others: Vendetta coverage of Tablighi Jamaat and Kumbh Mela during the COVID-19 pandemic in India. *Media Asia*, *48*(4), 346–351. doi:10.1080/01296612.2021.1946282

Konstantinidis, C., Mystridis, C., Tsagkalidou, E., & Rizopoulou, E. (2021). The Religious Tourism as a Competitive Advantage of the Prefecture of Pieria, Greece. *Journal of Tourism Hospitality Management*, *9*(3), 173–181. doi:10.17265/2328-2169/2021.03.005

Lin, L.-P., & Hsieh, W.-K. (2022). Exploring how perceived resilience and restoration affected the wellbeing of Matsu pilgrims during COVID-19. *Tourism Management*, *90*, 104473. doi:10.1016/j.tourman.2021.104473

Manhas, P. S., & Balakrishnan Nair, B. (2020). Strategic role of religious tourism in recuperating the Indian tourism sector post-covid-19. *International Journal of Religious Tourism Pilgrimage*, *8*(7), 6. doi:10.21427/ka25-fq52

Manhotra, D. (2017). Local residents cash in on Amarnath yatra. *Tribune India*. https://www.tribuneindia.com/news/archive/j-k/local-residents-cash-in-on-amarnath-yatra-434197

Mosier, W., Elhadary, T., Elhaty, I. A., & Safaei, M. (2020). Crisis management and the impact of pandemics on religious tourism. *International Journal of Religious Tourism Pilgrimage*, *8*(7), 3. doi:10.21427/6808-ct83

NCAER. (2003). *Domestic Tourism Survey: 2002–2003*. National Council of Applied Economic Research and Ministry of Tourism and Culture, Government of India.

Papazoglou, A. S., Moysidis, D. V., Tsagkaris, C., Dorosh, M., Karagiannidis, E., & Mazin, R. (2021). Spiritual Health and the COVID-19 Pandemic: Impacts on Orthodox Christianity Devotion Practices, Rituals, and Religious Pilgrimages. *Journal of Religion and Health*, *60*(5), 3217–3229. doi:10.100710943-021-01347-x PMID:34263390

Rashid, A. G. (2018). Religious tourism – a review of the literature. *Journal of Hospitality and Tourism Insights*, *1*(2), 150–167. doi:10.1108/JHTI-10-2017-0007

Reddy, S.S. (2006*). E-Governance in temple administration: Tirumala Tirupati Devasthanams.*

Sharma, A., Hassan, A., & Mohanty, P. (Eds.). (2022). *COVID-19 and the Tourism Industry: Sustainability, Resilience and New Directions*. Routledge. doi:10.4324/9781003207467

Sharma, S., & Sharma, S. (2015). Understanding Pithoragarh's cultural jewels: Its fairs and festivals. *Worldwide Hospitality and Tourism Themes*, *7*(4), 331–346. doi:10.1108/WHATT-03-2015-0007

Shinde, K. (2012). Policy, planning, and management for religious tourism in Indian pilgrimage sites. *Journal of Policy Research in Tourism, Leisure & Events*, *4*(3), 277–301. doi:10.1080/19407963.2012.726107

Shinde, K. A. (2010). Entrepreneurship and indigenous entrepreneurs in religious tourism in India. *International Journal of Tourism Research*, *12*(5), 523–535.

Shinde, K. A. (2011). This Is Religious Environment. *Space and Culture*, *14*(4), 448–463. doi:10.1177/1206331211412280

Suarja, I. K., Sudiarta, M., Armoni, N. L. E., Sutarma, I. G. P., & Jendra, I. W. (2023). Spiritual Tourism: Self-Healing through Melukat Ritual in Bali. *International Journal of Multicultural Multireligious Understanding*, *10*(3), 47–55.

Tharoor, I. (January 15, 2013). The world of the Kumbh Mela: Inside the largest single gathering of humanity. *Time Magazine*. https://world.time.com/2013/01/15/the-kumbh-mela-inside-the-worlds-single-largest-gathering-ofhumanity/

Thommandru, A., Espinoza-Maguiña, M., Ramirez-Asis, E., Ray, S., Naved, M., & Guzman-Avalos, M. (2023). Role of tourism and hospitality business in economic development. *Materials Today: Proceedings*, *80*, 2901–2904. doi:10.1016/j.matpr.2021.07.059

Times, T. E. (2019). Amarnath yatra crosses last year's figure of 2.85 lakh pilgrims. *Economic Times*. https://economictimes.indiatimes.com/news/politics-and-nation/amarnath-yatra-crosses-last-years-figure-of-2-85-lakh-pilgrims/articleshow/70345998.cms

Timothy, D. J. (2013). Religious views of the environment: Sanctification of nature and implications for tourism. In A. Holden & D. Fennell (Eds.), *The Routledge handbook of the tourism and the environment* (pp. 31–42). Routledge.

Tyagi, T., & Dhar, R. L. (2014). Factors affecting health of the police officials: Mediating role of job stress. *Policing*, *37*(3), 649–664. doi:10.1108/PIJPSM-12-2013-0128

UNWTO. (2011). *Religious tourism in Asia and the Pacific*. Author.

Verma, M., & Sarangi, P. (2019). Modeling attributes of religious tourism: A study of Kumbh Mela, India. *Journal of Convention & Event Tourism*, *20*(4), 296–324. Advance online publication. doi:10.1080/15470148.2019.1652124

Verma, M., Verma, A., & Khurana, S. (2018). Influence of travel motivation and demographic factors on tourist participation in World's largest mass religious gathering – The Kumbh Mela, Prabandhan. *Indian Journal of Management*, *11*(8), 7. doi:10.17010/pijom/2018/v11i8/130641

Compilation of References

Abbas, A. F., Jusoh, A., Mas'od, A., Alsharif, A. H., & Ali, J. (2022). Bibliometrix analysis of information sharing in social media. *Cogent Business and Management*, *9*(1), 2016556. doi:10.1080/23311975.2021.2016556

Abbasian, S., Onn, G., & Arnautovic, D. (2020). Overtourism in Dubrovnik in the eyes of local tourism employees: A qualitative study. *Cogent Social Sciences*, *6*(1), 6. doi:10.1080/23311886.2020.1775944

Abbate, C. S., & Di Nuovo, S. (2013). Motivation and personality traits for choosing religious tourism: A research on the case of Medjugorje. *Current Issues in Tourism*, *16*(5), 501–506. doi:10.1080/13683500.2012.749844

Abdulhanaa. (2021). A Review of Islamic Economic Law on Religious Tourism Arrangements in South Sulawesi. *Samarah: Jurnal Hukum Keluarga dan Hukum Islam*, *5*(1), 450-470. http://dx.doi.org/ doi:10.22373/sjhk.v5i1.9088

Adetunji, J. (2022, February 9). How Lourdes became a byword for hope. *The Conversation*. https://theconversation.com/how-lourdes-became-a-byword-for-hope-174574

Ahmad, A., Rahman, M. A., Rehman, F. U., Lbath, A., Afyouni, I., Khelil, A., & Wahiddin, M. R. (2014). A framework for crowd-sourced data collection and context-aware services in Hajj and Umrah. In *2014 IEEE/ACS 11th International Conference on Computer Systems and Applications (AICCSA)* (pp. 405-412). IEEE. 10.1109/AICCSA.2014.7073227

Ahmad, A., Rahman, M. A., Ridza Wahiddin, M., Ur Rehman, F., Khelil, A., & Lbath, A. (2018). Context-aware services based on spatio-temporal zoning and crowdsourcing. *Behaviour & Information Technology*, *37*(7), 736–760. doi:10.1080/0144929X.2018.1476586

Ahmed, Q. A., & Memish, Z. A. (2019). From the "Madding Crowd" to mass gatherings – Religion, sport, culture and public health. *Travel Medicine and Infectious Disease*, *28*, 91–97. doi:10.1016/j.tmaid.2018.06.001 PMID:29879514

Ainsworth, M. D. S., & Bowlby, J. (1991). An ethological approach to personality development. *The American Psychologist*, *46*(4), 331–341. doi:10.1037/0003-066X.46.4.333

Alam, S. S., & Choudhury, S. R. (2020). Water and Sanitation Services in Tourist Places: A Review of Policies and Strategies. *Journal of Cleaner Production*, *253*, 119926. doi:10.1016/j.jclepro.2019.119926

Ali, M. S., & Zaman, S. U. (2019). Water and Sanitation Services in Tourist Places: A Review of Innovative Solutions and Best Practices. *Journal of Environmental Management*, *250*, 109523. doi:10.1016/j.jenvman.2019.109523

Al-Makhadmah, I. M. (2020). The role of virtual museum in promoting religious tourism in Jordan. *Geo Journal of Tourism and Geosites*, *28*(1), 268–274. doi:10.30892/gtg.28121-468

Almuhrzi, H. M., & Alsawafi, A. M. (2017). Muslim perspectives on spiritual and religious travel beyond Hajj: Toward understanding motivations for Umrah travel in Oman. *Tourism Management Perspectives*, *24*(1), 235–242. doi:10.1016/j.tmp.2017.07.016

Alnsour, Y., & Al-Abdallah, G. (2020). The impact of service quality on the satisfaction of Hajj pilgrims: An empirical study in Makkah. *Journal of Tourism and Hospitality Management*, *8*(1), 12–25.

Al-Shaer, H., Algharabat, R., & Altarawneh, I. (2020). Service quality of sanitation facilities in Hajj pilgrimage: A case study of Mina camps. *Journal of Islamic Marketing*, *11*(3), 768–781.

Amaro, S., Antunes, A., & Henriques, C. (2018). A closer look at Santiago de Compostela's pilgrims through the lens of motivations. *Tourism Management*, *64*, 271–280. doi:10.1016/j.tourman.2017.09.007

Ambrosio, V. (2015). *Sacred pilgrimage and tourism as secular pilgrimage. Religious Tourism and Pilgrimage Management–An International Perspective*. CABI.

Amore, A., Falk, M., & Adie, B. A. (2020). One visitor too many: Assessing the degree of overtourism in established European urban destinations. *International Journal of Tourism Cities*, *1*(6), 117–137. doi:10.1108/IJTC-09-2019-0152

ARC (Alliance of religions and Conservation). (2011). Microsoft Word - ARC pilgrimage statistics 155m 11-12-19. doc. ARC.

Aria, M., & Cuccurullo, C. (2017). bibliometrix: An R-tool for comprehensive science mapping analysis. *Journal of Informetrics*, *11*(4), 959–975. doi:10.1016/j.joi.2017.08.007

Arokiaraj, D., Ganeshkumar, C., & Paul, P. V. (2020). Innovative management system for environmental sustainability practices among Indian auto-component manufacturers. *International Journal of Business Innovation and Research*, *23*(2), 168–182. doi:10.1504/IJBIR.2020.110095

Ashworth, G. J., & Tunbridge, J. E. (2000). The tourist city. In *Advances in Tourism Research,* (pp. 53–81). Science Direct. https://www.sciencedirect.com/science/article/abs/pii/B9780080436753500077

Atout France/ European Travel Commission. (2022). *Restarting tourism for the better: Performance of European tourism before, during & after COVID-19*. ETC. https://etc-corporate.org/uploads/2022/03/2022_Performance-of-European-tourism-before-during-and-beyond the Covid-19-pandemic.pdf

Badone, E. (2008). Pilgrimage, tourism and The Da Vinci Code at Les-Saintes-Maries-de-la-Mer, France. *Culture and Religion*, *9*(1), 23–44. doi:10.1080/14755610801954847

Balakrishnan Nair, B. (2020). Strategic role of religious tourism in recuperating the Indian tourism sector post-covid-19. *International Journal of Religious Tourism and Pilgrimage*, *8*(7), 6.

Bansiddhi, P., Brown, J. L., Thitaram, C., Punyapornwithaya, V., & Nganvongpanit, K. (2020). Elephant Tourism in Thailand: A Review of Animal Welfare Practices and Needs. *Journal of Applied Animal Welfare Science*, *23*(2), 164–177. doi:10.1080/10888705.2019.1569522 PMID:30689409

Barber, P. (2001). Mito, religione e conoscenza: la mappa del mondo medievale. In C. Pirovano (Ed.), Segni e sogni della terra. Il disegno del mondo dal mito di Atlante alla geografia delle reti (pp. 48–79). De Agostini.

Barber, R. (1993). *Pilgrimages*. The Boydell Press.

Barik, S. (2020). Odisha temple priests to resort to unusual protest. *The Hindu*. https://www.thehindu.com/news/national/other-states/odisha-temple-priests-to-resort-to-unusual-protest/article32947907.ece

Barja, I., Silvan, G., Rosellini, S., Pinerio, A., Gonzalez, A., Camacho, L., & Illera, J. C. (2007). Stress physiological responses to tourist pressure in a wild population of European pine marten. *The Journal of Steroid Biochemistry and Molecular Biology*, *104*(3-5), 136–142. doi:10.1016/j.jsbmb.2007.03.008 PMID:17467268

Bausch, T., Schröder, T., & Gunya, A. (2020). Pilgrimage as a field of tourism development: The case of Kavala (Northern Greece) and its institutional and cultural challenges. *International Journal of Religious Tourism and Pilgrimage*, *8*(3), 111–129.

Bayih, B. E. (2018). Potentials and Challenges of Religious Tourism Development in Lalibela, Ethiopia. *African Journal of Hospitality, Tourism and Leisure*, *7*(4).

BBC. (2020). Coronavirus: Search for hundreds of people after Delhi prayer meeting. *BBC*. https://www.bbc.com/news/world-asia-india-52104753

Bec, A., Moyle, B., Timms, K., Schaffer, V., Skavronskaya, L., & Little, Ch. (2019). Management of immersive heritage tourism experiences: A conceptual model. *Tourism Management*, *72*, 117–120. doi:10.1016/j.tourman.2018.10.033

Behrens, K. (2009). TonyYengeni's Ritual Slaughter: Animal Anti-Cruelty vs. Culture. *South African Journal of Philosophy*, *28*(3), 271–289. doi:10.4314ajpem.v28i3.47859

Belk, R., & Yeh, J. H.-Y. (2011). Tourist photographs: Signs of self. *International Journal of Culture, Tourism and Hospitality Research*, *5*(4), 345–353. doi:10.1108/17506181111174628

Benyo, G. (2023). *Calvary above the city*. Rest Buda.: https://pestbuda.hu/cikk/20221105_golgota_a_varos_felett_20_eve_ujult_meg_pesthidegkut_kalvariaja

Berger, P., & Luckmam, T. (1967). *The Social construction of reality: A treatise in the sociology of knowledge*.

Bertella, G. (2014). The Co-creation of Animal-based Tourism Experience. *Tourism Recreation Research*, *39*(1), 115–125. https://www.trrworld.org/. doi:10.1080/02508281.2014.11081330

Bhat, F. A. (2018). Challenges and sustainable tourism development in Ladakh region. *Journal of Tourism & Hospitality (Los Angeles, Calif.)*, *7*(3), 1–7.

Bhatt, G. P. (2004). The Skanda Purana. Motilal Banarasidass Publishing House.

Bhattacharya, A., & Ghosh, A. (2016). Economic impact of tourism: A case study of Ladakh. *Luyou Xuekan*, *31*(10), 38–45.

Bhat, Z. A. (2013). Tourism industry and pilgrimage tourism in Jammu and Kashmir: Prospects and challenges. *International Journal of Research In Management & Technology*, *2*, 105–113.

Bhava, I. (2022). *India Tourism Statistics at a Glance 2022*. Incredible India.

Bhonsale, M. (2019). Religious Tourism as Soft Power: Strengthening India's Outreach to Southeast Asia. *ORF Special Report No. 97*. Observer Research Foundation. SR 97-new text (orfonline.org)

Bhowmik, P. (2021). Heritage Tourism: A bibliometric review. *Anatolia*, *32*(3), 387–403. doi:10.1080/13032917.2021.1875250

Bilska-Wodecka, E. (2000). *Typology of European calvaries*. RUJ. https://ruj.uj.edu.pl/xmlui/bitstream/handle/item/77831/bilska-wodecka_typology_of_european_calvaries_2000.pdf?sequence=1&isAllowed=y

Biswas, D., Chatterjee, S., & Sultana, P. (2021). Stigma and fear during COVID-19: Essentializing religion in an Indian context. *Humanities & Social Sciences Communications*, *8*(1), 130. doi:10.105741599-021-00808-8

Blackwell, R. (2007). Motivations for religious tourism, pilgrimage, festivals and events. *Religious tourism and pilgrimage festivals management: An international perspective*, 35-47.

Blackwell, R. (2010). Motivation for pilgrimage: Using theory to explore motivations. *Scripta Instituti Donneriani Aboensis*, *22*, 24–37. doi:10.30674cripta.67360

Blau Nautica. (2023). *Underwater Museum.* Blue Nautica. https://www.blue-nautica.com/diving/underwater-museum/

Boaz, D. N. (2019). The "Abhorrent" Practice of Animal Sacrifice and Religious Discrimination in the Global South. *Religions, 10*(3), 160. Advance online publication. doi:10.3390/rel10030160

Boonchuay, T., & Pinit, W. (2018). Improving Water and Sanitation Services in Tourist Places: A Case Study of Phuket, Thailand. *Journal of Environmental Management, 209*, 65–73. doi:10.1016/j.jenvman.2017.12.003

Bouchon, F., & Rauscher, M. (2019). Cities and tourism, a love and hate story; towards a conceptual framework for urban overtourism management. *International Journal of Tourism Cities, 4*(5), 598–619. doi:10.1108/IJTC-06-2019-0080

Bowlby, J. (1958). The nature of the child's tie to his mother. *The International Journal of Psycho-Analysis, 39*, 350–373. PMID:13610508

Bowman, M. S., & Pezzullo, P. C. (2009). What's so 'dark' about 'dark tourism'?: Death, tours, and performance. *Tourist Studies, 9*(3), 187–202. doi:10.1177/1468797610382699

Braslavsky, C., & Fernǹdez, A. L. (n.d.). *A Educaô Secundr̀ia: mudanà ou imutabilidade? - Brasḯa: UNESCO, 2002. United Nations Educational Scientific Cultural Organisation (UNESCO).* Brazil Chapter.

Braun, V., & Clarke, V. (2006). Using thematic analysis in psychology. *Qualitative Research in Psychology, 3*(2), 77–101. doi:10.1191/1478088706qp063oa

Brdesee, H., Corbitt, B., & Pittayachawan, S. (2013). Barriers and motivations affecting Information Systems usage by Hajj–Umrah religious tourism operators in Saudi Arabia. *AJIS. Australasian Journal of Information Systems, 18*(1). doi:10.3127/ajis.v18i1.809

British Broadcasting Corporation [BBC] (Director). (2019, October 24). *Kartarpur corridor: India and Pakistan sign deal on Sikh temple project.* BBC. https://www.bbc.com/news/world-asia-india 50138275

Bryman, A., & Bell, E. (2015). *Business research methods* (4th ed.). Oxford University Press.

Burkart, A. J., & Medlik, S. (1974). *Tourism: past, present and future* (2nd ed.). William Heinemann Ltd.

Butler, M. (2008). *Turismo de cruceros. Situación actual y tendencias.* Organización Mundial del Turismo.

Buzinde, C. N., Kalavar, J. M., Kohli, N., & Manuel-Navarrete, D. (2014). Emic understandings of Kumbh Mela pilgrimage experiences. *Annals of Tourism Research, 49*, 1–18. doi:10.1016/j.annals.2014.08.001

Byrne, D. (2016). Heritage Corridors: Transnational flows and the built environment of migration. *Journal of Ethnic and Migration Studies, 42*(14), 2360–2378. https://doi.com/10.1080/1369183X.2016.1205850. doi:10.1080/1369183X.2016.1205805

Čábyová, Ľ. (2016). Pilgrimage tourism in Slovakia, the potential for development. *European Journal of Science and Theology, 12*(1), 231–244.

Calvaries in Budapest (2023). Budapest. *Csendhegyek* blogspot.com/search/label/Budapest

Campbell, C. K. (1999). *An approach to research in recreational geography.* Department of Geography, University of British Columbia.

Cape Town Declaration on Responsible Tourism. (2002). *International Conference on Responsible Tourism in Destinations*, Cape Town, South Africa.

Carrigan, A. (2014). Dark tourism and postcolonial studies: Critical intersections. *Postcolonial Studies, 17*(3), 236–250. doi:10.1080/13688790.2014.993425

Castro, M. P., & Zermeño, M. G. G. (2020). Being an entrepreneur post-COVID-19–resilience in times of crisis: A systematic literature review. *Journal of Entrepreneurship in Emerging Economies*, *13*(4), 721–746. doi:10.1108/JEEE-07-2020-0246

Ceesay, A. M., Jallow, O. G., & Njie, M. N. (2019). Sustainable Water and Sanitation Services for Tourism Development: A Case Study of the Gambia. *Sustainable Development*, *27*(2), 329–338. doi:10.1002d.1885

Census. (2011). Religion. *Census India.* https://censusindia.gov.in/Census_and_You/religion.aspx

Chadha. Neeru and Choudhary. Pooja (2016), Nature, Problems and Prospects of Rural Tourism in Punjab: An Analysis. *Pacific Business Review International*, *1*(1), 46-48.

Chadwick, I. C., & Raver, J. L. (2020). Psychological resilience and its downstream effects for business survival in nascent entrepreneurship. *Entrepreneurship Theory and Practice*, *44*(2), 233–255. doi:10.1177/1042258718801597

Charles. (2023). Personal Communication. [Discussion].

Chatterjee, A., Bajpai, R., Gerdes, M. W., & Gerdes, M. (2021). *Analyze the Impact of Healthy Behavior on Weight Change with a Mathematical Model using the Harris-Benedict Equations. ("Analyze the Impact of Healthy Behavior on Weight Change with a ...").* Research Square. doi:10.21203/rs.3.rs-584141/v1

Chatterjee, K. (2022). Modern Tourism Development through Sustainable Tourism Approach: A Case study of Pilgrimage Tourism in Varanasi. *The International Journal of Creative Research Thoughts*, *10*(5), 361–372.

Cheer, J. M. (2023). Host Communities and Pilgrimage Tourism. In R. N. Progano, J. M. Cheer, & X. M. Santos (Eds.), *Perspectives on Asian Tourism* (pp. 229–233). Springer.

Cheer, J. M., Belhassen, Y., & Kujawa, J. (2017). The search for spirituality in tourism: Toward a conceptual framework for spiritual tourism. *Tourism Management Perspectives*, *24*, 252–256. doi:10.1016/j.tmp.2017.07.018

Cheng, T.-M., Chang, S., & Dai, Z. (2015). The mediating effect of situation involvement of novelty seeking and satisfaction for Yanshuei fireworks festival in Taiwan. *International Journal of Culture, Tourism and Hospitality Research*, *9*(2), 154–167. doi:10.1108/IJCTHR-10-2014-0084

China Discovery. (2022). *Wong Tai Sin Temple – An active Taoist Temple*. China Discovery. https://www.chinadiscovery.com/hong-kong/wong-tai-sin-temple.html

Chiriyankandath, J. (2000). 'Creating a secular state in a religious country': The debate in the Indian constituent assembly. *Commonwealth and Comparative Politics*, *38*(2), 1–24. doi:10.1080/14662040008447816

Chistyakova, O. V. (2020). Ethnic Tourism as Knowing Other. *RUDN Journal of Philosophy*, *24*(4), 720–729. doi:10.22363/2313-2302-2020-24-4-720-729

Chistyakov, D. (2020). The influence of Mass Media on consumer culture: Religious tourism. *Logos*, 140–147. doi:10.24101/logos.2020.58

Choe, J. Y., & O'Regan, M. (2015). Religious tourism experiences in south east. *Asia.*

Chon, K. S. (2018). Sustainable community-based tourism in Ladakh: A destination life cycle perspective. *Journal of Sustainable Tourism*, *26*(5), 753–771.

Clow, K. E., & Beisel, J. L. (1995). Managing consumer expectations of low-margin, high-volume services. *Journal of Services Marketing*, *9*(1), 33–46. doi:10.1108/08876049510079871

Cohen, E. (1992). Pilgrimage and Tourism: Convergence and Divergence. In A. Morinis (Ed.), *Sacred Journeys: The Anthropology of Pilgrimage* (pp. 47–61). Greenwood Press.

Cohen, E. H. (2011). Educational dark tourism at an in populo site: The Holocaust Museum in Jerusalem. *Annals of Tourism Research*, *38*(1), 193–209. doi:10.1016/j.annals.2010.08.003

Cohen, E., & Cohen, S. (2014). A mobilities approach to tourism from emerging world regions. *Current Issues in Tourism*, *18*(1), 11–43. doi:10.1080/13683500.2014.898617

Coleman, Simon, & Eade, J. (Eds.). (2004). *Reframing Pilgrimage Cultures in Motion* (1st ed.). Routledge.

Coleman, S. M. (2023). Pilgrimage Religion. In *Britannica*. Encyclopedia Britannica.

Collins-Kreiner, N., & Wall, G. (2015). Tourism and Religion: Spiritual Journeys and Their Consequences. In Business Media Dordrecht. doi:10.1007/978-94-017-9376-6_34

Collins-Kreiner, N. (2007). Graves as attractions: Pilgrimage-tourism to Jewish holy graves in Israel. *Journal of Cultural Geography*, *24*(1), 67–89. doi:10.1080/08873630709478217

Collins-Kreiner, N. (2010). Current Jewish pilgrimage tourism: Modes and models of development. *Tourism: An International Interdisciplinary Journal*, *58*(3), 259–270.

Collins-Kreiner, N. (2010). Geographers and pilgrimages: Changing concepts in pilgrimage tourism research. *Tijdschrift voor Economische en Sociale Geografie*, *101*(4), 437–448. doi:10.1111/j.1467-9663.2009.00561.x

Collins-Kreiner, N. (2016). Dark tourism as/is pilgrimage. *Current Issues in Tourism*, *19*(12), 1185–1189. doi:10.1080/13683500.2015.1078299

Collins-Kreiner, N. (2016). The lifecycle of concepts: The case of 'Pilgrimage Tourism'. *Tourism Geographies*, *18*(3), 322–334. doi:10.1080/14616688.2016.1155077

Collins-Kreiner, N. (2016). The lifecycle of concepts: The case of 'Pilgrimage Tourism'. *Tourism Geographies: An International Journal of Tourism Space. Place and Environment.*, *18*(3), 322–334.

Collins-Kreiner, N. (2018). Pilgrimage-tourism: Common themes in different religions. *International Journal of Religious Tourism and Pilgrimage*, *6*(1), 8–17.

Collins-Kreiner, N. (2020). Pilgrimage tourism-past, present and future rejuvenation: A perspective article. *Tourism Review*, *75*(1), 145–148. doi:10.1108/TR-04-2019-0130

Collins-Kreiner, N. (Ed.). (2006). *Christian tourism to the Holy Land: Pilgrimage during security crisis*. Ashgate Publishing, Ltd.

Conference lodge. (2022). *Culture recreation*. Conference Lodge. http://www.conferencelodge.hk/en/discovery/culture-recreation.html

Conzen, M. P., & Wulfestieg, B. M. (2001). Metropolitan Chicago's Regional Cultural Park: Assessing the Development of the Illinois & Michigan Canal National Heritage Corridor. *The Journal of Geography*, *100*(3), 111–117. doi:10.1080/00221340108978426

Creswell, J. W. (2013). Steps in conducting a scholarly mixed methods study.

Creswell, J. W., Shope, R., Plano Clark, V. L., & Green, D. O. (2006). How interpretive qualitative research extends mixed methods research. *Research in the Schools*, *13*(1), 1–11.

Cronin, J. J. Jr, & Taylor, S. A. (1992). Measuring service quality: A reexamination and extension. *Journal of Marketing*, *56*(3), 55–68. doi:10.1177/002224299205600304

Csapo, J. (2012). The role and importance of cultural tourism in modern tourism industry. *Strategies for tourism industry-micro and macro perspectives, 10*, 201-212.

Curtin, P. A., & Gaither, T. K. (2007). Global public relations and the circuit of culture. Thousand Oaks, CA: sage. doi:10.4135/9781452224817.n3

Dadamuhamedov, A. (2019). The role of information and communications technologies in pilgrimage tourism in Uzbekistan. *The Light of Islam*, *2019*(1), 17.

Dafni, A., & Böck, B. (2019). Medicinal plants of the Bible—Revisited. *Journal of Ethnobiology and Ethnomedicine*, *15*(1), 57. doi:10.118613002-019-0338-8 PMID:31775790

Dar, H. (2020). Hindu Religious Motivations in Kashmir Valley. *International Journal of Religious Tourism and Pilgrimage*, *8*(3), 1–14.

Datrika, V. M. R., & David, A. (2022). Refined Model of Maslow's Needs Theory in Internet Era. *Organization and Human Capital Development (ORCADEV)*.

Dauda Goni, M., Hasan, H., Naing, N. N., Wan-Arfah, N., Zeiny Deris, Z., Nor Arifin, W., & Abubakar Baaba, A. (2019). Assessment of knowledge, attitude and practice towards prevention of respiratory tract infections among Hajj and Umrah Pilgrims from Malaysia in 2018. *International Journal of Environmental Research and Public Health*, *16*(22), 4569. doi:10.3390/ijerph16224569 PMID:31752224

David, A. (2022). *Influences of Demographic Variables on the Adoption of Green Products by Consumers*.

David, A., Nagarjuna, K., Mohammed, M., & Sundar, J. (2019). Determinant Factors of Environmental Responsibility for the Passenger Car Users. *International Journal of Innovative Technology and Exploring Engineering*.

David, A. (2020). Consumer purchasing process of organic food product: An empirical analysis. [QAS]. *Journal of Management System-Quality Access to Success*, *21*(177), 128–132.

David, A., Kumar, C. G., & Paul, P. V. (2022). Blockchain technology in the food supply chain: Empirical analysis. [IJISSCM]. *International Journal of Information Systems and Supply Chain Management*, *15*(3), 1–12. doi:10.4018/IJISSCM.290014

Davie, G. (2000). *Religion in Modern Europe: A Memory Mutates*. Oxford University Press.

De Ascaniis, S., & Cantoni, L. (2017). Online visit opinions about attractions of the religious heritage: An argumentative approach. *CHURCH, COMMUNICATION AND CULTURE*, *2*(2), 179–202. doi:10.1080/23753234.2017.1350585

Desai, M. (2017). *Banaras Reconstructed. Architecture and Sacred Space in a Hindu Holy City*. University of Washington Press.

Dewan, M. (2020). An Ethnographic Inquiry Towards 'Hare-Rama Hare-Krishna Religious Tourists Food Consumption Habits While on A Holiday in India. [JOGHAT]. *Journal of Gastronomy Hospitality and Travel*, *3*(1), 30–41. doi:10.33083/joghat.2020.28

Di Giovine, M. A., & Elsner, J. (2016). Pilgrimage tourism. In J. Jafari & H. Xiao (Eds.), *Encyclopedia of Tourism*. Springer., doi:10.1007/978-3-319-01384-8_146

Dichoso, T. J. (2015). Lourdes: A uniquely Catholic Approach to Medicine. *The Linacre Quarterly, 82*(1). Catholic Medical Association. https://www.doi.org/10.1179/2050854914Y.0000000034

Digance, J. (2003). Pilgrimage at contested sites. *Annals of Tourism Research*, *30*(1), 143–159. doi:10.1016/S0160-7383(02)00028-2

Digance, J. (2006). Religious and Secular Pilgrimage. In D. J. Timothy & D. H. Olsen (Eds.), *Tourism, Religion and Spiritual Journeys* (pp. 36–48). Routledge.

Dikshit, R. (2018, May 8). Soon, access Kashi Vishwanath temple directly from ghats. *The times of India*. https://timesofindia.indiatimes.com/city/varanasi/soon-access-kashi-vishwanath-temple-directly-from-ghats/articleshow/64072869.cms

Dikshit, R. (2022). 7.35 crore visit Kashi Vishwanath Dham in opening year. *The Times of India*. https://timesofindia.indiatimes.com/city/varanasi/7-35-crore-visit-kashi-vishwanath-dham-in-opening-year/articleshow/96211539.cms

Divya, A. (2021, December 14). Explained: What is changing at the ancient Kashi Vishwanath temple complex? *Indian Express*. https://indianexpress.com/article/explained/kashi-vishwanath-temple-narendra-modi-varanasi-7669709/

Donlon, J. G., Donlon, J. H., & Agrusa, J. (2010). Cultural tourism, camel wrestling, and the tourism 'bubble' in Turkey. Anatolia. *International Journal of Tourism and Hospitality Research*, *21*(1), 29–39.

Du Cros, H., & McKercher, B. (2014). *Cultural tourism*. Routledge. doi:10.4324/9780203790601

Dube, K., Nhamo, G., & Chikodzi, D. (2021). COVID-19 cripples global restaurant and hospitality industry. *Current Issues in Tourism*, *24*(11), 1487–1490. doi:10.1080/13683500.2020.1773416

Duffy, R., & Moore, L. (2011). Global regulations and local practices: The politics and governance of animal welfare in elephant tourism. *Journal of Sustainable Tourism*, *19*(4), 589–604. doi:10.1080/09669582.2011.566927

Durán-Sánchez, A., Álvarez-García, J., del Río-Rama, M. D. L. C., & Oliveira, C. (2018). Religious tourism and pilgrimage: Bibliometric overview. *Religions*, *9*(9), 1–15. doi:10.3390/rel9090249

Durgamohan, M. (2015). Cultural and economic significance of Hampi festival. *Worldwide Hospitality and Tourism Themes*, *7*(4), 377–392. doi:10.1108/WHATT-03-2015-0013

Durnali, M., Orakci, Ş., & Khalili, T. (2023). Fostering creative thinking skills to burst the effect of emotional intelligence on entrepreneurial skills. *Thinking Skills and Creativity*, *47*, 101200. doi:10.1016/j.tsc.2022.101200

Duval, L. (2022). Personal Communication [Interview].

Dwivedi, O. P., & Tiwari, B. H. (1987). *Environmental Crisis and Hindu Religion*. Gitanjali Publishing House.

Dwivedi, R. (2021). Infrastructure planning for sustainable pilgrimage tourism: A case of Uttarakhand, India. *Journal of Tourism Planning and Development*, *18*(3), 317–336.

Dwyer, L., & Kim, C. (2003). Destination competitiveness: Determinants and indicators. *Current Issues in Tourism*, *6*(5), 369–414. doi:10.1080/13683500308667962

Ebrahim, S. H., & Memish, Z. A. (2020). Saudi Arabia's drastic measures to curb the COVID-19 outbreak: Temporary suspension of the Umrah pilgrimage. *Journal of Travel Medicine*, *27*(3), taaa029. doi:10.1093/jtm/taaa029 PMID:32109274

Eck, D. L. (1982). *Banaras. The City of Light*. Penguin India.

Eickelman, D. F., & Piscatori, J. P. (1990). *Muslim Travellers: Pilgrimage, Migration, and the Religious Imagination* (D. F. Eickelman & J. P. Piscatori, Eds.). Routledge.

Elazhary, H. (2017). Cloud-based context-aware mobile applications and framework for Hajj and umrah management. *International Journal of Computer Trends and Technology*, *47*(2), 106–115. doi:10.14445/22312803/IJCTT-V47P115

Enrico, B., Nuccio, M., & Durio, A. (2021). Proximity tourism and cultural amenities: Evidence from a regional museum card. *Tourism Economics*, *27*(1), 187–204. doi:10.1177/1354816619890230

Erica, A., Randall, M., & Marta, T. (2000). *Values and Heritage Conservation.* The Getty Conservation Institute.

Erschbamer, M. (2019). Taming of supernatural entities and animal sacrifice, The synthesis of Tibetan Buddhism and local shamanistic traditions in Northern Sikkim (India). *Études mongoles et sibériennes, centrasiatiques et tibétaines, 50.* . doi:10.4000/emscat.3915

Fahey, F. (2002). Pilgrims or Tourists? *The Furrow*, *53*(4), 213–218.

Farmaki, A., & Antoniou, K. (2017). Politicising dark tourism sites: Evidence from Cyprus. *Worldwide Hospitality and Tourism Themes*, *9*(2), 175–186. doi:10.1108/WHATT-08-2016-0041

Faverio, M., Nortay, J., Diamant, J., & Smith, G. A. (2023). *Online Religious Services Appeal to Many Americans, but Going in Person Remains More Popular.*

Fennel, D. A. (2014). Exploring the boundaries of a new moral order for tourism's global code of ethics: An opinion piece on the position of animals in the tourism industry. *Journal of Sustainable Tourism*, *22*(7), 983–996. doi:10.1080/09669582.2014.918137

Fennell, D. A. (2012). Tourism, animals, and utilitarianism. *Tourism Recreation Research*, *37*(3), 239–249. doi:10.1080/02508281.2012.11081712

Fennell, D. A. (2013). Tourism and Animal Welfare. *Tourism Recreation Research*, *38*(3), 325–340. doi:10.1080/02508281.2013.11081757

Fennell, D. A., & Sheppard, V. (2020). Tourism, animals and the scales of justice. *Journal of Sustainable Tourism*, *29*(3), 314–335. doi:10.1080/09669582.2020.1768263

FICCI. (2012). Diverse Beliefs: Tourism of Faith Religious tourism gains ground. FICCI. https://ficci.in/spdocument/20207/diverse-beliefs-tourism-of-faith.pdf

Field, A. (2014). Discovering statistics using IBM SPSS statistics. *Sage (Atlanta, Ga.).*

Findeasy. (2023). Population and more. https://www.findeasy.in/christian-population-in-india/

Fleetwood, S. (1995). *Collection of Tourism Expenditure Statistics.*

Formichi, C., & O'Connor, P. (2015). Introduction: Overlooked religions in Hong Kong. *Asian Anthropology*, *14*(1), 3–7. doi:10.1080/1683478X.2015.1025594

Fornell, C., & Larcker, D. (1981). Evaluating Structural Equation Models with Unobservable Variables and Measurement Error. *JMR, Journal of Marketing Research*, *18*(1), 39–50. doi:10.1177/002224378101800104

Fourati, J., Issaoui, B., & Zidi, K. (2017). Literature Review of Crowd Management: A Hajj Case Study. In ICINCO, (1) (pp. 346-351). doi:10.5220/0006472103460351

Francois, B., Sternberg, E., & Fee, E. (2012). The Lourdes medical cures revisited. *Journal of the History of Medicine and Allied Sciences*, *69*(1), 135–162. doi:10.1093/jhmas/jrs041 PMID:22843835

Freitag, S. (1989). Introduction: The History and Political Economy of Bañaras. In *Cultural and Power in Banaras* (pp. 1–22). University of California Press., doi:10.1525/9780520313392-005

Friedrich, M., Stone, P. R., & Rukesha, P. (2018). Dark tourism, difficult heritage, and memorialisation: A case of the Rwandan genocide. The Palgrave handbook of dark tourism studies, 261-289. Palgrave.

Fuller, C. J. (2004). The camphor flame: popular Hinduism and society in India (Revised and Expanded ed.), Princeton University Press.

Gabor, M. T. (2016). Faith-based tourism. *Business Mirror*. http://www.businessmirror.com.ph/faith-based-tourism/

Gairola, S. U. (2020). Protection of environment through green pilgrimage. *International Journal of Ecology and Environmental Sciences*, *2*(3), 199–203.

Galav, P., Jain, A., Katewa, S. S., & Nag, A. (2010). Animal healthcare practices by livestock owners at Puskar animal fairs, Rajasthan. *Indian Journal of Traditional Knowledge*, *9*(3), 581–584.

Ganai, I. A. (2017). Pilgrimage tourism development in Jammu and Kashmir: Issues and challenges. *International Journal of Research in Social Sciences*, *7*(1), 261–272.

Ganeshkumar, C., David, A., & Jebasingh, D. R. (2022). Digital transformation: artificial intelligence based product benefits and problems of Agritech industry. In *Agri-Food 4.0*. Emerald Publishing Limited. doi:10.1108/S1877-636120220000027010

Ganeshkumar, C., David, A., Sankar, J. G., & Saginala, M. (2023). Application of Drone Technology in Agriculture: A Predictive Forecasting of Pest and Disease Incidence. In *Applying Drone Technologies and Robotics for Agricultural Sustainability* (pp. 50–81). IGI Global.

Ganeshkumar, C., Prabhu, M., Reddy, P. S., & David, A. (2020). Value chain analysis of Indian edible mushrooms. *International Journal of Technology*, *11*(3), 599–607. doi:10.14716/ijtech.v11i3.3979

Ganeshkumar, C., Sankar, J. G., & David, A. (2023). Adoption of Big Data Analytics: Determinants and Performances Among Food Industries. [IJBIR]. *International Journal of Business Intelligence Research*, *14*(1), 1–17. doi:10.4018/IJBIR.317419

Garwal, D., Satish, Y., Paul, M., & David, A. (2020). A Preliminary Study of Job Satisfaction among Women Employees in Banking Sector of Delhi NCR, Sonepat of Haryana, India. *International Journal of Management*, *11*(10).

Gawthrop, M. (2020). Advising Hajj and Umrah travellers in general practice. *Practice Nursing*, *31*(4), 156–161. doi:10.12968/pnur.2020.31.4.156

Gedecho, E. K. (2014). Challenges of religious tourism development: The case of Gishen Mariam, Ethiopia. *American Journal of Tourism Research*, *3*(2), 42–57.

Gesler, W. (1996). Lourdes: Healing in a place of pilgrimage. *Health & Place*, *2*(2), 95–105. https://www.doi.org/10.1016/1353-8292(96)00004-4. doi:10.1016/1353-8292(96)00004-4

Gladstone, D. L. (2013). *From pilgrimage to package tour: Travel and tourism in the third world*. Routledge. doi:10.4324/9780203689363

Goni, M. D., Naing, N. N., Hasan, H., Wan-Arfah, N., Deris, Z. Z., Arifin, W. N., Hussin, T. M. A. R., Abdulrahman, A. S., Baaba, A. A., & Arshad, M. R. (2020). Development and validation of knowledge, attitude and practice questionnaire for prevention of respiratory tract infections among Malaysian Hajj pilgrims. *BMC Public Health*, *20*(1), 1–10. doi:10.118612889-020-8269-9 PMID:32114986

Goodwin, H. (2016). *Responsible tourism: using tourism for sustainable development*. Goodfellow. doi:10.23912/978-1-910158-84-5-3101

Goodwin, H. (2021). *Responsible Tourism in India. Government of India Institute of Secretariat Training & Management*. ISTM.

Google Maps. (n.d.) [Poondi]. https://www.google.com/maps/place/Poondi,+Tamil+Nadu+613105/@16.4673358,68.613474,3755947m/data=!3m1!1e3!4m6!3m5!1s0x3baaebe35ef12321:0x81de0812de77a633!8m2!3d10.8655294!4d78.9411821!16s%2Fg%2F1vf99t15?entry=ttu

Google Maps. (n.d.) [Velankanni]. https://www.google.com/maps/place/Velankanni,+Tamil+Nadu+611111/@10.9260104,80.9059162,2319868m/data=!3m1!1e3!4m6!3m5!1s0x3a556ed46f2b695d:0xb2fb9d52ccf6410e!8m2!3d10.6819262!4d79.8436796!16zL20vMDR2bjUz?entry=ttu

GoogleMaps. (n.d.) [Lourdes]. https://www.google.com/maps/place/65100+Lourdes,+France/@41.4860998,2.2709957,1414979m/data=!3m1!1e3!4m6!3m5!1s0xd57d3f0e00a9549:0x406f69c2f3d7c20!8m2!3d43.091463!4d-0.045726!16zL20vMDE4cjN3?entry=ttu

Government of Hong Kong. (2016). *Hong Kong: The Facts Religion and Custom.* Government of Hong Kong. https://www.gov.hk/en/about/abouthk/factsheets/docs/religion.pdf

Government of Hong Kong. (2021). Religion and Custom. Government of Hong Kong. https://www.yearbook.gov.hk/2021/en/pdf/E21.pdf

Government of Hong Kong. (2022). *Hong Kong: The Facts Religion and Custom.* Government of Hong Kong. https://www.gov.hk/en/about/abouthk/factsheets/docs/religion.pdf

Grace Fellowship Church. (2020). *Who were on Golgotha during the crucifixion of Jesus?* Grace Fellowship Church. https://www.graceomaha.org/blog/2020/03/19/who-were-on-golgotha-during-the-crucifixion-of-jesus

Graham, B. (2000). The past in place: Historical geographies of identity. *Modern historical geographies,* 70-99.

Graham, B. (2002). Heritage as Knowledge: Capital of Culture. *Urban Studies (Edinburgh, Scotland), 39*(56), 1003–1017. doi:10.1080/00420980220128426

Gray, P. B., & Young, S. M. (2011). Human–pet dynamics in cross-cultural perspective. *Anthropos, 24*(1), 17–30.

Griffin, K. (2018). The Importance of Religious Tourism and Pilgrimage: Reflecting on definitions, motives and data. *The international journal of religious tourism and pilgrimage, 5*(3), 2-9.

Griffin, K. A. (2007). The globalization of pilgrimage tourism? Some thoughts from Ireland. *Religious tourism and pilgrimage management: An international perspective*, 15-34.

Griffin, K., & Raj, R. (2018). The Importance of Religious Tourism and Pilgrimage: Reflecting on definitions, motives and data. *International Journal of Religious Tourism and Pilgrimage*, *5*(3), ii–ix. doi:10.21427/D7242Z

Griffin, K., & Raj, R. (2018). The importance of religious tourism and pilgrimage: Reflecting on definitions, motives and data. *The International Journal of Religious Tourism and Pilgrimage*, *5*(3), 2–9.

Grönroos, C. (1984). A service quality model and its marketing implications. *European Journal of Marketing*, *18*(4), 36–44. doi:10.1108/EUM0000000004784

Grover, S., Sahoo, S., Mehra, A., Avasthi, A., Tripathi, A., Subramanyan, A., Pattojoshi, A., Rao, G. P., Saha, G., Mishra, K. K., Chakraborty, K., Rao, N. P., Vaishnav, M., Singh, O. P., Dalal, P. K., Chadda, R. K., Gupta, R., Gautam, S., Sarkar, S., ... Janardran Reddy, Y. C. (2020). Psychological impact of COVID-19 lockdown: An online survey from India. *Indian Journal of Psychiatry*, *62*(4), 354–362. doi:10.4103/psychiatry.IndianJPsychiatry_427_20 PMID:33165368

Gunn, C. A. (1994) Tourism Planning, New York, Taylor and Francis

Gupta, B. (2007). Pilgrimage Tourism: A Resource for Sustainable Tourism Development. In M. Chaudhary, K.K. Kamra, S.S. Boora, R. Bhushan, M. Chand and R.H. Taxak (Eds.), New Delhi. Identification of Tourism Circuits across India: Ministry of Tourism, Government of India, Interim Report, Haryana. Anmol Publication.

Gupta, V. (1999). Sustainable tourism: learning from Indian religious traditions. *International Journal of Contemporary Hospitality Management, 11,* 91–95. //. doi:10.1108/09596119910250751

Gupta, S. K. (1998). Tourism in Garhwal Himalaya: Strategy for Sustainable Development. In D. S. Bhardwaj, O. P. Kandari, M. Chaudhary, & K. K. Kamra (Eds.), *Domestic Tourism in India* (pp. 66–75). Indus Publishing Company.

Gurung, R. (2016). *Marketing Religious Tourism Destination: Case study: The birth place of Gautama Buddha*. Centria University of Applied Sciences.

Habibi, A., Mousavi, M., Jamali, S. M., & Ale Ebrahim, N. (2022). A bibliometric study of medical tourism. *Anatolia, 33*(3), 415–425. doi:10.1080/13032917.2021.1954042

Hair, J. F., Black, W. C., Babin, B. J., Anderson, R. E., & Tatham, R. L. (2006). *Multivariate data analysis* (6th ed.).

Hair, J. F., Sarstedt, M., Hopkins, L., & Kuppelwieser, V. G. (2014). Partial least squares structural equation modeling (PLS-SEM): An emerging tool in business research. *European Business Review*, *26*(2), 106–121. doi:10.1108/EBR-10-2013-0128

Hair, J., Hollingsworth, C. L., Randolph, A. B., & Chong, A. (2017). An updated and expanded assessment of PLS-SEM in information systems research. *Industrial Management & Data Systems*, *117*(3), 442–458. doi:10.1108/IMDS-04-2016-0130

Hall, C. M. (1992). *Tourism Planning: Policies, Processes and Relationships*. Prentice Hall.

Hall, C. M. (2006). Buddhism, tourism and the middle way. In *Tourism, religion and spiritual journeys* (pp. 172–185). Routledge.

Hall, C. M. (2010). *Tourism and regional development: New pathways*. Routledge.

Handriana, T., Yulianti, P., & Kurniawati, M. (2019). Exploration of pilgrimage tourism in Indonesia. *Journal of Islamic Marketing*.

Hao, X., Zhang, X., & Mai, P. (2020). Anime pilgrimage tourists' behavioral patterns and their homologous affection: Taking anime film your name. As an example. *Luyou Xuekan*, *35*(1), 95–108.

Hartmann, R. (2014). Dark tourism, thanatourism, and dissonance in heritage tourism management: New directions in contemporary tourism research. *Journal of Heritage Tourism*, *9*(2), 166–182. doi:10.1080/1743873X.2013.807266

Harvey, D. (2020). The condition of postmodernity. In *The New social theory reader* (pp. 235–242). Routledge. doi:10.4324/9781003060963-38

Hasana, U., Swain, S. K., & George, B. (2022). A bibliometric analysis of ecotourism: A safeguard strategy in protected areas. *Regional Sustainability*, *3*(1), 27–40. doi:10.1016/j.regsus.2022.03.001

Hassan, S. C. (2016). Influence of Culture on the Food Choice of Students of the University of Bedfordshire. [Dissertation, University of Bedfordshire (UK)].

Haverila, M. J., & Haverila, K. C. (2018). Examination of customer-centric measures among different types of customers in the context of major Canadian ski resort. *Asia Pacific Journal of Marketing and Logistics*, *30*(2), 438–459. doi:10.1108/APJML-05-2017-0096

Henderson, J. (2000). Attracting tourists to Singapore's Chinatown: A case study in conservation and promotion. *Tourism Management, 21*(5), 525–534. doi:10.1016/S0261-5177(99)00108-9

Hermawan, B., Salim, U., Rohman, F., & Rahayu, M. (2019). Making Borobudur a Buddhist Religious Tourist Destination: An Effort to Preserve Buddhist Temples in Indonesia. *International Review of Social Research, 9*(1), 71–77. doi:10.2478/irsr-2019-0008

Higgins-Desbiolles, F. (2016). Sustaining spirit: A review and analysis of an urban Indigenous Australian cultural festival. *Journal of Sustainable Tourism, 24*(8–9), 1280–1297. doi:10.1080/09669582.2016.1149184

Hillmer-Pegram, K. (2016). Integrating indigenous values with capitalism through tourism: Alaskan experiences and outstanding issues. *Journal of Sustainable Tourism, 24*(8–9), 1194–1210. doi:10.1080/09669582.2016.1182536

Hindustan Times Correspondent. (2022, September 24). Complete Vindhya Corridor in six months: Yogi. *Hindustan Times*. https://www.hindustantimes.com/cities/lucknow-news/complete-vindhya-corridor-in-six-months-yogi-101664044828857.html

Hoang, V. T., Gautret, P., Memish, Z. A., & Al-Tawfiq, J. A. (2020). Hajj and Umrah Mass Gatherings and COVID-19 Infection. *Current Tropical Medicine Reports, 7*(4), 1–8. doi:10.100740475-020-00218-x PMID:33169095

Hole, Y., Khedkar, E. B., & Pawar, S. (2019). The significance of pilgrimage tourism to sustainable development with special reference to the Indian context. *African Journal of Hospitality, Tourism and Leisure, 8*(3), 1–12.

Holst, A., & Martens, P. (2016). *Determinants of animal protection policy: a cross country empirical study*. Politics and Animals.

Hong Kong Tourism Board. (n.d.) *Wong Tai Sin: zen in the city*. Hong Kong Tourism Board. https://www.discoverhongkong.com/in/explore/neighbourhoods/wong-tai-sin/zen-in-the-city.html

Honggang, X., Dan, Z., & Jigang, B. (2016). Sustainability and nature-based mass tourism: Lessons from China's approach to the Huangshan Scenic Park. *Journal of Sustainable Tourism, 24*(2), 182–202. doi:10.1080/09669582.2015.1071381

Hooper, G., & Lennon, J. J. (Eds.). (2016). *Dark tourism: Practice and interpretation*. Routledge. doi:10.4324/9781315575865

Hopler, W. (2020, August 27). The Virgin Mary's Apparitions and Miracles in Vailankanni, India. *Learn Religions*. https://www.learnreligions.com/virgin-marys-miracles-in-vailankanni-india-124597

https://www.nios.ac.in/media/documents/tourism_337_courseE/3
37_Tourism_Eng/337_Tourism_Eng_L8.pdf

Huang, K., Pearce, P., Guo, Q., & Shen, S. (2020). Visitors' spiritual values and relevant influencing factors in religious tourism destinations. *International Journal of Tourism Research, 22*(3), 314–324. doi:10.1002/jtr.2337

Hungarian Central Statistical Office. (n.d.). [Data set]. Hungarian Central Statistical Office. https://www.ksh.hu/?lang=en

Hungarian Tourism Agency (MTÜ). (n.d.). *Home*. MTU. https://mtu.gov.hu/

Hung, K., Wang, S., & Tang, C. (2015). Understanding the normative expectations of customers toward Buddhism-themed hotels: A revisit of service quality. *International Journal of Contemporary Hospitality Management, 27*(7), 1409–1441. doi:10.1108/IJCHM-12-2012-0264

Hunziker, W., & Krapf, K. (1942). Grundriss der allgemeinen Fremdenverkehrslehre. *Polygraph*.

Hyde, K. F., & Harman, S. (2011). Motives for a secular pilgrimage to the Gallipoli battlefields. *Tourism Management, 32*(6), 1343–1351. doi:10.1016/j.tourman.2011.01.008

Illiyas, F. T., Mani, S. K., Pradeepkumar, A. P., & Mohan, K. (2013). Human stampedes during religious festivals: A comparative review of mass gathering emergencies in India. *International Journal of Disaster Risk Reduction*, *5*, 10–18. doi:10.1016/j.ijdrr.2013.09.003

India Code [IC]. (1983). The Uttar Pradesh Sri Kashi Vishwanath Temple Act, 1983. In *Digital Repository of State Act*. India Code. https://www.indiacode.nic.in/bitstream/123456789/12272/1/Kashi1983up.pdf

India Tourism Statistics. (2022). *Ministry of Tourism Report 2022*. Market Research Division, Ministry of Tourism, Government of India. https://www.tourism.gov.in/sites/default/files/2022-09/India%20Tourism%20Statistics%202022%20%28English%29.pdf

Indo-Asian news service [IANS]. (2019, March 8). *PM Modi Lays Foundation Stone Of Kashi Vishwanath Temple Corridor*. NDTV. https://www.ndtv.com/india-news/pm-modi-lays-foundation-stone-of-kashi-vishwanath-temple-corridor-2004557

Jackson, R. H., & Hudman, L. (1995). Pilgrimage tourism and English cathedrals: The role of religion in travel. *Tourism Review*.

James, P., & Mandaville, P. (2010). Globalization and Culture, Vol. 2: Globalizing Religions. Sage.

Jeganathan, G. S., David, A., & Ganesh Kumar, C. (2022). Adaptation of Blockchain Technology in HRM. *Korea Review of International Studies*, 10-22.

Jeganathan, G. S., David, A., & Ilangovan, K. (2022). Determination of Hospitality Services Quality and Customer Satisfaction–a Holserv Approach. In *AU Virtual International Conference on Entrepreneurship & Sustainability in Digital Era under the theme of" Challenges of Organizational & Business Management in Dynamic Digital Dimension* (*Vol. 3*, No. 1, pp. 325-334).

Jeganathan, G. S., & Srinivasulu, Y. (2014). Growth Drivers of Tourism Industry in India. *International Journal of Business and Management*, *2*(11), 100.

Jena, A., Yadav, R. A., & Rambarki, R. (2021). Demonizing the others: Vendetta coverage of Tablighi Jamaat and Kumbh Mela during the COVID-19 pandemic in India. *Media Asia*, *48*(4), 346–351. doi:10.1080/01296612.2021.1946282

Jina, M. S. (2017). Ladakh: A case study of sustainability in tourism development. In L. D. Singh & S. Raj (Eds.), *Sustainable Tourism in Rural Ladakh: Strategy for Development* (pp. 15–33). Springer.

Jongmeewasin, S. (2016). Religious tourism, pilgrimage and cultural tourism. In *6th National and International Graduate Study Conference: Creative Multi-Disciplinary Studies for Sustainable Development & the 1st National and International Conference on Creative Multi-Disciplinary Studies for Sustainable Development* (NCMSD & ICMSD 2016) (pp. 394-95).

Joseph. (2023). Personal Communication. [Interview].

Jover, J., & Díaz-Parra, I. (2022). Who is the city for? Overtourism, lifestyle migration and social sustainability. *Tourism Geographies*, *1*(24), 9–32. doi:10.1080/14616688.2020.1713878

Judák, V., Petrikovičová, L., & Akimjak, A. (2022). Religious Tourism on the Example of Nation Pilgrim Places in Slovakia (Patronages of the Virgin Mary). *Journal of Education Culture and Society*, *13*(2), 615–630. doi:10.15503/jecs2022.2.615.630

Jyotsna, J. H., & Prakash Sai, L. (2022). Modelling pilgrim-tourist experience in Hindu religious destinations: An Interactive Qualitative Analysis. *Journal of Tourism and Cultural Change*, 1–19. doi:10.1080/14766825.2022.2095914

Kaczmarek, J., Stasiak, A., & Włodarczyk, B. (2001). The Tourist Product and its Evolution, *Turyzm/Tourism, 11*(1), 7-23. doi:10.18778/0867-5856.11.1.02

Kang, J. (2005). Suitability analysis of heritage corridor in the rapidly urbanising region: A case study of Taizhou City. Geographical Research, Pecking University, Beijing.

Karar, A. (2010). Impact of pilgrim tourism at Haridwar. *Anthropologist, 12*(2), 99–105. doi:10.1080/09720073.2010.11891138

Kaul, H., & Gupta, S. (2009). Sustainable Tourism in India. *Worldwide Hospitality and Tourism Themes, 1*(1), 12–18. doi:10.1108/17554210910949841

Kim, H., Yilmaz, S., & Ahn, S. (2019). Motivational Landscape and Evolving Identity of a Route-Based Religious Tourism Space: A Case of Camino de Santiago. *Sustainability, 11*(13), 3547, 1-20. doi:10.3390/su11133547

Kim, B., Kim, S., & King, B. (2020). Religious tourism studies: Evolution, progress, and future prospects. *Tourism Recreation Research, 45*(2), 185–203. doi:10.1080/02508281.2019.1664084

Kim, D. H., Park, J. H., & Kim, H. (2020). Service quality, satisfaction, and behavioral intention in pilgrimage tourism. *Sustainability, 12*(13), 5479.

Kirmani, S. (2018, November 29). *Kartarpur corridor: A road to peace between India and Pakistan?* BBC news online. https://www.bbc.com/news/world-asia-46382657

Klimiuk, J., & Moriarty, K. J. (2021). The Lourdes Pilgrimage and the impact on Pilgrim quality of life. *Journal of Religion and Health, 60*(6), 3775–3787. doi:10.100710943-021-01398-0 PMID:34505260

Kliot, N., & Collins-Kreiner, N. (2000). Pilgrimage tourism in the Holy Land: The behavioural characteristics of Christian pilgrims. *GeoJournal, 50*(1), 55–67. https://link.springer.com/article/10.1023/A:1007154929681. doi:10.1023/A:1007154929681

Konstantinidis, C., Mystridis, C., Tsagkalidou, E., & Rizopoulou, E. (2021). The Religious Tourism as a Competitive Advantage of the Prefecture of Pieria, Greece. *Journal of Tourism Hospitality Management, 9*(3), 173–181. doi:10.17265/2328-2169/2021.03.005

Korber, S., & McNaughton, R. B. (2017). Resilience and entrepreneurship: A systematic literature review. *International Journal of Entrepreneurial Behaviour & Research, 24*(7), 1129–1154. doi:10.1108/IJEBR-10-2016-0356

Korstanje, M. (2016). *The Rise of Thana Capitalism and Tourism.* Routledge. doi:10.4324/9781315457482

Korstanje, M. E. (2020). The dark tourist: Consuming dark spaces in the periphery. In M. Korstanje & H. Seraphin (Eds.), *Tourism, terrorism and security*. Emerald Group. doi:10.1108/978-1-83867-905-720201009

Korstanje, M. E. (2022). COVID-19 and Its Effects in the Global Transportation System: Is This the End of Tourism, at Least as We Know It? In *Challenges and Opportunities for Transportation Services in the Post-COVID-19 Era* (pp. 212–228). IGI Global. doi:10.4018/978-1-7998-8840-6.ch011

Koshim, A. G., Sergeyeva, A. M., Saparov, K. T., Berdibayeva, S. K., & Assylbekova, A. A. (2020). Underground Mosques of Mangystau as the objects of Religious Tourism. *Geo Journal of Tourism and Geosites, 34*(1), 33–41. doi:10.30892/gtg.34105-616

Kumar, P., Kumar, A., & Sankar, J. G. (2020). Service quality of health tourism in India: A structural equation modelling approach. *Solid State Technology*, 1187–1191.

Kumar, R. (Ed.). (2003). *Essays on Indian music*. Discovery Publishing House.

Kumar, S. (2020). Sustainable pilgrimage tourism development in Uttarakhand Himalayas: A community perspective. *Journal of Heritage Tourism, 15*(1), 1–17.

Kumar, S. A. V., & Singh, S. K. (2019). Assessing the Quality of Sanitation Services in Tourist Places: A Case Study of Goa, India. *Journal of Environmental Management*, *233*, 605–612. doi:10.1016/j.jenvman.2018.12.062 PMID:30605791

Kuznets, S. (1955). Economic growth and income inequality. *The American Economic Review*, *45*, 1–28.

Ladhari, R. (2008). Alternative measures of service quality: A review. *Managing Service Quality*, *18*(1), 65–86. doi:10.1108/09604520810842849

Lankford, S. V., Dieser, R. B., & Walker, G. J. (2005). Self-construal and pilgrimage travel. *Annals of Tourism Research*, *32*(3), 802–804. doi:10.1016/j.annals.2004.10.005

Lash, S. M., & Urry, J. (1993). *Economies of signs and space* (Vol. 26). Sage.

Lazzaretti, V. (2021). The boundary within: Demolitions, dream projects and the negotiation of Hinduness in Banaras. In *Spaces of Religion in Urban South Asia* (pp. 87–99). Routledge.

Lee, D. (2017). HEALTHQUAL: A multi-item scale for assessing healthcare service quality. *Service Business*, *11*(3), 491–516. doi:10.100711628-016-0317-2

Leiper, N. (1979). The framework of tourism: Towards a definition of tourism, tourist, and the tourist industry. *Annals of Tourism Research*, *6*(4), 390–407. doi:10.1016/0160-7383(79)90003-3

Leon, P. (2023). Personal Communication. [Discussion].

Lerner, P., & Rabello, A. M. (2007). The Prohibition of Ritual Slaughtering (Kosher Shechita and Halal) and Freedom of Religion of Minorities. *The Journal of Law and Religion*, *22*(1), 1–62. doi:10.1017/S0748081400003210

Lewis, B. R., & Mitchell, V. W. (1990). Defining and measuring the quality of customer service. *Marketing Intelligence & Planning*, *8*(6), 11–17. doi:10.1108/EUM0000000001086

Light, D. (2017). Progress in dark tourism and thanatourism research: An uneasy relationship with heritage tourism. *Tourism Management*, *61*, 275–301. doi:10.1016/j.tourman.2017.01.011

Lin, L. H., Ho, Y. L., & Lin, W. H. E. (2013). Confucian and Taoist work values: An exploratory study of the Chinese transformational leadership behavior. *Journal of Business Ethics*, *113*(1), 91–103. doi:10.100710551-012-1284-8

Lin, L.-P., & Hsieh, W.-K. (2022). Exploring how perceived resilience and restoration affected the wellbeing of Matsu pilgrims during COVID-19. *Tourism Management*, *90*, 104473. doi:10.1016/j.tourman.2021.104473

Linnenluecke, M. K. (2017). Resilience in business and management research: A review of influential publications and a research agenda. *International Journal of Management Reviews*, *19*(1), 4–30. doi:10.1111/ijmr.12076

Liro, J. (2021). Visitors' motivations and behaviours at pilgrimage centres: Push and pull perspectives. *Journal of Heritage Tourism*, *16*(1), 79–99. doi:10.1080/1743873X.2020.1761820

Liu, T. (2003). A Nameless but Active Religion: An Anthropologist's View of Local Religion in Hong Kong and Macau. *The China Quarterly*, *174*, 373–394. doi:10.1017/S0009443903000238

Lopez, L. (2013). How long does the pilgrimage tourism experience to Santiago de Compostela last? *The international journal of religious tourism and pilgrimage, 1*(1), 1-14.

Lopez, L. (2014). How Long Does the Pilgrimage Tourism Experience to Santiago de Compostela Last? *International Journal of Religious Tourism and Pilgrimage*, *1*(1), 1–14. doi:10.21427/D7C133

Lourdes, T. V. (2023). *TV Lourdes – Le Sanctuaire de Lourdes en direct.* [Video]. YouTube. https://www.youtube.com/watch?v=Zz2lqOMiBeo

Lubbe, B. A., Preez, E. A., Douglas, A., & Wessels, F. F. (2017). The impact of rhino poaching on tourist experiences and future visitation to National Parks in South Africa. *Current Issues in Tourism*, *22*(1), 8–15. doi:10.1080/13683500.2017.1343807

Luz, N. (2020). Pilgrimage and religious tourism in Islam. *Annals of Tourism Research*, *82*, 102915. doi:10.1016/j.annals.2020.102915

Lynch, J. J. (1977). *The broken heart: The medical consequences of loneliness*. Basic Books.

Lynch, J. J. (2000). *A cry unheard: New insights into the medical consequences of loneliness*. Bancroft Press.

Ly, T. P., & Tan, X. (2023). Temple keepers in religious tourism development: A case in Macao. *Journal of Heritage Tourism*, *18*(1), 67–83. doi:10.1080/1743873X.2022.2134019

Madaan, G., Swapna, H. R., Kumar, A., Singh, A., & David, A. (2021). Enactment of sustainable technovations on healthcare sectors. *Asia Pacific Journal of Health Management*, *16*(3), 184–192. doi:10.24083/apjhm.v16i3.989

Makrum, M. (2020). Hajj, its islamic perspective, management and role of supply chain in Indonesian perspectives. [IJSCM]. *International Journal of Supply Chain Management*, *9*(1), 794–804.

Ma, L., & Tsui, A. S. (2015). Traditional Chinese philosophies and contemporary leadership. *The Leadership Quarterly*, *26*(1), 13–24. doi:10.1016/j.leaqua.2014.11.008

Malchrowicz-M, E., Munsters, W., Korzeniewska-Nowakowska, P., & Gravelle, F. (2020). Controversial Animal Tourism Considered from a Cultural Perspective. *Tourism / Turyzm, 30,* 21-30. . doi:10.18778/0867-5856.30.1.14

Manhas, P. S., & Balakrishnan Nair, B. (2020). Strategic role of religious tourism in recuperating the Indian tourism sector post-covid-19. *International Journal of Religious Tourism Pilgrimage*, *8*(7), 6. doi:10.21427/ka25-fq52

Manhotra, D. (2017). Local residents cash in on Amarnath yatra. *Tribune India*. https://www.tribuneindia.com/news/archive/j-k/local-residents-cash-in-on-amarnath-yatra-434197

Margry, P. J. (2008). Secular Pilgrimage: A contradiction in Terms? In P. J. Margry (Ed.), *Shrines and Pilgrimages in the Modern Word*. Amsterdam University Press.

Marine-Roig, E. (2015). Religious tourism versus secular pilgrimage: The basilica of La Sagrada Família. *The international journal of religious tourism and pilgrimage, 3*(1), 25-37.

Marmot, M. (2017). Inclusion health: Addressing the causes of the causes. *Lancet*, *391*(10117), 186–188. doi:10.1016/S0140-6736(17)32848-9 PMID:29137870

Martinho, V. J., & Nunes, J. (2020). Camino de Santiago: The routes in the region of Viseu. *Pasos (El Sauzal)*, *18*(3), 357–370. doi:10.25145/j.pasos.2020.18.025

Maslow, A. H. (1943). A theory of human motivation. *Psychological Review*, *50*(4), 370–396. doi:10.1037/h0054346

Mazumdar, S., & Mazumdar, S. (2004). Religion and Place Attachment: A Study of Sacred Places. *Journal of Environmental Psychology*, *24*(3), 385–397. doi:10.1016/j.jenvp.2004.08.005

McKercher, B., & du Cros, H. (2002). *Cultural tourism: The partnership between tourism and cultural heritage management*. The Howarth Hospitality Press. [Google Scholar]

Mehdi, A., Chaudhry, D., Tomar, P., Mehdi, A., Chaudhry, D., & Tomar, P. (2019). Sectoral Perspectives. Freedoms, Fragility and Job Creation: Perspectives from Jammu and Kashmir, India, 85-110. doi:10.1007/978-981-13-1220-5_4

Memish, Z. A., Steffen, R., White, P., Dar, O., Azhar, E. I., Sharma, A., & Zumla, A. (2019). Mass gatherings medicine: Public health issues arising from mass gathering religious and sporting events. *Lancet*, *393*(10185), 2073–2084. doi:10.1016/S0140-6736(19)30501-X PMID:31106753

Metti, M. S. (2011). *Jerusalem - the most powerful brand in history.*

Meyer, D. (2004). *Tourism routes and gateways: key issues for the development of tourism routes and gateways and their potential for Pro-Poor Tourism.* [Discussion Paper, Overseas Development Institute].

Ministry of Tourism. (2018). *India Tourism Statistics 2018.* Ministry of Tourism.

Mishra, M., & Gupta, S. (2021). Impact of service quality on pilgrim loyalty: A case study of Vaishno Devi shrine. *Tourism Management Perspectives*, *39*, 100857.

Misra, R. K., & Behera, B. (2018). Socio-cultural impacts of Kumbh Mela on host communities: An analysis of Prayagraj (Allahabad). *Journal of Tourism and Cultural Change*, *16*(4), 345–363.

Mittal, R., & Sinha, P. (2022). Framework for a resilient religious tourism supply chain for mitigating post-pandemic risk. *International Hospitality Review*, *36*(2), 322–339. doi:10.1108/IHR-09-2020-0053

Mkenda, M. A., Mwambusi, A. M., & Msoffe, E. P. (2020). Sustainable Water and Sanitation Services for Tourist Places: A Case Study of Zanzibar, Tanzania. *Journal of Environmental Management*, *269*, 110792. doi:10.1016/j.jenvman.2020.110792

Mohanty, G. N., & Mohanty, S. N. (2004). Living with Snakes: The Lifestyle of the Snake Charmers (SapuaKela. *Adivasi*, *44*(12), 24–40.

Molina-Azorin, J. F. (2016). *Mixed methods research: An opportunity to improve our studies and our research skills.*

Moodie, D. (2019). On blood, power, and public interest: The concealment of Hindu sacrificial rites under Indian law. *The Journal of Law and Religion*, *34*(2), 1–18. doi:10.1017/jlr.2019.24

Morinis, A. (1992). *Sacred journeys: The anthropology of pilgrimage.* Greenwood Press.

Morris, M. C. (2013). Improved animal welfare is more related to income equality than it is to income. *Journal of Applied Animal Science*, *16*, 272–293. doi:10.1080/10888705.2013.768921 PMID:23795689

Mosier, W., Elhadary, T., Elhaty, I. A., & Safaei, M. (2020). Crisis management and the impact of pandemics on religious tourism. *International Journal of Religious Tourism Pilgrimage*, *8*(7), 3. doi:10.21427/6808-ct83

Motichandra. (1962). *Kashi Ka Itihas.* Mumbai: Hindi Granth Ratnakar Pvt. Ltd.

Mu, Z., Li, H., Jian, W., Ji, L., Yan, J., & Xiting, L. (2007). Religious tourism and cultural pilgrimage. *Religious Tourism and Pilgrimage Festivals Management*, *14*, 98.

Nandagopal. (2023). Personal Communication. [Discussion].

Narayanan, Y. (2013). Religion and Sustainable Development: Analysing the Connections. *Sustainable Development (Bradford)*, *21*(2), 131–139. doi:10.1002d.1557

Nash, D., Akeroyd, A. V., Bodine, J. J., Cohen, E., Dann, G., Hermans, D., Jafari, J., Kemper, R. V., LaFlamme, A. G., Manning, F., Noronha, R., Pi-Sunyer, O., Smith, V. L., Stoffle, R. W., Thurot, J. M., Watson-Gegeo, K. A., & Wilson, D. (1981). Tourism as an anthropological subject. *Current Anthropology*, *22*(5), 461–468. doi:10.1086/202722

Nasser, N., Anan, M., Awad, M. F. C., Bin-Abbas, H., & Karim, L. (2017). An expert crowd monitoring and management framework for Hajj. In *2017 International Conference on Wireless Networks and Mobile Communications (WINCOM)* (pp. 1-8). IEEE.10.1109/WINCOM.2017.8238202

Naude, R., Kruger, S., De Beer, L. T., Saayman, M., & Jonker, C. (2016). The relationship between personality types and leisure time activities amongst Casino employees' workplace expectations. *SA Journal of Human Resource Management*, *14*(1), 1–11. doi:10.4102ajhrm.v14i1.761

Navruz-zoda, Z. (2020). Evaluation of Holy Places of the Regions for the Development of Pilgrimage Tourism. *Indonesian Journal of Law and Economics Review*, *6*, 10–21070. doi:10.21070/ijler.2020.V6.493

NCAER. (2003). *Domestic Tourism Survey: 2002–2003*. National Council of Applied Economic Research and Ministry of Tourism and Culture, Government of India.

Nilsson, M., & Tesfahuney, M. (2018). The post-secular tourist: Re-thinking pilgrimage tourism. *Tourist Studies*, *18*(2), 159–176. doi:10.1177/1468797617723467

Nkwanyana, S., & Njiyela, T. (2021). The Sanctity of Emmaus Pilgrimage Heritage Site: The Case of Abbot Francis Pfanner. *Pharos Journal of Theology*, *6*(102), 1–12. doi:10.46222/pharosjot.102.024

Nolan, M. L., & Nolan, S. (1992). Religious sites as tourism attractions in Europe. *Annals of Tourism Research*, *19*(1), 68–78. doi:10.1016/0160-7383(92)90107-Z

Noor, T. H., Atlam, E. S., Malki, Z. S., Alzighaibi, A. R., Hashim, H., & El-Agamy, R. (2020). Android/iPhone Mobile Application for Quick Response Pilgrims Campaign Locator. *Journal of Computer and Communications*, *8*(05), 1–13. doi:10.4236/jcc.2020.85001

Norman, A. (2012). The Varieties of the Spiritual Tourist Experience. In *Literature & Aesthetics, 22*(1). https://laparisienne.wordpress.com/2009/04/30/buddy-om/

Nyaupane, G. P., Timothy, D. J., & Poudel, S. (2015). Understanding tourists in religious destinations: A social distance perspective. *Tourism Management*, *48*(1), 343–353. doi:10.1016/j.tourman.2014.12.009

O'Regan, M., Choe, J., & Di Giovine, M. (2019). Reframing and reconceptualising gambling tourism in Macau as a Chinese pilgrimage. *Tourism Geographies*, *21*(3), 508–528. doi:10.1080/14616688.2018.1545248

Olsen, D. H., & Esplin, S. C. (2020). The Role of Religious Leaders in Religious Heritage Tourism Development: The Case of the Church of Jesus Christ of Latter-Day Saints. *Religions*, *11*(256), 1–22. doi:10.3390/rel11050256

Olsen, D. H., & Korstanje, M. E. (Eds.). (2019). *Dark tourism and pilgrimage*. CABI.

Olsen, D. H., & Timothy, D. J. (1999). Tourism 2000. *Selling the Millennium in Tourism Management*, *20*(4), 389–392.

Olsen, D., & Timothy, D. (2006). Tourism and religious journeys. In *Tourism, religion and spiritual journeys* (1st ed., pp. 1–21). Routledge.

Owaidah, A., Olaru, D., Bennamoun, M., Sohel, F., & Khan, N. (2019). Review of modelling and simulating crowds at mass gathering events: Hajj as a case study. *Journal of Artificial Societies and Social Simulation*, *22*(2), 9. doi:10.18564/jasss.3997

Ozturk, I., Aslan, A., & Altinoz, B. (2021). Investigating the nexus between CO2 emissions, economic growth, energy consumption and pilgrimage tourism in Saudi Arabia. *Economic Research-Ekonomska Istraživanja*, 1-16.

Padmapriya, K. S. (2011, December 20). *Poondi Madha Basilica*. Aalayam Kanden (Temples I saw). www.aalayamkanden.blogspot.com/2011/12/poondi-madha-basilica.html

Page, S. J., & Hall, C. M. (2014). *The geography of tourism and recreation: Environment, place and space*. Routledge.

Palni, L. (2019). Tibet: The Sacred Landscape in Transition. The State of Ecology of the Tibetan Plateau.

Pandey, S., & Shukla, R. (2022). Contribution And Importance of Dietetics in Ayurveda. *World Journal of Pharmaceutical Research, 11*(11), 1069. https://doi.org/ doi:10.20959/wjpr202211-25127

Pandya, S. P. (2017). Food, faith, social service and institution building: the Annamrita programme of the Hare Krishna movement in India. (Description: Food, faith, social service and institution building). *Religion State & Society, 45*(1), 4–22. doi:10.1080/09637494.2017.1281598

Papazoglou, A. S., Moysidis, D. V., Tsagkaris, C., Dorosh, M., Karagiannidis, E., & Mazin, R. (2021). Spiritual Health and the COVID-19 Pandemic: Impacts on Orthodox Christianity Devotion Practices, Rituals, and Religious Pilgrimages. *Journal of Religion and Health, 60*(5), 3217–3229. doi:10.100710943-021-01347-x PMID:34263390

Parasuraman, A., Zeithaml, V. A., & Berry, L. (1988). *SERVQUAL: A multiple-item scale for measuring consumer perceptions of service quality, 64*(1), 12-40. APA.

Parasuraman, A., Zeithaml, V. A., & Berry, L. L. (1985). A conceptual model of service quality and its implications for future research. *Journal of Marketing, 49*(4), 41–50. doi:10.1177/002224298504900403

Park, H. Y. (2013). *Heritage Tourism*. Routledge., doi:10.4324/9781315882093

Parne, M. D., Chandrika, K. G., & David, A. (2021). *General Adjustments and Work-Satisfaction of Indian Expatriates*.

Patange, P., Srinithivihahshini, N. D., & Mahajan, D. M. (2013). Pilgrimage and the environment: Challenges in a pilgrimage centre in Maharashtra, India. *International Journal of Environmental Sciences, 3*(6), 2269–2277.

Pathak, S., & Goltz, S. (2021). An emotional intelligence model of entrepreneurial coping strategies. *International Journal of Entrepreneurial Behaviour & Research, 27*(4), 911–943. doi:10.1108/IJEBR-01-2020-0017

Patwardhan, V., Ribeiro, M. A., Payini, V., Woosnam, K. M., Mallya, J., & Gopalakrishnan, P. (2020). Visitors' place attachment and destination loyalty: Examining the roles of emotional solidarity and perceived safety. *Journal of Travel Research, 59*(1), 3–21. doi:10.1177/0047287518824157

Paul, J., Lim, W. M., O'Cass, A., Hao, A. W., & Bresciani, S. (2021). Scientific procedures and rationales for systematic literature reviews (SPAR-4-SLR). *International Journal of Consumer Studies, 45*(4), 1–16. doi:10.1111/ijcs.12695

Pearce, D. (1997). The roles of the public sector in conservation and tourism planning. *Tourism and heritage management*, 88-100.

Pearce, J. (1997). Commentary by John Pearce. *Bulletin de l'Institut Pasteur, 95*(3), 182. doi:10.1016/S0020-2452(97)81374-7

Pearce, P. L. (1982). *The social psychology of tourist behavior: International Series in Experimental Social Psychology* (1st ed.). Pergamon Press.

Pearce, P. L. (1985). A Systematic Comparison of Travel-Related Roles. *Human Relations, 38*(11), 1001–1011. doi:10.1177/001872678503801101

Pearson, M., & Sullivan, S. (2013). *Looking after heritage places: The basics of heritage planning for managers, landowners and administrators*. Melbourne Univ. Publishing.

Pécsek, B. (2021). The transformation of the Budapest Fiumei Road Cemetery into staycation. *Változó világ, változó turizmus*, 323-336.

Pécsek, B. (2015). City Cemeteries as Cultural Attractions: Towards an Understanding of Foreign Visitors. *DETUROPE Cent. Eur. J. Reg. Dev. Tour*, *7*(1), 44–61. doi:10.32725/det.2015.004

Pécsek, B. (2018). *A lassú turizmus mint a városi turisztikai desztinációk fejlesztésének fenntartható alternatívája*. Szent Istvan University of Agriculture and Life Sciences.

Pereiro, X. (2016). A review of indigenous tourism in Latin America: Reflections on an anthropological study of Guna Tourism (Panama). *Journal of Sustainable Tourism*, *24*(8–9), 1121–1138. doi:10.1080/09669582.2016.1189924

Pew Research Center Forum on Religion & Public Life. (2012). *The Global Religious Landscape*.

Plate, S. B. (2009). The Varieties of Contemporary Pilgrimage. *Cross Currents*, *59*(3), 260–267. doi:10.1353/cro.2009.a782452

Podoynitsyna, K., Van der Bij, H., & Song, M. (2012). The role of mixed emotions in the risk perception of novice and serial entrepreneurs. *Entrepreneurship Theory and Practice*, *36*(1), 115–140. doi:10.1111/j.1540-6520.2011.00476.x

Poondi Madha Basilica. (2023) *Our Basilica*. Poondi Madha Basilica. https://www.poondimadhabasilica.org/about/

Potts, T. J. (2012). 'Dark tourism'and the 'kitschification'of 9/11. *Tourist Studies*, *12*(3), 232–249. doi:10.1177/1468797612461083

Pourtaheri, M., Rahmani, K., & Ahmadi, H. (2012). Impacts of religious and pilgrimage tourism in rural areas: The case of Iran. *Journal of Geography and Geology*, *4*(3), 122. doi:10.5539/jgg.v4n3p122

Pradhan, R. R., & Mishra, S. K. (2019). Water and Sanitation Services in Tourist Places: A Review of Challenges and Best Practices. *Environmental Science and Pollution Research International*, *26*(23), 23367–23384. doi:10.100711356-019-05722-6

Prasad, N. (2019, June 13). Kashi Vishwanath Temple Corridor's final DPR ready! PM Narendra Modi's dream project is every pilgrim's joy. *Financial Express*. https://financialexpress.com. https://www.financialexpress.com/lifestyle/travel-tourism/kashi-vishwanath-temple-corridors-final-dpr-ready-pm-narendra-modis-dream-project-is-every-pilgrims-joy/1606684

Prasetio, A. P., Tai, T. D., Opulencia, M. J. C., Abbas, M., El-Ebiary, Y. A. B., Abbas, S. F., Bykanova, O., Samal, A., & Iswanto, A. H. (2022). Impact of the COVID-19 pandemic on religious tourism amongst Muslims in Iraq. *Hervormde Teologiese Studies*, *78*(4), 1–6. doi:10.4102/hts.v78i4.7565

Pratama, S., Arthana, W., & Kusuma, N. (2020). Assessing Water and Sanitation Service Quality in Tourist Places: A Case Study of Bali, Indonesia. *Journal of Environmental Management*, *261*, 110242. doi:10.1016/j.jenvman.2020.110242

Progano, R. N., & Santos, X. M. (2023). Introduction to Host Communities and Pilgrimage Tourism: Asia and Beyond. In *Host Communities and Pilgrimage Tourism: Asia and Beyond* (pp. 3–14). Springer Nature Singapore. doi:10.1007/978-981-19-9677-1_1

Psyminds. (2020). Kau Chim- A Chinese Fortune-Telling Practice. *Psy Minds*. https://psy-minds.com/kau-chim-fortune-telling/

Raj, R., Griffin, K., & Blackwell, R. (2015). Motivations for religious tourism, pilgrimage, festivals and events. In Religious tourism and pilgrimage management: An international perspective (pp. 103-117). Wallingford UK. doi:10.1079/9781780645230.0103

Raja, J. (2017). Religious Tourism and Pilgrimage in Non-religious Country: The Czech Republic. In Raj, R., Griffin, K., & ProQuest. Conflicts, religion and culture in tourism (pp. 57-81). CABI Religious tourism and Pilgrimage series.

Raj, R. (2012). Religious tourist's motivation for visiting religious sites'. *International Journal of Tourism Policy*, *4*(2), 95–105. doi:10.1504/IJTP.2012.048997

Raj, R., & Griffin, K. A. (Eds.). (2015). *Religious tourism and pilgrimage management: An international perspective*. Cabi. doi:10.1079/9781780645230.0000

Ramakrishnan, S. (2019, August 30). Annai Velankanni Festival: All you need to know about the 11-day event. *India Express*. https://indianexpress.com/article/cities/chennai/annai-velankanni-festival-all-you-need-to-know-about-the-event-5951112

Ramasamy, R., David, A., Ariyaputhiri, S., & Kumar, P. (2021). The Socio-Cultural impacts of Ethnic Festival Events on Local Community. International Journal of Humanities, Law &. *Social Science*, *8*(3), 91–99.

Ranaweerage, E., Ranjeewa, A. D. G., & Sugimoto, K. (2015). Tourism-induced disturbance of wildlife in protected areas: A case study of free ranging elephants in Sri Lanka. *Global Ecology and Conservation*, *4*, 625–631. doi:10.1016/j.gecco.2015.10.013

Rangaswami, C. (2017). Ramanathaswamy temple. In C. M. Rangaswami (Ed.), Ramanathaswamy temple (pp. 201–207). Bharti publications.

Rani, W. M., Wan, Tamjes, M.S., & Wahab, M.H. (2018). Governance of Heritage Conservation: Overview on Malaysian Practice. *The Journal of Social Sciences Research*. 847-851. https://ideas.repec.org/a/arp/tjssrr/2018p847-851.html

Rao, N., & Suresh, K. T. (2013). *Domestic tourism in India, The Native Tourist*. Routledge.

Rashid, A. G. (2018). Religious tourism – a review of the literature. *Journal of Hospitality and Tourism Insights*, *1*(2), 150–167. doi:10.1108/JHTI-10-2017-0007

Rather, A. Y. (2022). KASHMIR TOURISM: POSSIBILITIES AND CHALLENGES. International Journal of Economic, Business, Accounting [IJEBAS]. *Agriculture Management and Sharia Administration*, *2*(3), 291–298.

Ravi, S., David, A., & Imaduddin, M. (2018). Controlling & calibrating vehicle-related issues using RFID technology. *International Journal of Mechanical and Production Engineering Research and Development*, *8*(2), 1125–1132. doi:10.24247/ijmperdapr2018130

Reddy, S.S. (2006*). E-Governance in temple administration: Tirumala Tirupati Devasthanams.*

Reddy, M. V., Boyd, S. W., & Nica, M. (2020). Towards a post-conflict tourism recovery framework. *Annals of Tourism Research*, *84*, 102940. doi:10.1016/j.annals.2020.102940 PMID:32367898

Reed, A. (2014). *Pilgrimage tourism of diaspora Africans to Ghana*. Routledge. doi:10.4324/9781315771304

Remenyik, B., Barcza, A., Csapó, J., Szabó, B., Fodor, G., & Dávid, L. D. (2021). Overtourism in Budapest: Analysis of Spatial Process and Suggested Solutions. *Reg. Stat.*, *11*(3), 179–197. doi:10.15196/RS110303

Richards, G. (1996, ed.). Cultural Tourism in Europe. CAB International, Wallingford.

Richards, G., & Munsters, W. (Eds.). (2010). *Cultural tourism research methods*. Cabi. doi:10.1079/9781845935184.0000

Robinson, P., Heitmann, S., & Dieke, P. (Eds.). (2011). *Research themes for tourism*. CABI. doi:10.1079/9781845936846.0000

Roche, E. (2019, October 24). India and Pakistan signed a historic agreement to open the Kartarpur corridor. *Livemint*. https://www.livemint.com/politics/policy/india-pakistan-sign-pact-to-enable-indian-pilgrims-visit-kartarpur-11571901789231.html

Rodríguez-Antón, J. M., Alonso-Almeida, M. D., Klassen, R., & Vázquez-Casielles, R. (2020). Is there a path to destination sustainability? A complexity theory approach to tourism resilience. *Journal of Sustainable Tourism, 28*(7), 972–991.

Rodriguez, J. (2017). Cleaning Up Bodhgaya: Conflicts over Development and the Worlding of Buddhism. *City & Society, 29*(1), 59–81. doi:10.1111/ciso.12112

Romanelli, M., Gazzola, P., Grechi, D., & Pollice, F. (2021). Towards a sustainability-oriented religious tourism. *Systems Research and Behavioral Science, 38*(3), 386–396. doi:10.1002res.2791

Rotherham, I. D. (2007). Sustaining Tourism Infrastructures for Religious Tourists and Pilgrims within the U.K. In R. Raj & N. D. Morpeth (Eds.), *Religious Tourism and Pilgrimage Festivals Management* (pp. 64–77). CABI. doi:10.1079/9781845932251.0064

Roy, M. (2020). *Tourism and Its Environmental Implication: A Case Study OfKaziranga National Park.* JISU Journal of Multidisciplinary Research. https://www.researchgate.net/publication/344225631

Russel, P. (1999). Religious Travel in the New Millennium in Travel and Tourism. *Analyst (London), 5*, 39–68.

Ruttanavisanon, W., & Agmapisarn, C. (2022). Religious tourism in Thailand-A review of Thai literature from 2012-2021. *Social Science Asia, 8*(1), 12–25.

Saad, M., See, T. P., Adil, M. A. M., & Kassim, N. M. (2015). Spiritual Intelligence on Leadership Effectiveness and Food-hygiene Practices in Public Institutions. *Procedia: Social and Behavioral Sciences, 201*, 146–155. doi:10.1016/j.sbspro.2015.08.161

Saidovna, V. F. (2022). The Usage of Pilgrimage Tourism Terms in the English Language. *International journal of inclusive and sustainable education, 1*(4), 66-70.

Salazar, A. M. (2021, July 12) *8 key findings about Christians in India.* Pew Research. https://www.pewresearch.org/short-reads/2021/07/12/8-key-findings-about-christians-in-india/#:~:text=Some%20say%20Christianity%20in%20India,2.4%25%20of%20India's%20massive%20population

Salazar, N. B. (2012). Community-based cultural tourism: Issues, threats and opportunities. *Journal of Sustainable Tourism, 20*(1), 9–22. doi:10.1080/09669582.2011.596279

Saleh, F. I. M., & Karia, N. (2020). Benchmarks for INGOs' effective responses during COVID-19 pandemic. *Benchmarking, 27*(10), 2863–2886. doi:10.1108/BIJ-04-2020-0157

Salem, M. A., Al-Shaer, H., & Alqurashi, F. (2019). Measuring the quality of water services in pilgrim's accommodation facilities in Mecca, Saudi Arabia. *Environmental Science and Pollution Research International, 26*(28), 28967–28978.

San Filippo, M. (2001). The Religious Niche. *Travel Weekly, 60*(18), 12.

Sankar, J. G., Valan, P., & Siranjeevi, M. S. (2020). Various Models to Evaluate Quality in the Service Industry. In Digital Transformation and Innovative Services for Business and Learning (pp. 181-194). IGI Global.

Sankar, J. G., Rajeswari, M., & Dwarak, M. V. S. (2020). To Measure the Service Quality of Film and Television Technology. *Solid State Technology*, 1192–1195.

Sather-Wagstaff, J. (2016). *Heritage that hurts: Tourists in the memoryscapes of September 11.* Routledge. doi:10.4324/9781315427539

Savishinsky, J. S. (1983). Pet ideas: The domestication of animals, human behavior, and human emotions. In A. H. Katcher & A. M. Beck (Eds.), *New perspectives on our lives with companion animals* (pp. 112–131). University of Pennsylvania Press.

Saxena, G., & Sharma, K. (2019). Developing tourism policy framework for sustainable tourism in Ladakh, India. *Journal of Tourism and Cultural Change*, *17*(4), 422–442.

Schramm, K. (2004). Coming Home to the Motherland: Pilgrimage Tourism in Ghana. In S. Coleman & J. Eade (Eds.), Reframing Pilgrimage: Cultures in Motion (pp. 133–149). Routledge.

Schramm, K. (2004). Coming home to the motherland: Pilgrimage tourism in Ghana. In *Reframing pilgrimage* (pp. 143–159). Routledge.

Scorrano, P., Luštický, M., & Passafaro, P. (2022). The impact of religious tourism on local development: The case of Ostuni in Italy. *Local Economy*, *37*(1), 50–70.

Seaton, A. V. (1996). Guided by the dark: From thanatopsis to thanatourism. *International Journal of Heritage Studies*, *2*(4), 234–244. doi:10.1080/13527259608722178

Sébastia, B. (2004). Like an Indian God, Hinduisation of the cult of Saint Anthony of Padua in Tamil Nadu. *Auxlieuxd'Etre*.

Sehgal, S. (2018, February 6) *Velankanni Church*. Sky Scanner. https://www.skyscanner.co.in/news/velankanni-church

Senbeto, D. L. (2023). The Greener, the Better? Probing Green Innovation in Pilgrimage Tourism Destinations. *Journal of Hospitality & Tourism Research (Washington, D.C.)*. doi:10.1177/10963480231151674

Serpell, J. (1987). Pet keeping in non-western societies: Some popular misconceptions. *Anthrozoos*, *1*(3), 166–174. doi:10.2752/089279388787058443

Sestino, A., Pino, G., & Guido, G. (2023). The effect of fervid attachment to religious rites on tourism: Evidence from the Holy Week in Southern Italy. *Journal of Cultural Heritage Management and Sustainable Development*, *13*(2), 201–218. doi:10.1108/JCHMSD-01-2021-0010

Settis, S. (2014). *Se Venezia muore*. Giulio Einaudi.

Shackley, M. (1996). Community impact of the camel safari industry in Jaisalmer, Rajasthan. *Tourism Management*, *17*(3), 213–218. doi:10.1016/0261-5177(96)00008-8

Shaheer, I., Carr, N., & Insch, A. (2021). *Spatial distribution of participation in boycott calls: a study of tourism destination boycotts associated with animal abuse*. Anatolia, An International Journal of Tourism and Hospitality Research.

Shahi, A., & Pizam, A. (2008). Towards an ethical framework for animal-based attractions. *International Journal of Contemporary Hospitality Management*, *20*(6), 679–693. doi:10.1108/09596110810892236

Shao Zhen, T., Hassan, S. H., & Mohamad, O. (2019). Buddhists' Motivations for Revisiting Religious Temples in Malaysia. *European Journal of Business and Management Research*, *4*(4). doi:10.24018/ejbmr.2019.4.4.81

Sharma, A. (2015). Smart City plan of Varanasi identifies traffic congestion and encroachment as major problems. *The Economics Times*. https://economictimes.indiatimes.com/news/economy/infrastructure/smart-city-plan-of-varanasi-identifies-traffic-congestion-and-encroachment-as-major-problems/articleshow/50226443.cms?from=mdr

Sharma, P. (2022, October 9). Explained: The Rs 3,200 crore Shree Jagannatha Temple corridor project. *The Week*. https://www.theweek.in/theweek/cover/2022/09/30/explained-the-rs-3-200-crore-shree-jagannatha-temple-corridor-project.html

Sharma, A., Hassan, A., & Mohanty, P. (Eds.). (2022). *COVID-19 and the Tourism Industry: Sustainability, Resilience and New Directions*. Routledge. doi:10.4324/9781003207467

Sharma, B., & Dyer, P. (2009). An investigation of differences in the measurement of satisfaction and evaluation of perceived service quality for religious and non-religious tourists. *Journal of Vacation Marketing*, *15*(4), 278–293.

Sharma, S., & Sharma, S. (2015). Understanding Pithoragarh's cultural jewels: Its fairs and festivals. *Worldwide Hospitality and Tourism Themes*, *7*(4), 331–346. doi:10.1108/WHATT-03-2015-0007

Sharpley, R. (2008). *Tourism and development in the developing world.* Routledge.

Sharpley, R., & Sundaram, P. (2015). *Pilgrimage and tourism: Sacred journeys as tourist attractions.* Channel View Publications.

Shastri, B. (1994). *KalikapuraneMurtivinirdesah.* Motilal Banarsidass Publications.

Shastri, J. L. (1970). *The Siva Puraṇa.* Motilal Banarsidass.

Shekhar, S., & Valeri, M. (2022). Evolving Themes in Dark Tourism Research: A Review Study. *Tourism (Zagreb)*, *70*(4), 624–641. doi:10.37741/t.70.4.6

Sheldon, P. J. (2020). Designing tourism experiences for inner transformation. *Annals of Tourism Research*, *83*, 102935. doi:10.1016/j.annals.2020.102935

Shinde, K. A. (2007). Case study 3: Religious tourism in the Indian Himalayas. In The Encyclopaedia of Tourism and Recreation in Marine Environments.

Shinde, K. A., & Olsen, D. H. (2023). Reframing the Intersections of Pilgrimage, Religious Tourism, and Sustainability. *Sustainability*, *15*(1), 461, 1-13. doi:10.3390/su15010461

Shinde, K. (2012). Policy, planning, and management for religious tourism in Indian pilgrimage sites. *Journal of Policy Research in Tourism, Leisure & Events*, *4*(3), 277–301. doi:10.1080/19407963.2012.726107

Shinde, K. A. (2010). Entrepreneurship and indigenous entrepreneurs in religious tourism in India. *International Journal of Tourism Research*, *12*(5), 523–535.

Shinde, K. A. (2011). This Is Religious Environment. *Space and Culture*, *14*(4), 448–463. doi:10.1177/1206331211412280

Shrestha, S., Kandel, S. B., & Adhikari, S. R. (2018). Sanitation Service Quality in Tourist Places: A Case Study of Kathmandu, Nepal. *Journal of Environmental Management*, *210*, 1–8. doi:10.1016/j.jenvman.2018.01.069

Shri Kashi Viswanath Temple Trust. (n.d.). *Home.* Shri Kashi Viswanath Temple Trust. https://www.shrikashivishwanath.org/

Shuo, Y., Ryan, C., & Liu, G. (2009). Taoism, temples and tourists: The case of Mazu pilgrimage tourism. *Tourism Management*, *30*(4), 581–588. doi:10.1016/j.tourman.2008.08.008

Sigala, M., & Leslie, D. (2016). *International cultural tourism: management, implications, and cases.* Routledge.

Sik Sik Yuen. (2022). *Cultural Affairs. Wong Tai Sin Beleif.* Sik Sik Yuen. https://www2.siksikyuen.org.hk/en/index

Singh, B. (2020). *A Critical Analysis of Dissonant Heritage and Dark Tourism in India: The Case of the 'Wall of Truth'Memorial* [Doctoral dissertation, University of Central Lancashire].

Singh, R. P. (2011). Varanasi, India's cultural heritage city: contestation, conservation & planning. *Heritagescapes and cultural landscapes*, 205-254.

Singh, L. (2008). Role of tourism in the economy of Jammu and Kashmir. *The Indian Journal of Labour Economics : the Quarterly Journal of the Indian Society of Labour Economics*, *51*(4), 689.

Singh, N., & Kaur, N. (2016). Punjab tourism: A SWOT analysis. *International Journal of Multidisciplinary Research and Development*, *3*(1), 131–133.

Singh, R. P., & Singh, R. (2009). Pilgrimage and tourism in India: A study of Banaras. *Annals of the National Association of Geographers, India, 29*(1), 1–14.

Singh, S. (2004). Religion, Heritage and Travel: Case References from the Indian Himalayas. *Current Issues in Tourism, 7*(1), 44–65. doi:10.1080/13683500408667972

Singh, S. (2006). Secular pilgrimages and sacred tourism in the Indian Himalayas. *GeoJournal, 64*(3), 215–223. doi:10.100710708-005-5649-8

Singh, S. (2011). *Religious Tourism in India - A Mix, a Fix or an Abiding Concern for Planners?* Religious Tourism in Asia and the Pacific. UNWTO Publication.

Singh, S. V., & Ranjan, R. (2019). Online travel portal and their effect on travel agency: A study on outbound visitors of Varanasi. [IJRAR]. *International Journal of Research and Analytical Reviews, 6*(2), 387–393.

Sinha, A., Harkness, T., & Harkness, T. (2009). Views of the Taj—Figure in the Landscape. *Landscape Journal, 28*(2), 198–217. https://www.jstor.org/stable/43323846. doi:10.3368/lj.28.2.198

Smith, V. L. (1992). Introduction: The quest in Guest. *Annals of Tourism Research, 19*(1), 1–17. doi:10.1016/0160-7383(92)90103-V

Smørvik, K. K. (2020). Why enter the church on holiday? Tourist encounters with the Basilica of Santa Maria in Trastevere, Rome. *Journal of Heritage Tourism, 16*(3), 337–348. doi:10.1080/1743873X.2020.1807557

Sofield, T. H. (2006). Border tourism in Thailand: Kaosan Road, Bangkok as a transit hub for backpackers. In Tourism in Southeast Asia: Challenges and new directions (pp. 151-175).

Sofield, T., & Li, S. (2011). Tourism governance and sustainable national development in China: A macro-level synthesis. *Journal of Sustainable Tourism, 19*(4–5), 501–534. doi:10.1080/09669582.2011.571693

South, N., & Wyatt, T. (2011). Comparing illicit trades in wildlife and drugs: An exploratory study. *Deviant Behavior, 32*(6), 538–561. doi:10.1080/01639625.2010.483162

Spode, H. (1998). Geschichte der Tourismuswissenschaft. In G. Haedrich (Ed.), Tourismus-management: Tourismus-marketing Und Fremdenverkehrsplanung (German). de Druyter. doi:10.1515/9783110815993.911

Spreng, R. A., & McKoy, R. D. (1996). *An empirical examination of a model of perceived.*

Stations of the Cross. (n.d.) Britannica. online. https://www.britannica.com/topic/Stations-of-the-Cross

Štefko, R., Kiráľová, A., & Mudrík, M. (2015). Strategic marketing communication in pilgrimage tourism. *Procedia: Social and Behavioral Sciences, 175*, 423–430. doi:10.1016/j.sbspro.2015.01.1219

Stibich, M. (2022). *What Is Religion? The Psychology of Why People Believe.* Verywell Mind.

Stone, P. R. (2006). A dark tourism spectrum: Towards a typology of death and macabre related tourist sites, attractions and exhibitions. *Tourism: An International Interdisciplinary Journal, 54*(2), 145–160.

Stone, P. R. (2012). Dark tourism and significant other death: Towards a model of mortality mediation. *Annals of Tourism Research, 39*(3), 1565–1587. doi:10.1016/j.annals.2012.04.007

Stone, P., & Sharpley, R. (2008). Consuming dark tourism: A thanatological perspective. *Annals of Tourism Research, 35*(2), 574–595. doi:10.1016/j.annals.2008.02.003

Storytrails. (2021, April 13) *When did Christianity come to India? The story of St. Thomas, St. Thomas Basilica, Chennai.* [Video]. YouTube. https://www.youtube.com/watch?v=I9M0wx5N24k&t=2s

Suarja, I. K., Sudiarta, M., Armoni, N. L. E., Sutarma, I. G. P., & Jendra, I. W. (2023). Spiritual Tourism: Self-Healing through Melukat Ritual in Bali. *International Journal of Multicultural Multireligious Understanding*, *10*(3), 47–55.

Suhartanto, D., & Triyuni, N. N. (2016). Tourist loyalty to religious destination: The role of satisfaction and destination image. *Tourism (Zagreb)*, *64*(2), 146–157.

Suzanne, L. C. (2002), Animals and the Concept of Dignity: Critical Reflections on a Circus Performance. *Ethics and the Environment, 7*(2), 104-126. https://www.jstor.org/stable/40339038

Swain, M. B. (1989). Developing ethnic tourism in Yunnan, China: Shilin Sani. *Tourism Recreation Research*, *14*(1), 33–39. doi:10.1080/02508281.1989.11014541

Swarbrooke, J. (2012). *Sustainable tourism management*. CABI.

Sweet, J. (2023). *22 top-rated tourist attractions in Budapest.* Planetware. https://www.planetware.com/tourist-attractions-/budapest-h-ps-bud.htm

Syamala, G., & Kakoti, S. (2016). A Study on Religious Tourism Potential and Possibilities with reference to Shirdi A Place of Religious Tourism. *Online International Interdisciplinary Research Journal : An International Multidisciplinary Journal*, *6*(3).

Szilágyi, I. (1975). Kálváriák. Budapest: Corvina kiadó.

Tambo, E., El-Dessouky, A. G., Khater, E. I., & Xianonng, Z. (2020). Enhanced surveillance and response approaches for pilgrims and local Saudi populations against emerging Nipah, Zika and Ebola viral diseases outbreaks threats. *Journal of Infection and Public Health*, *13*(5), 674–678. doi:10.1016/j.jiph.2020.01.313 PMID:32265162

Tarifa-Fernández, J., Carmona-Moreno, E., & Sánchez-Fernández, R. (2022). An attempt to clarify what deserves to remain dark: A long look back. *Tourism and Hospitality Research*, 14673584221110358. doi:10.1177/14673584221110358

Teo, P., & Chang, T. C. (2012). 8 Singapore: tourism development in a planned context. In *Tourism in South and Southeast Asia: Issues and cases* (pp. 117–128). Routledge.

Terzidou, M., Scarles, C., & Saunders, M. N. K. (2018). The complexities of religious tourism motivations: Sacred places, vows and visions. *Annals of Tourism Research*, *70*, 54–65. doi:10.1016/j.annals.2018.02.011

Tharoor, I. (January 15, 2013). The world of the Kumbh Mela: Inside the largest single gathering of humanity. *Time Magazine*. https://world.time.com/2013/01/15/the-kumbh-mela-inside-the-worlds-single-largest-gathering-ofhumanity/

The Catholic Company. (2023). *Who spread the Stations of the Cross throughout the world?* Get Fed. https://getfed.catholiccompany.com/who-spread-the-stations-of-the-cross-throughout-the-world/

The Pew Forum on Religion & Public Life. (2012). *Global Religion Landscape*. Pew Research. https://assets.pewresearch.org/wp-content/uploads/sites/11/2014/01/global-religion-full.pdf

Thomas, W. (2009, August 2) Our Lady of Health, Velankanni, India. *Catholic Voice*. https://web.archive.org/web/20110213110628/http://www.catholicvoice.ie/features/2009/08/02/our_lady_of_health_velankanni_india

Thommandru, A., Espinoza-Maguiña, M., Ramirez-Asis, E., Ray, S., Naved, M., & Guzman-Avalos, M. (2023). Role of tourism and hospitality business in economic development. *Materials Today: Proceedings*, *80*, 2901–2904. doi:10.1016/j.matpr.2021.07.059

Thummula, E., Yadav, R. K., & David, A. (2019). A cost-effective technique to avoid communication and computation overhead in vehicle insurance database for online record monitoring. [IJMPERD]. *International Journal of Mechanical and Production Engineering Research and Development*, *9*(2), 711–722.

Tian, J., Li, T., Chen, R., Yang, K., Li, P., & Wen, S. (2023). Converting idol worship into destination loyalty: A study of "idol pilgrimage tour" in China. *Journal of Research in Interactive Marketing*. doi:10.1108/JRIM-01-2023-0020

Timeout. (2023). *Things to do.* Timeout. https://www.timeout.com/budapest/things-to-do/best-attractions-in-budapest

Times Now. (2022, October 19). After Kashi and Mahakal Lok, BJP to build Banke Bihari Corridor: All you need to know. *Times Now.* https://www.timesnownews.com/india/after-kashi-and-mahakal-lok-bjp-to-build-banke-bihari-corridor-all-you-need-to-know-article-94969557

Times of India Report [TOI]. (2022, October 11). All you need to know about Ujjain's Mahakal Lok Corridor. *Times of India.* https://timesofindia.indiatimes.com/city/indore/all-you-need-to-know-about-ujjains-mahakal-lok-corridor/articleshow/94776634.cms

Times, T. E. (2019). Amarnath yatra crosses last year's figure of 2.85 lakh pilgrims. *Economic Times.* https://economictimes.indiatimes.com/news/politics-and-nation/amarnath-yatra-crosses-last-years-figure-of-2-85-lakh-pilgrims/articleshow/70345998.cms

Timothy, D. J. (2000). Sacred Journeys: Religious heritage and tourism. *Tourism Recreation Research, 27*(2), 3–6. doi:10.1080/02508281.2002.11081214

Timothy, D. J. (2013). Religious views of the environment: Sanctification of nature and implications for tourism. In A. Holden & D. Fennell (Eds.), *The Routledge handbook of the tourism and the environment* (pp. 31–42). Routledge.

Timothy, D. J., & Boyd, S. W. (2006). Heritage tourism in the 21st century: Valued traditions and new perspectives. *Journal of Heritage Tourism, 1*(1), 1–16. doi:10.1080/17438730608668462

Touropia. (2023). *Tourist Attractions in Budapest.* Touropia. https://www.touropia.com/tourist-attractions-in-budapest/

Tozser, A. (2020). An examination of the Csíksomlyó shrine based on the tourist life cycle curve. *Geo Journal of Tourism and Geosites, 32*(4), 1304–1309. doi:10.30892/gtg.32417-573

Tripadvisor. (2023). *Budapest, Central Hungary.* Tripadvisor. https://www.tripadvisor.com/Attraction_Products-g274887-a_contentId.40525991899+761525408-Budapest_Central_Hungary.html

Tul, H. (2015). Kali's Tongue: Shame, Disgust, and the Rejection of Blood and Violence in Vedic and Hindu Thought, *International Journal of Hindu Studies, 19*, 301–332. . doi:10.1007/s11407-015-9181-2

Tumarkin, M. M. (2005). *Traumascapes: The power and fate of places transformed by tragedy.* Melbourne Univ. Publishing.

Tunbridge, J. E., & Ashworth, G. J. (1996). Dissonant heritage. *The Management of the Past as a Resource in Conflict.*

Tunbridge, J. E., & Ashworth, G. J. (2016). Is all tourism dark? In *Dark Tourism* (pp. 12–25). Routledge.

Turnbull, C. (1992). Postscript: Anthropology as Pilgrimage: Anthropologist as Pilgrim. In A. Morinis (Ed.), *Sacred Journeys: The Anthropology of Pilgrimage* (pp. 257–274). Greenwood Press.

Turner, V., & Turner, E. (1978). *Image and Pilgrimage in Christian Culture: Anthropological perspectives.* Basil Blackwell.

Tyagi, T., & Dhar, R. L. (2014). Factors affecting health of the police officials: Mediating role of job stress. *Policing, 37*(3), 649–664. doi:10.1108/PIJPSM-12-2013-0128

Tylor, E. B. (1871). *Primitive Culture: Researches Into the Development of Mythology, Philosophy, Religion, Art, and Custom* (Vol. 1). John Murray.

Tzanelli, R. (2016). *Thanatourism and cinematic representations of risk: Screening the end of tourism.* Routledge. doi:10.4324/9781315624105

Ugli, V. A. A. (2023). The importance and advantages of the development of pilgrimage tourism in Uzbekistan. *Galaxy International Interdisciplinary Research Journal, 11*(5), 304–307.

United Nations Educatiional, Scientific and Cultural Organisation [UNESCO]. (2003). *Tradition of Vedic chanting.* UNESCO. https://ich.unesco.org/en/RL/tradition-of-vedic-chanting-00062

United Nations Educational, Scientific and Cultural Organization report [UNESCO]. (n.d.). Index of development of a multidimensional framework for heritage sustainability. In *CDIS Methodology Manual V7*. United Nations Educational, Scientific and Cultural Organization. https://en.unesco.org/creativity/sites/creativity/files/cdis/heritage_dimension.pdf

United Nations World Tourism Organisation. (n.d.). *Report*: *Tourism and Culture*. UN WTO. https://www.unwto.org/tourism-and-culture

University of Alberta Office of Sustainability. (n.d.). *What is Sustainability?* University of Alberta.

UNWTO. (2008). *GLOSSARY OF TOURISM TERMS*. UN, WTO.

UNWTO. (2011). *Religious tourism in Asia and the Pacific*. Author.

UNWTO. (2017). *Annual report.* UNWTO. https://www.unwto.org/global/publication/unwto-annual-report-2017

Urban, H. B. (2016). Desire, Blood and Power– Georges Bataille and the Study of Hindu Tantra in Northeast India. *RevistaCientíficaArbitrada de la Fundación MenteClara, 1*(3), 26–48. doi:10.32351/rca.v1.3.22

Urban, H. B. (2019). 'The Cradle of Tantra': Modern transformations of a Tantric Center in Northeast India from Nationalist Symbol to Tourist Destination. *South Asia, 42*(2), 256–277. doi:10.1080/00856401.2019.1570609

Vahdat-Nejad, H., Asani, E., Mahmoodian, Z., & Mohseni, M. H. (2019). Context-aware computing for mobile crowd sensing: A survey. *Future Generation Computer Systems, 99*, 321–332. doi:10.1016/j.future.2019.04.052

Valencia, M. A., Santos, L. R., & Chavez, M. A. (2018). Water and Sanitation Services in Tourist Places: A Case Study of Cancun, Mexico. *Journal of Environmental Management, 217*, 180–189.

Valsan, B. (2012, September 9) Goa to Vailankanni: A journey of faith. *Times of India.* https://timesofindia.indiatimes.com/city/goa-to-vailankanni-a-journey-of-faith/articleshow/16317746.cms

Van Eck, N. J., & Waltman, L. (2010). Software survey: VOSviewer, a computer program for bibliometric mapping. *Scientometrics, 84*(2), 523–538. doi:10.100711192-009-0146-3 PMID:20585380

Vande Berg, T., & Kniss, F. (2008). ISKCON and immigrants: The rise, decline, and rise again of new religious movement. *The Sociological Quarterly, 49*(1), 79–104. doi:10.1111/j.1533-8525.2007.00107.x

Velmurugan, T. A., & Sankar, J. G. (2017). A comparative study on motivation theory with Maslow's hierarchy theory and two factor theory in organization. *Indo-Iranian Journal of Scientific Research, 1*(1), 204–208.

Venkatachalam, V. B., & Patwardhan, V. (2011). Sustainable Religious Tourism Destination Management: The Case of Udupi, India. In Manhas, P.S.(Ed.). Sustainable Destination Excellance: Innovations in Alternative Tourism, pp. 44-53. SHTM, University of Jammu, Jammu & Kashmir.

Venu, V., & Goodwin, H. (2008, March). *The Kerala Declaration on responsible tourism.* In Incredible India, Second International Conference on Responsible Tourism in Destinations, Kerala.

Verma, L. (2021, December 13). PM Modi to inaugurate Kashi Vishwanath corridor in Varanasi today. *Indianexpress.com.* https://indianexpress.com/article/india/pm-modi-to-inaugurate-kashi-vishwanath-corridor-in-varanasi-7669259/

Verma, M., & Sarangi, P. (2019). Modeling attributes of religious tourism: A study of Kumbh Mela, India. *Journal of Convention & Event Tourism*, *20*(4), 296–324. Advance online publication. doi:10.1080/15470148.2019.1652124

Verma, M., Verma, A., & Khurana, S. (2018). Influence of travel motivation and demographic factors on tourist participation in World's largest mass religious gathering – The Kumbh Mela, Prabandhan. *Indian Journal of Management*, *11*(8), 7. doi:10.17010/pijom/2018/v11i8/130641

Vetitnev, A., Chigarev, D., & Afanasiev, O. (2021). Overtourism and Tourismophobia at a Russian resort. *European Journal of Tourism Research*, *28*, 2812. doi:10.54055/ejtr.v28i.2263

Vidhya, K., & Selvam, V. (2019). Factors affecting the perceptions of pilgrim tourists in selection of Accommodation and Transportation: A Case Study of Golden temple, Vellore city, Tamil Nadu, India. *Humanities & Social Sciences Reviews*, *7*(5), 1254–1261. doi:10.18510/hssr.2019.75164

Vijayanand, S. (2012). Socio-economic impacts in pilgrimage tourism. *International Journal of Multidisciplinary Research*, *2*(1), 329–343.

Vila, T. D., Darcy, S., & Gonzalez, E. A. (2015). Competing for the disability tourism market–a comparative exploration of the factors of accessible tourism competitiveness in Spain and Australia. *Tourism Management*, *47*, 261–272. doi:10.1016/j.tourman.2014.10.008

Walter, T. (1993). *Pilgrimage in Popular Culture* (I. Reader & T. Walter, Eds.; 1st ed.). Palgrave Macmillan.

Wang, K.-Y., Karim, A., & Yu, J. (2020). Religious Festival Marketing: Distinguishing between Devout Believers and Tourists. *Religions*, *11*(8), 413. doi:10.3390/rel11080413

Wang, S., Blasco, D., Hamzah, A., & Verschuuren, B. (2023). "Tourists and 'philosophers': Nature as a medium to consciousness and transcendence in spiritual tourism." ("Annals of Tourism Research | Vol 99, March 2023 - ScienceDirect"). *Annals of Tourism Research*, *99*, 103543. doi:10.1016/j.annals.2023.103543

Wei, Y., & Jian, K. (2005). Soundscape and Sound Preferences in Urban Squares: A Case Study in Sheffield. *Journal of Urban Design*, *10*(1), 61–80. doi:10.1080/13574800500062395

White, L., & Frew, E. (2013). *Dark tourism and place identity: Managing and interpreting dark places*. Routledge. doi:10.4324/9780203134900

Wiedmann, F., & Salama, A. M. (2021). Urban Form, Evolution, and Liveability of Cities in the Arabian Peninsula: A Case Study of Doha, Qatar. In Contemporary Architecture and Urbanism in the Middle East (pp. 47-68). University Press of Florida.

Winter, C. (2020). A review of research into animal ethics in tourism: Launching the annals of tourism research curated collection on animal ethics in tourism. *Annals of Tourism Research*, *84*(10), 29–89. doi:10.1016/j.annals.2020.102989

Wong, C. U. I., Ryan, C., & McIntosh, A. (2013). The Monasteries of Putuoshan, China: Sites of secular or religious tourism? *Journal of Travel & Tourism Marketing*, *30*(6), 577–594. doi:10.1080/10548408.2013.810996

World Commission on Environment and Development. (1987). *Our common future*. Oxford University Press.

World Health Organization (WHO). (2020). *Key planning recommendations for mass gatherings in the context of COVID-19: interim guidance, May 29th 2020 (No. WHO/2019-nCoV/POE_mass gathering/2020.3)*. World Health Organization.

www.google.com/images

Xiang, L., Wang, D., Liang, X., Huang, L., & Trace, C. (2017). The new landscape of travel: A comprehensive analysis of smartphone apps. *Information Technology & Tourism*, *17*(1), 87–103.

Xudoyarov, A. (2019). A. (2019). Specific organizational and economic aspects of the development of pilgrimage tourism in uzbekistan. *The Light of Islam*, (4), 39.

Yadav, M., & Sharma, A. (2018). Prospects of religious tourism in India. *Journal of Tourism & Hospitality (Los Angeles, Calif.)*, *7*(3), 1–5.

Yamin, M. (2015). Secure and healthy Hajj management: a technological overview. *American Academic & Scholarly Research Journal, 7*(3).

Yousaf, S. (2021). Travel burnout: Exploring the return journeys of pilgrim-tourists amidst the COVID-19 pandemic. *Tourism Management*, *84*, 104285. doi:10.1016/j.tourman.2021.104285 PMID:36530602

Yuan, Y. H. E., & Wu, C. K. (2008). Relationships among experiential marketing, experiential value, and customer satisfaction. *Journal of Hospitality & Tourism Research (Washington, D.C.)*, *32*(3), 387–410. doi:10.1177/1096348008317392

Yu, X., Kim, N., Chen, C.-C., & Schwartz, Z. (2012). ARE YOU A TOURIST? TOURISM DEFINITION FROM THE TOURIST PERSPECTIVE. *Tourism Analysis*, *17*(4), 445–457. doi:10.3727/108354212X13473157390687

Zahari, R. (2016). Pilgrimage tourism management and development in Malaysia: A review. In W. Musa & J. H. Chin (Eds.), *Religion, Tourism and Sustainable Development in Malaysia* (pp. 115–128). Springer.

Zhang, L., & Xu, Y. (2016). Pathways to entrepreneurial intentions: A meta-analysis. *The International Entrepreneurship and Management Journal*, *12*(4), 659–685.

About the Contributors

Pankaj Kumar is an Assistant Professor at Mizoram University (A Central University), Mizoram, India. He is an academician, researcher and consultant in the area of travel, tourism and hospitality management with over 7 years (Post Ph.D.) of teaching and research experience. Dr. Kumar has formerly also worked with Amity University Noida, India, Himgiri ZEE University Dehradun, India and Arba Minch University, Ethiopia, Africa. Dr. Pankaj kumar completed his 'Master of Arts in Tourism Management' from prestigious Indian university 'Banaras Hindu University' Varanasi, India. He pursued his Ph.D from 'Centre for Mountain Tourism & Hospitality Studies', HNB Garhwal University (A Central University), Srinagar, Uttarakhand, India. He has published over a dozen research papers in Journals, 8 book chapters and edited two books. His research expertise encompasses diverse areas of tourism and hospitality such as Sustainable Tourism, Ecotourism, Community-based Tourism, Heritage Tourism, Rural Tourism and Tourism Business

Rajesh Ramasamy is an Assistant Professor at the Mizoram University in Aizawl, Mizoram, India, where he teaches in the Department of Tourism and Hospitality Management. He has an MBA in Tourism, a Ph.D. in Tourism Studies, and Postgraduate Diplomas in Statistical & Research Methods and Event Management. He has ten years of combined teaching and professional experience in the travel and tourism industry. Festival Tourism, Tourism Entrepreneurship, and Tourism Marketing are some of his research interests. He has authored an edited book, published a number of research papers in reputable national and international publications, and delivered a number of lectures at conferences and seminars.

***

Surjeet Kumar is currently an Assistant Professor in Tourism and Hotel Management at Kurukshetra University, Kurukshetra, he has nineteen years of experience in higher education. He has extensive experience of research in Tourism & Hotel Management in the form of books, articles, and research papers. His research on Event management, MICE tourism; religious gastronomy, food service quality; Indian cuisine attractiveness, wedding events, personality development etc. are widely appreciated. He has supervised more than 07 PhD and 05 M.Phil. scholars. He has developed / designed various courses in tourism & hospitality at university level. He has been in expert committees of universities and other government institutions. He is presently member of various academic and administrative committees at Kurukshetra University. He worked in other universities of Haryana state which includes Maharishi Markandeshwar University, Mullana, Ambala & Maharishi Dayanand University, Rohtak. he is socially very active, he remained programme officer of NSS (UTD) for three years and contributed positively to student - community engagement and their welfare, by holding various social events. He has also contributed to the development of heritage walks in Kurukshetra city for immersive cultural experiences.

Debasish Batabyal, PhD, is currently teaching at Amity University, Kolkata, India. Dr. Debasish regularly speaks at academic conferences in India and abroad on tourism and hospitality management, specifically on eTourism, ecotourism, destination development and management. He has authored two books and edited a book 'Managing Sustainable Tourism Resources'. He has 32 publications in various journals of repute and book chapters at national and international level. As a key member, he also worked with various research institutes and team members, including committee for the production of development reports by UNDP, District Gazetteer Development committee, Board of study, syllabus committee etc. and designed job oriented courses with industry-academia liaison.

Waseem Ahmad Bhat is a Ph.D. candidate in Tourism Management at the University of Kashmir, Srinagar. An accomplished scholar, he has qualified for the Junior Research Fellowship and contributed to the field with various published articles and conference presentations. Alongside his academic pursuits, Mr. Waseem brings practical industry experience as a former Vocational Trainer and Reservation Manager, blending practical insights with theoretical knowledge in his research

Namrata Chakrabortty is an Assistant Professor in Business Communication in JIS University, Kolkata. She has published research articles in internationally reputed journals (Scopus Index) as well as national reputed journals (UGC Care Journal). She also attended and presented paper in three national conferences .She has ten years of teaching experience with him along with great desire of research and contribution to the society.

Somnath Chatterjee is an Assistant Professor in the Department of Business Management, Aliah University, Kolkata, India. He has 16 years experience in teaching and research. He has contributed to many research journals and edited books of international repute. He is one of the recipients of SSTC fellowship from ILO and joined 11th Academy held in Madrid, Spain.

Pratim Chatterjee has been teaching Hospitality Management at the Department of Hospitality,Amity University, Kolkata, West Bengal. A graduate in Hotel and Hospitality Administration from IHM,Bhubaneswar(NCHMCT&AN) and Post-graduate in Business Management with specialization in marketing from ICFAI Business School. Mr. Chatterjee is currently pursuing his P.h.D program in Management. His areas of research interest include Food & Beverage management, Hospitality branding, Accommodation management, Tourist Destination and Hospitality & Tourism marketing policy. He has published research articles in internationally reputed journals (Scopus Index) as well as national reputed journals (UGC Care Journal).He also attended and presented paper in national and international conferences .He also serves as a reviewer in some reputed international journals He has fifteen years of total experience with him along with great desire of research and contribution to the society.

Rashmiranjan Choudhury is holding an academic position at Jain (Deemed to-be-University) as an Assistant professor with academic and research responsibilities. Dr. Choudhury has done a doctoral degree as UGC- Senior Research Fellow from the Dept. of Tourism and Hotel Management at North-Eastern Hill University (NEHU). He has presented several research papers in various conferences/seminars. In published research work he has 13 research articles in journals and book chapters to his credit.

Zubair Ahmad Dada is a senior Assistant Professor of Tourism Management (DDE) at the University of Kashmir. He has about eight years' experience in teaching, research and consultancy. He has been delivering his services in various Universities of national and regional importance that include University of Jammu, Central University of Jammu, Central University of Kashmir, and University of Kashmir. He has around 20 publications to his credit that have been published in various journals of national and international repute. He has been actively participating in various national and international research forums. He holds the editorship and membership of some top tier journals. He has many Ph.D. scholars under his supervision His research interests include Eco-tourism, entrepreneurship and venture creation, tourism, peace & conflict theory, tourism sociology, tourism anthropology, political economy of tourism, tourism and migration, travel journalism, tourism, cross border tourism, globalization and identity conflicts.

Arokiaraj David serves as an Associate Professor of Al Tareeqah Management Studies, UAE. He has successfully cleared the UGC-National Eligibility Test for Lectureship (NET) and has been honored with the Junior and Senior Research Fellow (JRF & SRF) titles. With over a decade of teaching and research experience, his areas of expertise include Green Marketing, Consumer Behavior, Product & Brand Management, Global Marketing, Retail Management, Strategic Management, Data Analytics, Sustainable Practices, Resources Management, Product Development, Environmental Responsibility, and General Management. Dr. David has authored more than 60 research articles, including 11 published in Scopus, 08 in ABDC, and 05 in Web of Science. He has also contributed to 07 book chapters, published 04 books, registered 07 patents, and received 01 copyright and 01 consultant project. He is proficient in various research tools and techniques such as SPSS, AMOS, E-Views, STATA, and PLS-SEM and has hands-on experience in both primary and secondary data analysis. Recently, Dr. David has been recognized with the Best Research Award, Young Academician Award, and Young Scientist Award.

Jeganathan Gomathi Sankar is a dedicated educationist who specializes in the field of marketing. He is a member of the BSSS Institute of Advanced Studies, where he teaches courses on various marketing subjects. Dr. Sankar has previously served as a full-time research scholar in the Department of International Business at Pondicherry University, as well as a faculty member at the Saveetha School of Management. Dr. Sankar's research interests are varied and include areas such as service quality, information diffusion, ICT, marketing technology, and AI. He has published his work in well-regarded journals that are indexed in Scopus and ABDC. Through his teaching and research, Dr. Sankar strives to contribute to the field of marketing and provide his students with a comprehensive understanding of the subject matter.

Shuaibu Chiroma Hassan is working as senior lecturer at Isa Kaita College of education, Dutsinma, Katsina State - Nigeria. He is the Students' Work Experience Scheme Coordinator; formerly HoD Home Economics. He is graduate of university of Bedfordshire (MSc International tourism and hospitality management, Higher National Diploma in Catering and Hotel Management from Kaduna Polytechnic, National Diploma in Catering and Hotel Management from Katsina Polytechnic, I have a Postgraduate Diploma in education from Usmanu Danfodiyo University, Sokoto, with Teachers' Registration Council of Nigeria. He has published some Book Chapters with Routledge, authored some papers in conferences and journals hospitality and tourism and co-authored a book on nutrition

Rashmi Jha is a UGC-JRF qualified Research Scholar at the Department of History of Art, Banaras Hindu University. She previously worked as a Research Assistant in an IoE Incentive Grant, B.H.U. (MoE, Govt. of India) sponsored project, "Developing a Digital Archive of the Art Heritage of Varanasi (Newly Discovered Group of Ancient Temples and Musical Traditions of Varanasi)". Her research interests are Indian vernacular (religious and non-religious) architecture, and ancient and contemporary craft traditions of Varanasi

M. Kalaivani is currently working as an Assistant professor in SRM Institute of Science and Technology, Vadapalani, Chennai. She has Seventeen years of experience in this field and she has high interest towards research activities. Her specialization in management includes Marketing Management. She did her Ph.D. in Bharathiyar University in the discipline of Marketing management.

Noorliza Karia is an Associate Professor of Logistics and Operations Management at the Universiti Sains Malaysia (USM). She has a Ph.D. in Logistics from the University of Hull, UK, an MBA from USM, a Bachelor of Science in Production and Operations Management from Denver, USA, and a Chartered Member of the Chartered Institute of Logistics and Transport, UK. Her interests are logistics, operations, and quality, and halal logistics, innovation, and sustainability. Her works appear in quality journals such as the International Journal of Production Economics, Production Planning & Control, Resources, Conservations, Recycling, Journal Production Cleaner, Nonprofit Management & Leadership, Knowledge Management Research & Practice, Asian Journal of Technology Innovation.

Maximiliano E. Korstanje is editor in chief of International Journal of Safety and Security in Tourism (UP Argentina) and Editor in Chief Emeritus of International Journal of Cyber Warfare and Terrorism (IGI-Global US). Korstanje is Senior Researchers in the Department of Economics at University of Palermo, Argentina. In 2015 he was awarded as Visiting Research Fellow at School of Sociology and Social Policy, University of Leeds, UK and the University of La Habana Cuba. In 2017 is elected as Foreign Faculty Member of AMIT, Mexican Academy in the study of Tourism, which is the most prominent institutions dedicated to tourism research in Mexico. He had a vast experience in editorial projects working as advisory member of Elsevier, Routledge, Springer, IGI global and Cambridge Scholar publishing. Korstanje had visited and given seminars in many important universities worldwide. He has also recently been selected to take part of the 2018 Albert Nelson Marquis Lifetime Achievement Award. a great distinction given by Marquis Who´s Who in the world.

Firas Izzat Mahmoud Saleh is a PhD holder at the School of Management in Universiti Sains Malaysia (USM). He has more than 23 years of industrial experience in the field of INGOs. Currently, he is the program quality director for CARE International in Jordan. He received his MA in Quality Management from University of Jordan, BSc in Computer Science from Princess Sumaya University for 6, Jordan. He has gained industrial experiences in management of international development and aid projects, PMOs establishment and management, quality management and strategic management in the international nongovernmental organizations industry as well as public and private sectors including Hajj management. He has PMP Certification since 2009. He has published his works in high-quality journals. His research interests center on quality management, nonprofit management, project management, service quality and supply chain management.

Priyakrushna Mohanty (PK) is an Assistant Professor at the Department of Business Administration (Tourism), Christ University, Bengaluru, India. He was a former U.G.C. Senior Research Fellow at the Department of Tourism Studies, Pondicherry University, India, from where he received his PhD in Tourism Studies. PK is also an awardee of the prestigious Travel Corporation (India) Gold Medal for his outstanding performance in Master's Degree in Tourism Studies from Pondicherry University, India. He also holds a Master's Degree in Finance along with three PG Diploma Degrees in Rural Development, Research Methodology, and Teaching Skills. PK served the Indian Railway Catering and Tourism Corporation Ltd. for two years, after which he was recruited as a Guest Faculty in the Dept. of Tourism Studies, Pondicherry University. He has published more than forty papers in both international and national journals and edited books. He has presented more than thirty papers in both international and national conferences to his name. Several national and international institutes have invited him as a guest speaker. He is the editor for 5 book projects with prestigious publishers like Routledge, Emerald, Springer Nature, and CABI. PK is passionate about academic areas of Research Methodology, Tourism Sustainability, Sustainable Livelihood, Events Tourism, and e-Tourism, along with Gender issues in Tourism Development.

Pécsek Brigitta works as a lecturer at Eszterházy Károly Catholic University, Eger. She is the Head of the MSc in Management and Leadership program in English, teaching a range of marketing and tourism courses for international students. Over the last decade she has performed impactful research in sustainable tourism, covering the breadth and depth of slow tourism in urban context. She has also explored the phenomenon of staycation and workcation. She is currently conducting research on establishing the link between biophilia and tourism motivation both in Hungary and in Japan. Her approach is always multidisciplinary in nature and broad in scope, taking advantage of her diverse background ranging from marketing and linguistics to education.

Reyaz Qureshi is a senior faculty for Tourism at Department of Tourism, Hospitality & Leisure Studies, University of Kashmir. He has his master's degree in Tourism Administration and Ph.D. in Tourism Studies from Centre for Mountain Tourism & Hospitality Studies, HNB Garhwal Central University Uttrakhand. He has served the Ministry of Tourism, Govt. of Jammu and Kashmir for about a decade. In 2010 he joined academics as a faculty in Tourism Management at University of Kashmir, Srinagar. His areas of interest are Tourism Planning and Development, Travel Agency Management and Tour Operation and Tourism Marketing.

Ashish Ranjan, Research Scholar, is currently pursuing PhD from Department of Tourism & Hotel Management, Kurukshetra University, Kurukshetra under the guidance of Dr. Surjeet Kumar on weddings destination competitiveness and attractiveness He is decorated with Gold Medal in Post graduation. He is a certified Bar Mixologist from Pernord Ricard USA. He has worked in hospitality Industry with Brands like The Oberois and Hilton.

Rajeev Ranjan is an Assistant Professor, currently working with the Mangalayatan University Aligarh(Uttar Pradesh). He has six years of working experience in industry and academics. He has published several research papers in National and International Journals. He is UGC- JRF & SRF in Tourism Administration & Management, pursued his undergraduate from Jamia Millia Islamia (New Delhi) and Post graduate and Doctorate from Banaras Hindu University. He has keen interest in Heritage Management, Cultural Management, Destination Management as well as Policy and Planning of Indian Tourism & Hospitality.

Swati Shah is a Ph.D in Tourism with a specialization in the area of customer satisfaction, and employee satisfaction in the service Industry. She holds an experience of approximately 10 years in academics. In her academic career, she has worked with reputed institutions like Amity University, Noida, Kalindi College Delhi University, CVS Delhi University, and Kumaun University Nainital. She has been actively involved with students from teaching, training, placements, and designing the course curriculum. In her academic career, she has also participated in research workshops, seminars, and conferences. Some of her research work is published in journals of international repute.

Gautam Shandilya is an academician with an experience of more than 20 years in industry and academia. He has several research papers, book and book chapters to his credit.

Aditya Sharma is a doctoral student at the Department of Tourism and Hospitality Management at Mizoram University, Aizawl. His research interests include homestay tourism, sustainability, social media and bibliometry.

Amrik Singh is working as Professor and Head in the School of Hotel Management and Tourism at Lovely Professional University, Punjab, India. He has published many quality research papers indexed in Scopus or Web of Science. He has authored 2 edited books and 1 textbook. He also participated and acted as a resource person in various national and international conferences, seminars, research workshops, and industry talks. His area of research interest is Hospitality Management, HRM, Sustainability, Higher Education and Tourism etc. He is also associated as editorial and reviewer of various quality indexed journals.

Sujay Vikram Singh is a Senior Research Fellow currently pursuing a doctorate from Banaras Hindu University Graduated and post-graduated from IHM Lucknow.His research interest includes Hospitality, CRM, Service Marketing, Service Quality, and Systematic Literature reviews. Published papers in various handbooks and journals. His Recent journal publications include International Journal of Market Research (Sage publishing), ABDC-A,Scopus Journal of Global information management (IGI Publishing) Scopus, ABDC-A . Presented and published papers at various national and international conferences, seminars. Achieved best papers in various conferences (ihm Bhopal,subharati university)

Praveen Srivastava is an experienced academician with a demonstrated history of working in the education management industry for more than 20 years. He is skilled in Mendeley (Recipient of Mendeley Advisor Certification), MOOC Courses, Customer Relationship Management and ICT. Authored and edited three books and several journal articles. He is also recipient of "Award of Excellence" by IIT Mumbai for being amongst the top performer of the FDP on "Use of ICT in Education for Online and Blended learning" in 2016. In 2018, IIT Mumbai awarded him with the Top Performer badge in two-week Faculty Development Program (FDP 301X) on "Mentoring Educators in Educational Technology". He was a Mentor in the AICTE FDP (FDP101x and FDP201x) on Pedagogy for Online and Blended Teaching-Learning Process conducted by the IIT Mumbai from August 3, 2017, to September 7, 2017, and from September 14, 2017, to October 12, 2017 respectively.

Jigmet Stobdan (Pursuing Ph.D) is currently working as an assistant professor in the department of Commerce, University of Ladakh. The author is UGC-NET JRF, CAIIB fellow of IIBF. He has presented papers in various national and international conferences.

V. Suganya is currently working as an Assistant professor in SRM Institute of Science and Technology, Vadapalani, Chennai. She has Seven years of experience in this field and she has high interest towards research activities. Her specialization in management includes HR and Marketing Management. She did her Ph.d in Amet University in the discipline of HR under the topic of Work life balance among women police in Chennai and its impact on job satisfaction and psychological wellbeing.

Freeda Swarna M. has PG degrees in Tourism, Environment and a MPhil in Ecotourism. Her work with Dharthi NGO focuscs on Environment, Forests, Communities and Tourism. Her ability to work with Global Organizations has ensured a connect to the tourism industry. She is certified by the Global Council for Sustainable Tourism and has cleared the Sustainable Tourism Training Program. Her work with the communities in India in the domain of Tourism and Ecotourism has been appreciated. She has worked with various Forest Departments in States in India and the Tourism Departments. She has travelled to Europe and Africa and shared her knowledge. Within India, she is well travelled as well.

Atul Tripathi is a Professor and Head of the Department of History of Art and Tourism Management at Banaras Hindu University. Before this, he taught at the Department of History and Culture, Gujarat Vidyapeeth, Ahmedabad, as Lecturer and Reader, for over a decade. He specializes in the history of ancient and early mediaeval Indian architecture, sculpture and iconography. He has received several prestigious awards, completed a number of valuable research projects and is presently engaged in the IoE Incentive Grant, B.H.U. (MoE, Govt. of India) sponsored project entitled "Developing a Digital Archive of the Art Heritage of Varanasi (Newly Discovered Group of Ancient Temples and Musical Traditions of Varanasi)".

Po Man Tse holds a BA in Psychology from the University of British Columbia, MSc in Hotel and Tourism Management from The Hong Kong Polytechnic University. Research interests include consumer behavior, smart tourism, and hospitality education.

Index

www.ingramcontent.com/pod-product-compliance
Lightning Source LLC
LaVergne TN
LVHW080525300926
849162LV00007B/1537